CELL BIOLOGY RESEARCH PROGRESS

REGULATORY T CELLS

Cell Biology Research Progress

Additional books in this series can be found on Nova's website under the Series tab.

Additional E-books in this series can be found on Nova's website under the E-book tab.

Immunology and Immune System Disorders

Additional books in this series can be found on Nova's website under the Series tab.

Additional E-books in this series can be found on Nova's website under the E-book tab.

CELL BIOLOGY RESEARCH PROGRESS

REGULATORY T CELLS

REN S. HAYASHI
EDITOR

Nova Science Publishers, Inc.
New York

For permission to use material from this book please contact us:
Telephone 631-231-7269; Fax 631-231-8175
Web Site: http://www.novapublishers.com

Library of Congress Cataloging-in-Publication Data

Regulatory t cells / Ren S. Hayashi.
p. cm.
Includes index.
ISBN 978-1-61761-554-2 (hardcover)
1. T cells. 2. Immune response--Regulation. I. Hayashi, Ren S.
QR185.8.T2R443 2010
616.07'97--dc22

2010033116

Published by Nova Science Publishers, Inc. † New York

Contents

Preface

Regulatory T cells (sometimes known as suppressor T cells) are a specialized subpopulation of T cells that act to suppress activation of the immune system and thereby maintain immune system homeostasis and tolerance to self-antigens. This book presents topical research in the study of regulatory T cells, including the role of regulatory T cells in the treatment of multiple sclerosis and hepatocellular carcinoma; genetic engineering of antigen-specific regulatory T cells; regulatory T cells in arthritis, allergic disorders and immunomodulation; and the roles of immunoregulatory cells and the mechanisms of immune response.

Chapter 1- Research in the past decades has clearly demonstrated that chronic inflammation and autoimmune mechanisms play critical roles in the pathogeneses of many diseases including cardiovascular diseases and atherosclerosis, which are the leading causes of morbidity and mortality worldwide. However, several important issues remain to be addressed including the mechanisms underlying the regulation of autoimmune inflammation. It has been recently reported that pro-inflammatory interleukin-17 (IL-17)-producing $CD4^+$ T helper cells (Th17 cells) interplay with immunosuppressive $CD4^+CD25^{high}Foxp3^+$ regulatory T cells (Tregs) in controlling inflammatory and autoimmune diseases. Therefore, in this chapter, we will analyze the progress in the field by focusing on the following topics: *1)* the new paradigm of multiple T helper cell subsets; *2)* IL-17 family cytokines and Th17 cells; *3)* Effects of Th17/IL-17 on various inflammation and autoimmune diseases; *4)* the interaction of Tregs and Th17 cells in regulating autoimmune diseases and vascular inflammation. Regulation of autoimmunity and inflammation lies in the interplays of the different T helper subsets, therefore, better understanding of these subsets' interactions with one another would greatly improve our approaches in developing therapy to combat inflammatory and autoimmune diseases.

Chapter 2- Regulatory T-cells (Tregs) are key regulators of the immune system, however their role in infectious diseases remains rather controversial. Chronic persistent infections in humans such as parasites, viruses and (myco)bacteria all can result in the induction of both $CD4^+$ and $CD8^+$ Tregs. In these chronic persistent infections, Tregs may dampen inflammation to limit tissue damage, but they may also inhibit ensuing effector immunity, thereby impairing pathogen clearance. All of these may depend on timing and location of Treg responses, early during infections Tregs may be essential balancing the trafficking of effector cells between the sites of immune induction and infection whereas in later stages Tregs may hamper pathogen clearance. A remarkably increased frequency of Tregs has been observed at the site of infection, suggesting active local involvement. The vast majority of studies has focused on the role of (various subsets of) $CD4^+$ Treg cells. However, $CD8^+$ Treg cells play a prominent role as well, but surprisingly little is known about these cells, particularly in humans. The antigen specificity of such $CD8^+$ Tregs has been demonstrated for several pathogens and the genetic restriction of the T-cell responses can comprise classical as well as non-classical HLA class I family members. Here we review and discuss human $CD8^+$ Tregs and $CD8^+$ Treg subsets, with particular emphasis on their properties, effector mechanisms and their role in chronic, persistent infectious diseases.

Chapter 3- In the last years knowledge of regulatory T cells has expanded. They play a role in autoimmune disease but it is still unclear how they suppress auto-reactive T cells exactly. The possibilities to use regulatory T cells in the treatment of MS are discussed in this chapter. The focus will be on the function of HLA-G and $CD4^+CD25^+$ T cells expressing FoxP3 and CTLA-4.

Two types of Tregs with different ways of suppression are known. Adaptive Tregs are induced in the periphery and mainly secrete TGF-β and IL-10. These cytokines mediate suppression of T cells through induction of FoxP3. Natural Tregs are generated in the thymus and suppress T cells through HLA-G, CTLA-4 and FoxP3. HLA-G interrupts the signalling in APCs and T cells resulting in suppression. FoxP3 blocks the expression of IL-2 and consequently inhibits proliferation of T cells. The expression of FoxP3 is induced by TGF-β. TGF-β also induces $CD4^+CD25^-$ T cells to convert into $CD4^+CD25^+$ T cells which express FoxP3 and hence suppress T cell proliferation. The suppression mechanism of CTLA-4 is still unknown. It might be a positive feedback with TGF-β resulting in increased expression of FoxP3 or a block of co-stimulatory molecules on APCs causing reduced secretion of cytokines following reduced activation of T cells.

T_H17 cells also have an impact on MS. T_H17 cells are induced by IL-6 and TGF-β. These cytokines start the differentiation and activate the transcription factor ROR-γt that makes a T_H17 population expand. The presence of IL-6 inhibits the expression of FoxP3. Presence of TGF-β and absence of IL-6 induces FoxP3 which inhibits the expression of ROR-γt leading to the production of pro-inflammatory cytokines. Hence, FoxP3 and ROR-γt are of great influence and a disbalance between these factors will lead to a skewed balance in favour of T_H17 cells.

Consequently, increased levels of TGF-β will induce expression of FoxP3 and could normalise the situation.

The current treatment of MS is mainly symptomatic. Glatiramer acetate can promote activation of FoxP3, hence enhancing the function of Tregs. IFN-β up-regulates the expression of HLA-G. Unfortunately, these drugs do not cure MS, since glatiramer acetate and IFN-β are not able to skew the dysbalance back to normal.

Because TGF-β and HLA-G are able to induce the function of Tregs; the conclusion is that administration of TGF-β and HLA-G appears to be a future treatment of MS.

Chapter 4- The immune system requires a homeostatic equilibrium between the mechanisms that assure self-tolerance, those that control the capacity to mount life-long immunity to pathogenic microbes or toxins, and those that attenuate effector mechanisms from inducing immune pathology. A growing body of evidence suggests that an imbalance in this homeostasis plays a major role in the pathophysiology of autoimmune disease.

A great diversity of regulatory cells types are being progressively identified, such as tolerogenic DCs, $CD4^+CD25^+$ T_{Reg}, regulatory natural killer (rNK) and $CD8^+CD28^-$

suppressor T-lymphocytes, among others. Among the different types of immunoregulatory cells, $CD4^{+}CD25^{+}$ (IL-2 receptor alpha chain) regulatory T $CD4^{+}$ cells (T_{Reg}) play a pivotal role in the tolerance to self-antigens and tissue grafts, and in the suppression of autoimmune reactions. T_{Reg} exert modulatory functions upon the activity of self-aggressive T cells not deleted in the thymus and are one of the pillars for maintaining the immune system in homeostatic balance. Indeed, these cells modulate the intensity and quality of immune reactions through attenuation of the cytolytic activities of reactive immune cells. The outcome of different contributions of all these complex mechanisms under variable inflammatory conditions, the diversity of immune interactions between this constellation of cells conduct the evolution of the immune reaction. In this review, we summarize research characterizing the growing diversity of immunorregulatory cells subsets in the control of autoimmunity in rodents and humans. Here, we attempt to integrate the current experimental evidence on the major suppressive pathways of immunorregulation.

Chapter 5- Gastrointestinal mucosa is an important site for oral tolerance induction and allergic responses to exogenous innoxious antigens, such as food antigens. The failure or aberration of oral tolerance results in the development of allergic diseases, especially food allergy. Several factors, including mucosal barrier function, humoral factors, lymphocyte anergy, clonal deletion, and regulatory T cells, contribute to the development of oral tolerance. Accumulating clinical and experimental evidences have suggested that the decreased number and dysfunction of regulatory T cells are involved in the pathogenesis of food allergy and that the development of regulatory T cells is associated with outgrowing food allergy. Although the precise nature of regulatory T cells controlling oral tolerance to food Ags remains unknown, recent reports demonstrate that not only regulatory $CD4^{+}$ T cells but also subsets of $CD8^{+}$ T cells may function as Treg by inhibiting allergic responses. Increased understanding of the mechanisms of oral tolerance has made a paradigm shift of therapeutic strategies for food allergies from avoidance of foods to oral tolerance induction by augmentation of regulatory T cells. In this article, we review the roles of regulatory T cells in oral tolerance and the pathophysiology of food allergy, and discuss their therapeutic potential.

Chapter 6- The recognition and elimination of malignant cells by donor derived immune effector cells, the so-called graft-versus-leukemia effect (GVL), is key to the success of Allogeneic Hematopoietic Stem Cell Transplantation (AHSCT) in hematological malignancies. The ongoing challenge limiting the success of AHSCT is the maximization of this powerful GVL effect, while minimizing attack of the donor immune cells against the host normal tissue, the so-called graft-versus-host disease (GVHD). The role of CD4+CD25+FOXP3+ regulatory T cells (Treg) in the promotion of peripheral tolerance to self and thus the aversion of autoimmunity is now widely accepted. Furthermore, increasing evidence support their role in modulating alloresponses in the stem cell transplant setting, demonstrating that Tregs are critical players both in the engraftment of donor stem cells and the abrogation of allo-reactive responses that can result in GVHD. Importantly evidence from both murine models and human data suggests that Tregs dampen the alloresponse to host, thereby ameliorating the severity of GVHD without an apparent loss of useful GVL effect. This observation has sparked much research interest into both the underlying mechanisms by which Tregs mediate this effect and ways in which their function can be harnessed and translated into therapies to improve AHSCT outcomes. In this review we seek to summarize the existing data on the role of regulatory T cells in the stem cell transplant setting and examine how their manipulation might be incorporated as a useful technique to reduce GVHD following allogeneic stem cell transplantation.

Chapter 7- Hepatocellular Carcinoma (HCC) is the fifth most common cancer worldwide and also the third most common cause of cancer-related death. HCC arises most frequently in males with cirrhosis, which is most often a consequence of chronic hepatitis infection (HBV and HCV) or alcohol abuse. To date, the only effective approaches for patients with HCC are resection or liver transplantation.

Immunological mechanisms are important in the surveillance of malignancy and control of tumor progression. Tumor-Infiltrating Lymphocytes (TILs) have been described in HCC, and extensive infiltration has been associated with reduced tumor recurrence following resection. However continued tumor-growth, despite the presence of a lymphocytic infiltration, including tumor-specific T cells within and surrounding tumors, suggests a failure of immune control. Although, many mechanisms have been proposed for this attenuated immune response, it becomes evident that direct suppression of effector cells, supported by regulatory T cells could play a pivotal role in the suppression of immune response to tumors. Initially described in context of immune disorders such as inflammatory autoimmune pathologies, regulatory T lymphocytes are characterized by their capacity to inhibit T helper response. To date, several regulatory T cells are described, however CD4+CD25+ regulatory T cells and Tr1 sub-populations remain best characterized.

Currently, there is no evidence for direct implication of CD4+CD25+ regulatory T cells in the malignancy and control of HCC progress. However, recent studies showed that both regulatory T cells subpopulations and particularly Tr1 have been implicated in the modulation of the immune response during HCV chronic infection, supporting HCC progress.

Chapter 8- A strong relationship exists between regulatory T (Treg) cells and the development and progression of cancer. Increasingly, evidence suggests that Treg cells can migrate into tumors and suppress effective antitumor responses in the tumor microenvironment. The critical question that needs to be addressed is the mechanism of action of these Treg cells *in vivo*. In this chapter, we consider the nature and characteristics of Treg cells in the tumor microenvironment, including dynamic processes of Treg recruitment, activation and induction during tumor pathogenesis and their potential mechanism for multiple suppressive effects. Enhancing our understanding of the underlying molecular mechanisms will offer an insight into how Treg cell function can be manipulated more effectively for antitumor therapy.

Chapter 9- Adaptive immune responses are based on the cooperative harmony between $CD4^+$ and $CD8^+$ T cells. This principle applies not only to typical effector T cells but also to regulatory T cells. Studies of $CD4^+$ regulatory T cells have greatly advanced after the discovery of $CD4^+CD25^+Foxp3^+$ regulatory T cells (Treg), whereas progress in understanding the dynamics of $CD8^+$ regulatory T cells has been much slower. Although several $CD8^+$ regulatory T cells with different markers have been reported, these cells have not been studied as extensively as $CD4^+$ Treg. Among $CD8^+$ regulatory T cells, the $CD8^+CD122^+$ regulatory T cells in mice and their human counterparts, $CD8^+CXCR3^+$ cells, appear to be the most promising for further study. Following the initial discovery of $CD8^+CD122^+$ regulatory T cells, the same research group has clarified part of the molecular mechanism of these regulatory T cells and determined the important roles of these cells in certain autoimmune diseases by using experimental mouse models. In this review, we summarize the advantages and disadvantages of each $CD8^+$ regulatory T cell population in comparison with $CD4^+$ Treg and other $CD8^+$ regulatory T cells. Furthermore, we present experimental results showing the synergistic effect of $CD8^+CD122^+$ regulatory T cells and $CD4^+$ Treg and suggest that both $CD4^+$ and $CD8^+$ regulatory T cells be used for the treatment of immune diseases.

Chapter 10- Regulatory T cells (Treg) are crucial for suppression of autoimmune responses and employ a number of mechanisms for controlling the immune response. Treg are abundant at the site of inflammation in arthritis yet fail to control the pathological process. Recently the characterisation of surface markers which help define Treg in inflammation have aided investigations into why this failure of immune regulation occurs and has increased our understanding of Treg mechanisms of action. These investigations are complex due to overlap of Treg markers with those found on activated cells. There is increasing evidence that Treg in arthritis have defects in suppressive function. Here we discuss defects in Treg from patients with inflammatory arthritis, with the focus on adult rheumatoid arthritis (RA), and childhood arthritis (juvenile idiopathic arthritis JIA). We review evidence for how these Treg defects might be affected by treatment and possible novel therapeutic strategies.

Chapter 11- Regulatory T cells (Tregs) play a key role in the maintenance of immune homeostasis *in vivo*. The unique characteristics of Tregs endowed by foxp3, inability to produce IL-2 while constitutively expressing IL-2Rα (CD25), make themselves absolutely dependent on the responding cells for IL-2. When IL-2 is produced and secreted by the responder cells, it would bind preferentially to IL-2Rs on nearby Tregs, as the density of IL-2R on Tregs is higher than that on the responding cells. Therefore, Tregs start to proliferate earlier than the responding cells, and suppress the latter, resulting in an elegant loop of negative feedback. Furthermore, Tregs suppress the responder cells by lowering the production of IL-2 as well as by inducing earlier shedding of IL-2Rα from the responding cells. Meanwhile, the fate of secreted IL-2 and the mode of suppression are different according to the concentration of IL-2; if the stimulating signal is weak, a small amount of IL-2 is secreted from the responder cells and will bind preferentially to the IL-2Rs on Tregs. Therefore, Tregs will proliferate earlier and suppress the responder cells. If the stimulating signal is strong and the initially secreted IL-2 is enough, it would also bind to the IL-2Rs on the responder cells. In this case, the responder cells proliferate initially, but their late expansion is suppressed by Tregs. Either way, in the steady state, Tregs always win the responder cells in the competition for IL-2, hence the state of tolerance. However, in infection or inflammatory state, the suppressive activities of Tregs are suppressed (contra-suppression) by certain molecules secreted from the innate immune cells (i.e., dendritic cells), such as IL-6 or IL-21, resulting in the proliferation of the responder cells, hence immune response is elicited. In conclusion, regulation of IL-2 and IL-2Rα is an important mechanism of immune homeostasis maintained by Tregs.

Chapter 12- Foxp3 expression is required for the development and function of T regulatory cells. The last years intensive research focused on elucidating the transcriptional mechanisms behind Foxp3 expression have revealed a gene strictly regulated by epigenetic mechanisms, where multiple enhancer elements coordinates the expression of Foxp3. In this short communication article we will summarize the recent advances in studies of Foxp3 expression control.

Chapter 13- Regulatory T cells (Tregs) potently suppress a variety of innate and adoptive immune responses mediated by T effector cells (Teff). However, because of their low number (comprising only minor subset (4-10%) of the total CD4 T cell population) and their multiple specificities, large numbers of expanded antigen-specific Tregs are required in order to suppress an ongoing inflammatory response. This scarcity of Tregs greatly limits their clinical use against autoimmune inflammatory conditions. To overcome the scarcity of antigen-specific Tregs, several attempts have been made to redirect the specificity of Tregs to a pre-defined target antigen.

In this chapter, we review three approaches to endow Tregs with the required novel specificity; all rely on the ability to genetically engineer the recognition of Treg by their transduction with any receptor-gene of interest. The first approach utilizes a chimeric receptor whose specificity is comprised of a class-II-peptide complex; this receptor redirects the genetically modified Tregs against pathogenic antigen-specific effector T cells, resulting in suppression of an experimental model of multiple sclerosis, experimental allergic encephalomyelitis (EAE). The second approach makes use of TCR α and β chain gene transduction to generate MHC-1:peptide specific Tregs that were shown to suppress arthritis and rejection of a non-HLA-matched transplant. The third approach is based on an antibody-based chimeric receptor, which we have used to generate murine Tregs that suppress TNBS colitis in a murine model. In this review, we describe the activity of each of these modified Tregs, compare their reactivity to unmodified nTreg, and discuss their potential as well as obstacles towards future clinical application.

Chapter 14- The immune system protects the host from infection and exogenous stimuli using innate and acquired immune systems and is critical to maintaining health. These stimuli lead to type 1 helper T cell (Th1) and/or Th2 type cytokine secretion according to the character of the exogenous or endogenous stimuli. In addition, FoxP3^{+}(forkhead box P3^{+}) CD4^{+}CD25^{+}regulatory T cells: natural regulatory T cell (nTreg), IL-10 producing regulatory T cells (Tr1), and Th17 cells play a pivotal role in this immune balance triggering immune suppression and stimulation. Altogether they form a four-way balance.

In allergic diseases including atopic dermatitis (AD) and nasal allergy, it is understood that the cytokine profile comprises Th2 dominant, and this milieu gives rise to phenotypical abnormality and pathologic conditions including cellular infiltration and overproduction of IgE.

Chapter 15- Regulatory T cells (Tregs) are an important part of the complex adaptive system of the body that is able to respond to environmental challenges. Tregs maintain peripheral tolerance and play an important role in inflammatory reaction control. Several subsets of Tregs have been described. Naturally occurring CD4^{+}CD25^{+} Tregs are recognized as a major subset of immune cells maintaining immune self-tolerance in the periphery. Inhibitory potential of these cells against self-reacting T cells is linked to the thymic origin of Tregs. Another subtype of Tregs that is inducible. Such Tregs are generated in the periphery and exert their suppressive potential substantially in the form of anti-inflammatory activity. The latter plays an important role in cooperation of three principal anti-inflammatory mechanisms, which have been developed in the course of evolution: macrophages possessing the suppressive activity (most ancient), Tregs, and stress hormones (phylogenetically youngest). Normally, all three mechanisms of inflammation control are well equalized. However the balance may be disturbed by age due to repeated episodes of stress and HPA axis activation. Consequently, stress hormone release coupled with antigen overload results in Treg accumulation. In time HPA axis activation replaces by its depletion that is manifested by both basal cortisol level decrease and the reduction of stress-induced cortisol response. So, cortisol being in low concentration in the blood is not able any more to control inflammation and Tregs become a principal mechanism of anti-inflammatory machinary. Superfluous Treg accumulation results in development functional somatic syndromes, such as chronic fatigue syndrome, and in some patients, by reason of anticancer immunity suppression, to tumor growth. On the other hand, in childhood the lack of adequate antigen loading may lead to delay of Treg maturation. Allergy and asthma manifestation may be a consequence of such Treg insufficiency. Thus, both excess and deficiency in Treg number may be at the bottom of morbid conditions. The advances in modern pharmacology allow devising the methods of Treg level control.

In: Regulatory T Cells
Editor: Ren S. Hayashi, pp. 1-32

ISBN: 978-1-61761-554-2

Chapter 1

T Helper 17 Cells Interplay with CD4+CD25highFoxp3+ Regulatory T cells in Regulating Autoimmune Inflammation

Jietang Mai, Anthony Virtue, Hong Wang and Xiao-Feng Yang*

Department of Pharmacology, Cardiovascular Research Center, Temple University School of Medicine, Philadelphia, Pennsylvania, USA

1. Abstract

Research in the past decades has clearly demonstrated that chronic inflammation and autoimmune mechanisms play critical roles in the pathogeneses of many diseases including cardiovascular diseases and atherosclerosis, which are the leading causes of morbidity and mortality worldwide. However, several important issues remain to be addressed including the mechanisms underlying the regulation of autoimmune inflammation. It has been recently reported that pro-inflammatory interleukin-17 (IL-17)-producing $CD4^+$ T helper cells (Th17 cells) interplay with immunosuppressive $CD4^+CD25^{high}Foxp3^+$ regulatory T cells (Tregs) in controlling inflammatory and autoimmune diseases. Therefore, in this chapter, we will analyze the progress in the field by focusing on the following topics: *1)* the new paradigm of multiple T helper cell subsets; *2)* IL-17 family cytokines and Th17 cells; *3)* Effects of Th17/IL-17 on various inflammation and autoimmune diseases; *4)* the interaction of Tregs and Th17 cells in regulating autoimmune diseases and vascular inflammation. Regulation of autoimmunity and inflammation lies in the interplays of the different T helper subsets, therefore, better

* Corresponding author: Xiao-Feng Yang, MD, PhD, FAHA, Department of Pharmacology, Temple University School of Medicine, 3420 North Broad Street, Medical Research Building, Room 325, Philadelphia, PA 19140. Telephones: 215-707-5985; FAX: 215-707-7068; Email: xfyang@temple.edu

understanding of these subsets' interactions with one another would greatly improve our approaches in developing therapy to combat inflammatory and autoimmune diseases.

Keywords: Th17 cells, $CD4^+CD25^{high}Foxp3^+$ regulatory T cells (Tregs), inflammation, autoimmune, cardiovascular diseases, and atherosclerosis

2. Introduction

The pro-inflammatory cytokine, Interleukin-17 (IL-17), is primarily secreted from T cells, while a range of other cells such as mast cells, monocytes, macrophages, neutrophils, vascular smooth muscle cells, gamma delta T cells and natural killer T cells also contribute to the cytokine's production to a lesser extent [1-5]. IL-17 has been implicated in regulating various inflammatory and autoimmune diseases in human and mouse models. These diseases include rheumatoid arthritis [6, 7], systemic lupus erythematosus, encephalomyelitis [8], colitis [9, 10], tissue transport rejection [11], scleritis, uveitis [12], and diabetes [13, 14]. Among cytokines and more specifically, pro-inflammatory cytokines, IL-17 has spurred special attention due to the discovery of an IL-17-secreting subgroup of $CD4^+$ T helper cells (Th cells) appropriately named Th17 cells. This finding substantiates the emerging multi-subset paradigm with at least six subsets of T helper cells that have replaced the 20 year old Th1-Th2 paradigm [15]. These subsets consist of T helper cells (Th1), type 2 T helper cells (Th2), Th9 (IL-9 producing) [16-18], $CD4^+CD25^{high}Foxp3^+$ (Forkhead box P3 transcription factor) regulatory T cells (Tregs) [19, 20], follicular T helper cells (Tfh) [21, 22], and Th17 cells [15]. The new multi-subset model has substantially advanced our insight into the differentiation of functional $CD4^+$ T helper cell subsets and T cell regulation of inflammation and autoimmunity. This point is well demonstrated in the maturation of Tregs and Th17 cells, which arise in a mutually exclusive fashion, depending on whether their differentiation programs are activated in the presence of TGF-beta (transforming growth factor-beta) or TGF-beta plus pro-inflammatory cytokines such as IL-6 [23]. Even though significant progress has been made, the interplays between pro-inflammatory IL-17/Th17 and anti-inflammatory Tregs in orchestrating a variety of inflammation and autoimmune diseases remain to be elucidated. In this chapter, we will concentrate on the discussion of the current comprehension and advances in this front.

3. Comparison of the New Multi-Subset Paradigm to the old Th1-Th2 Paradigm

3.1. Imperfect Th1-Th2 Paradigm

Over 20 years ago, Mosmann and Coffman proposed the Th1-Th2 cell paradigm to explain how hosts are able to initiate T cell responses in order to eradicate the invasion of various pathogens [15]. Associated with the cell surface expression of chemokine receptors CCR5 (Chemokine C-C motif receptor 5) and CXCR3 (CXC-chemokine receptor 3), Th1 cells also are defined by the expression of cell surface molecules CD26 and lymphocyte activation gene-3. The primary cytokine secreted by Th1 cells is interferon-gamma (IFN-gamma), though they also produce IL-2, tumor necrosis factor-alpha (TNF-alpha), and lymphotoxin (LT). These cytokines are important in mediating delayed-type hypersensitivity reactions and macrophage activation. Functionally, Th1 cells play a key role in cellular immunity against virus and intracellular pathogens. Th2 cells differ from Th1 cells in a number of characteristics. They are identified by the expression of chemokine receptors CCR3 and CCR4 [24-26] and surface molecules CD62L, T1/ST2, and CD30 [27]. Moreover, IL-4 is the most important cytokine produced by Th2 cells although they also secrete IL-5, IL-9, IL-10, and IL-13 to a lesser extent. Differentially, Th2 cells play a critical role in humoral immunity, aid in B cell proliferation, and are important in allergic responses and protection against infection of helminthic parasites [15, 28].

Until recently, this paradigm has been the cornerstone to our perception of T cell responses. However, as several complicated pathological conditions which could not be illuminated by the basic Th1-Th2 model became evident, the replacement of the paradigm became apparent. A long held belief that supported the Th1-Th2 model was the idea that autoimmune diseases were primarily driven by the auto-reactivity of Th1 cells. This concept was supported by T cell passive transfer. For example, Th1 cells were thought to have a pathogenic function in the development of experimental autoimmune encephalomyelitis (EAE), which suggested that IFN-gamma producing Th1 cells have specificity for self-antigen and are pathogenically required for the induction of organ-specific autoimmunity in EAE. However, experimental data which examined the depletion of IFN-gamma gave contradictory results. IFN-gamma-deficient and IFN-gamma receptor-deficient mice and mice lacking other factors involved in Th1 lineage differentiation developed a more severe disease rather than being protected from the disease as predicted by the Th1-Th2 model [29, 30]. Furthermore, the relationship between IL-12 and IL-23 demonstrates another example of how Th1-Th2 paradigm fails to address autoimmunity. IL-12 and IL-23 both contain two subunits, a common p40 subunit, and a unique p35 (IL-12) or p19 (IL-23) subunit [31], respectively. It was shown that loss of the unique IL-23 subunit renders animals highly resistant to the development of autoimmunity and inflammation, whereas the lack of the IL-12 subunit did not. Since IL-12 is central in Th1 differentiation, this result suggests that Th1 cells are not required for the induction of autoimmune-mediated inflammation [32]. The described examples demonstrate the shortcomings of the Th1-Th2 paradigm and lead investigators to search for novel subsets and the proposition of a new multi-subset paradigm.

3.2. New $CD4^+$ Th Subsets – Tregs, Th17, Th9, Tfh, IL-10 IFN-Gamma-Secreting $CD4^+$ T cells and $CD4^+CD28^{null}$ T Cells

Six new $CD4^+$ Th subsets have been characterized since 1990, which have been characterized respectively in different extent:

3.2.1. Tregs

Tregs play an essential role in suppressing both innate and adaptive immune responses [33-35]. Naturally occurring Tregs are known to be central in modulating the activity of self-reactive cells. Additional interest in investigating the balance between autoimmunity and regulatory mechanisms in human autoimmune diseases has been spurred by the identification of Foxp3 as the critical determinant for Tregs maturation and function.

3.2.2. Th17

Th17 is a unique subset of $CD4^+$RORgamma $t^+$$IL23R^+$$CCR6^+$$CD161^+$ IL-17-producing T cells [36-38], which is distinct from Th1, Th2, Tfh [22], Th9 [16-18], and Treg cell subsets. These cells and their secreted effector cytokines are involved in mediating host defensive mechanisms against various infections such as the pathogenesis of various autoimmune diseases and extracellular bacterial infections.

3.2.3. Th9

Th9 cells in the presence of TGF-beta and IL-4 aptly secrete IL-9. Distinct from Th1, Th17, and $Foxp3^+$ Tregs, Th9 cells do not express any well-characterized transcription factors such as Th1 cell expressed T-box (T-bet), trans-acting Th2-cell-specific transcription factor GATA3, Th17-specific retinoid-related orphan receptor gamma t (RORgamma t), and Tregs-specific Foxp3. In fact, the specific transcription factor uniquely expressed by Th9 cells remains to be identified [16-18]. Despite this, findings support Th9 cells as a new subset of T helper cells include the fact that Th9 cells are neither anergic nor suppressive, and Th9 cells vigorously proliferate and enhance T cell proliferation.

3.2.4. Tfh

It took more than 30 years after the establishment in the 1960s of the prerequisite of T cell association for B cell production of antibody before the follicular $CD4^+$ T cell subset was characterized. Tfh cells maintain the unique ability to target and associate with B cell follicles in germinal centers which contributes to their expression of CXC-chemokine receptor 5 (CXCR5) and subsequently to induce antibody production [21, 22]. Of note, Tfh cells are the only $CD4^+$ T helper cells localized in the B cell follicles [21], whereas Th1, Th2, Th9, Th17, and Tregs are localized in the T cell zone and circulation. New findings clarify some aspects of the relationship between Th subsets: *a)* Tfh can be generated from Tregs, which challenges the current view that Tfh and Tregs are different subsets [39]; *b)* Tfh can produce IL-17 without expressing Th17-specific transcription factor RORgamma t [40, 41]; and *c)* inducible Tregs and Th17 cells arise in a mutually exclusive manner [23].

3.2.5. IL-10- IFN-gamma-secreting $CD4^+$ T cells

Initially reported in the early 1990s, IL-10-IFN-gamma-secreting $CD4^+$ T cells are suppressive T cells, which are able to inhibit cytotoxic T lymphocytes and are activated and liable in chronic infection. Despite having similar function, these $Foxp3^-$T-bet^+ cells are distinct from conventional Tregs. Furthermore, the relationships with other $CD4^+$ T cell subsets remain unclear [42].

3.2.6. $CD4^+CD28^{null}$ T cells

It has been postulated that $CD4^+CD28^{null}$ T cells are a unique subset of helper T cells which contribute to and have deleterious effects in atherosclerosis and coronary artery disease. However, it is unclear whether $CD4^+CD28^{null}$ T cells have common characteristics with Th1 and Th17 subsets [43].

4. Structural Characters of IL-17 Family Cytokines

Utilizing a subtractive hybridization method in rodents, IL-17 was initially cloned from an activated T cell hybridoma and originally named cytotoxic T-lymphocytes-associated antigen 8 (CTLA-8). This protein possesses a 57% homology with the 13th ORF of the T-lymphotropic virus, herpes virus Saimiri virus (initially HVS13, later vIL-17) [44]. However, the significance of this similarity is not completely understood. IL-17, also known as IL-17A, is the founding member of the IL-17 cytokine family containing IL-17A, IL-17B, IL-17C, IL-17D, IL-17E, and IL-17F. It consists of 155 amino acids and is secreted as a disulfide-linked homodimeric glycoprotein [45]. Cloned from a $CD4^+$ T cell library, cDNA from the human IL-17 cytokine was discovered to have 63% amino acid homology to mouse IL-17. As expected, IL-17 family members share varying degrees of structural similarity. With approximately 55% homology, IL-17 and IL-17F have the greatest structural likeness among the six cytokines within the IL-17 family. This may be attributed to the fact that only IL-17 and IL-17F are secreted by T cells, whereas the other members are produced by non-T cell tissues [46-49], Furthermore, IL-17B and IL-17C have approximately 28% amino acid homology with IL-17 [48], while IL-17E maintains 20% homology with IL-17, IL-17B, and IL-17C [49].

Using techniques such as immunoprecipitation followed by immunoblot, enzyme-linked immunosorbent assay (ELISA), and mass spectrometry, IL-17 and IL-17F have been found to exist either as homodimers or heterodimers with one another *in vitro* and in activated human $CD4^+$ T cell. All three forms demonstrate biological activity with IL-17F expressed at the highest concentration followed by IL17A/F heterodimer and minimal IL-17 detection in activated T cells from human donor [50]. IL-17A/F heterodimer has been found to be crucial to the regulation of neutrophil recruitment in the airways and activation of fibroblast and macrophages to produce other pro-inflammatory cytokines in the mouse [51, 52]. The 3-dimensional crystalline structure of IL-17F reveals that IL-17 family members form a canonical pseudo-cystine knot [47]. Several pertinent questions remain to be defined: *(a)* how the expressions of these IL-17 family members are regulated; and *(b)* how these IL-17 family members and other cytokines functionally interact in the regulation of inflammation and autoimmunity.

5. Th17 Cells, a New Proinflammatory and Autoimmune $CD4^+$ T Cell Subset

Despite the first member of IL-17 cytokine family being characterized over 15 years ago, only recently have the cells that distinctively secrete these cytokines been characterized. Predominately generated by activated $CD4^+$ T cells, IL-17 and IL-17F are additionally produced by an array of other cell types such as gamma delta T Cells, natural killer T cells (NKT), neutrophils, eosinophils, monocytes [53, 54], and vascular smooth muscle cells [4]. Further examination of cells that produce IL-17 and IL-17F led to the classification of a unique $CD4^+$ T cell subset named T helper 17 [55, 56]. In addition to IL-17 and IL-17F, Th17 cells also secrete other cytokines such as IL-21, IL-22, IL-26, and CCL20 [57-59].

As previously mentioned, Th17 is a distinctive subset of $CD4^+$ T cells, and is unique from Th1, Th2, Tfh [22], Th9, and Treg cell subsets[16-18]. Th17 cells and their effector cytokines are implicated in the pathogenesis of various autoimmune diseases and in mediating host defensive mechanisms against various infections, such as extracellular bacterial infections. Activation of T-cell receptors and co-stimulatory molecules are essential for the differentiation of naïve $CD4^+$ T cells into effector T helper cells. The necessity of distinct co-stimulatory molecules from those required by other T cell lineages [32, 58, 60-65], along with their implicated immune responses that are different from those of Th1, Th2, and Tregs, set Th17 cells apart from other subsets. For example, Th1 cells require IL-12 for development and primarily produce IFN-gamma, IL-2, and LT in order to orchestrate T cell-mediated immunity [66]. Th2 on the other hand, develops in the presence of IL-4 and produce IL-4, IL-5, and IL-13 to facilitate the humoral immune response [67]. Interestingly, the capacity to hinder the differentiation of Th17 cells has been attributed to cytokines and transcription factors produced by cells of other subsets. Both Th17 development and IL-17 expression is negatively regulated by IFN-gamma and IL-4 which are secreted by Th1 and Th2 cells respectively. Th17 cell development is also hindered by specific Th1 and Th2 transcriptional factors including STAT1 (Signal Transducer and Activator of Transcription), STAT4, T-bet, STAT6 and GATA3 [55, 68]. Furthermore, Th17 promoting cytokines maintain the ability to suppress the development of other T cell lineages as well [61]. This fact implies that the balance of the T cell lineage is regulated by the products of each subset, and that during an infection, the proper T helper cell activation is crucial for the host to effectively combat and control the infection. To further complicate the picture, it has been observed that a subset of Th17 cells can co-produce IFN-gamma and IL-17, suggesting that IFN-gamma may not always inhibit IL-17 production, but rather play a role in the pathogenic functions of Th17 [36, 69, 70]. The function of IFN-gamma in Th17-mediated pathogenesis remains to be fully resolved.

The cytokines and transcription factors involved in the generation, differentiation, and expansion of Th17 has been investigated to a greater extent recently. Despite this, there are still disagreements on the exact conditions needed for Th17 development, especially in humans. The factors that have been found to be involved in Th17 differentiation and development are TGF-beta, IL-6, IL-21, IL-23, and IL-1beta in humans and rodents with some variance among the species. The five transcription factors recognized in supporting the differentiation of Th17 from naïve T cells include: RORgamma t, RORalpha, STAT3, interferon-regulatory factor-4 (IRF4) [71], and the activator protein (AP)-1 B cell-activating transcription factor (Batf) [1, 72]. Additionally, specific for Th17 cell development, the transcriptional factor RORgamma t is up-regulated by STAT3 [73-75].

Secreted by an assortment of cells, the cytokine TGF-beta is essential in the development, homeostasis, and tolerance of T cells. The cytokine acts as a vital differentiation factor in the development of Tregs which inhibit inflammation and autoimmunity by counteracting the effects of other T helper cells. Produced by $Foxp3^+$ Tregs, TGF-beta inhibits Th1 cell differentiation [76] and is considered crucial in Th17 development in mice and humans as supported by several reports [62, 77, 78]. However, contradicting reports that TGF-beta may not be necessary for Th17 lineage commitment in humans exist. Resolution of this controversy regarding which factors drive Th17 cell development in humans remains elusive

[79-81]. In mice, TGF-beta and IL-6 are important in driving the development of Th17 in naïve T cells while IL-21 and IL-23 are critical in maintaining and expanding Th17 cells [58, 61]. Studies carried out *in vitro* have demonstrated that TGF-beta stimulation of T cells supports the development of Tregs [82]. However, Th17 differentiation is seen with the addition of pro-inflammatory cytokines such as IL-6 in the presence of TGF-beta (Figure 1) while the ability of IL-6 and TGF-beta to induce Th17 cells is amplified when IL-1beta and TNF-alpha are used as co-stimulators [60, 64]. TGF-beta is also important in making Th17 responsive to IL-23, a cytokine important in Th17 cell differentiation [83], through the induction of IL-17 and IL-23 receptor (IL-23R) in naïve T cells. IL-23 stabilizes the expansion of Th17 cells, and the IL-23R is induced in T cells after stimulation with IL-6 plus TGF-beta [62]. However, high concentrations of TGF-beta have been shown to inhibit the expression of IL-23R on T cells, implying that TGF-beta has a biphasic function in T cell development[1]. On a transcriptional level, TGF-beta up-regulates RORgamma t expression on T cells, hinders the expression of IL-17 in naïve T cells, and stimulates Treg differentiation. This inhibition is mitigated only with the addition of pro-inflammatory cytokines such as IL-6 [84]. The presence of IL-6 increases the expression of IL-21 leading to an observed up-regulation of IL-23R levels, though the induction of IL-23R is nearly abolished in the absence of IL-21R. Noteworthy is the fact that IL-21 is generated by Th17 cells, and can regulate Th17 cells [63] and IL-17 production via positive feedback. Furthermore, Treg differentiation is found to suppress TGF-beta-induced expression of Foxp3 [57]. IL-6, IL-21, and IL-23 all have the ability to independently induce RORgamma t expression, which is dependent on STAT3. This is demonstrated by the complete loss of IL-17 induction in T cells lacking STAT3 in the presence of IL-6 or IL-21 with TGF-beta [65]. However, they are not sufficient to induce Th17 differentiation alone since Th17 expression is only seen when TGF-beta is also present. This finding suggests that RORgamma t alone is not enough to induce Th17 differentiation, indicating that additional factors induced by TGF-beta are required to drive Th17 proliferation. Thus taken together, IL-6, IL-21, and IL-23 act in conjunction with TGF-beta to induce RORgamma t-dependent differentiation of Th17 cells, while STAT3 is also activated downstream of IL-6, IL-21, and IL-23. STAT3 and RORgamma t work together to cause maximal IL-17 differentiation. Also noteworthy, mouse and human Th17 cell differentiation is selectively inhibited by the small molecule halofuginone (HF) through activation of a cytoprotective signaling pathway, the amino acid starvation response [85]. In addition, it has been shown that the effect of IL-1beta is pre-programmed via IL-1RI in, IL-1 receptor I $(IL\text{-}1RI)^+$ memory Th17 cells. These cells have amplified gene expression of IL17, RORC (RAR-related orphan receptor C), and IRF4 even before T-cell antigen receptor triggering. Although Th17 cells from human umbilical cord blood did not express IL-1RI, the cytokines IL-7, IL-15, and TGF-beta up-regulated IL-1RI levels on naïve Th17 cells. This fact indicates that $IL\text{-}1RI^+$ naïve Th17 cells develop in the periphery [86]. Recently, Th17 is shown to also secrete IL-9 and IL-9 may also be an important molecule in Th17 differentiation. IL-9 plus TGF-β induced Th17 cell differentiation from naïve T cell. Unlike IL-6, which has a negative effect on Tregs in the presence of TGF-β, IL-9 in the presence of TGF-β drives the differentiation of Th17 and enhances the function of Tregs[87].

The ability of IL-9 to induce Th17 differentiation is probably attributed to the activation of STAT3 which IL-9 is reported to induce [88]. Interestingly, IL-23 is shown to have a negative effect on IL-9 secretion by Th17 cells[87].

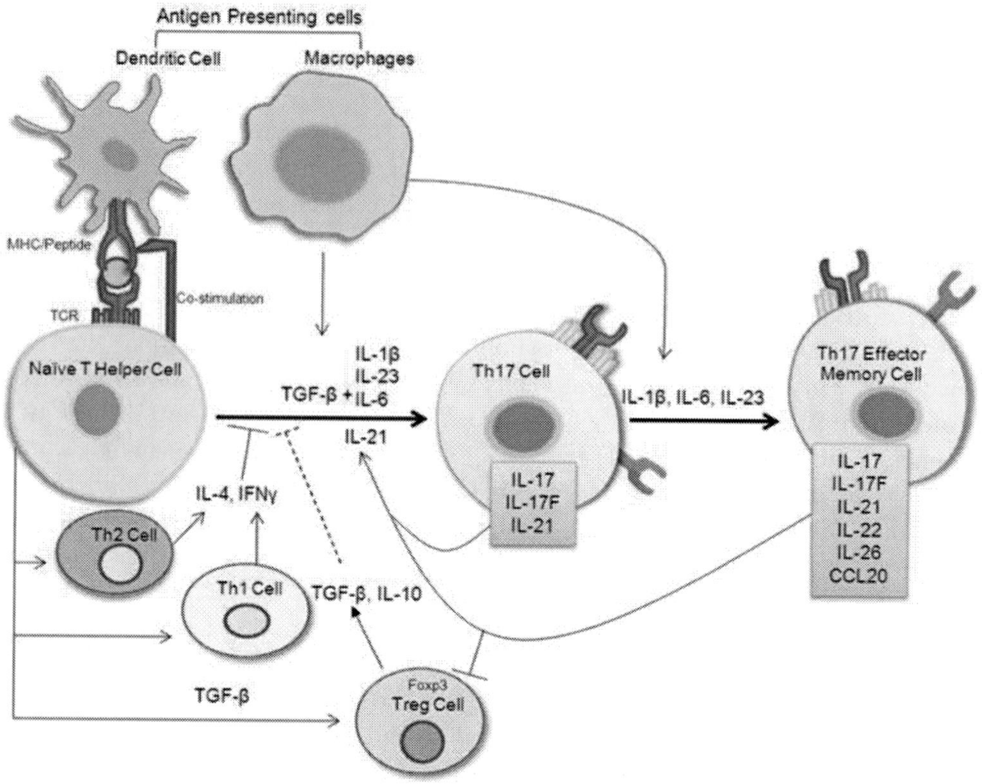

Figure 1. The Differentiation of Th17 Cells. Upon activation by antigen presenting cells, naïve T helper cells can differentiate into different T helper subsets. In the presence of TGF-β and pro-inflammatory cytokines such as IL-6, IL-23, or IL-1β, naïve T cell differentiates into Th17 cells. The differentiation of Th17 is inhibited by hallmark cytokines of other T helper subsets. Pro-inflammatory cytokines can induce further effector molecules in committed Th17 cells to establish their terminally differentiated effector phenotype

6. Wide Spectrum Effects of IL-17 on Various Cell Types and Tissues

The effects of IL-17 cytokine family on the production other chemokine and cytokine widely varies among cell types. The profile of cytokines secreted is contingent on cell type and the IL-17 cytokine(s) involved. We will outline several examples in the following text.

6.1. Synovial Stromal Cells

Original investigations of IL-17 revealed that the cytokine induces IL-6 and IL-8 generation in a dose-dependent manner in human foreskin fibroblasts while enhanced surface expression of intracellular adhesion molecule-1 (ICAM-1) is seen with IL-17 incubation [45]. In primary culture of synovial fluid, an analogous effect is witnessed. Again, IL-17 induces secretion of IL-6 and IL-8; however, levels of prostaglandin E2 (PGE2) and granulocyte colony-stimulating factor (G-CSF) are also increased in stromal cells which lead to the hematopoiesis of neutrophils. The recruitment of neutrophils by IL-6 and IL-8 allows for the immune response to fight off infections [89].

6.2. Fibroblasts

IL-17F stimulates the production of cytokines such as IL-6, IL-8, and G-CSF, and contributes to cartilage matrix turn over regulation via induction of matrix metalloproteinases (MMPs) [47]. The ability of IL-17 to affect mRNA levels has been shown in the fibroblastoid L929 cell line on monocyte chemotactic protein-1 (MCP-1), macrophage-inflammatory protein-2 (MIP-2), and tissue inhibitor of metalloproteinases (TIMP). This finding has been supported in mouse embryonic fibroblast (MEF) where granulocyte chemotatic protein-2 (GCP-2), MMP-3, MMP-9, and MMP-13 mRNAs are induced with IL-17 stimulation [90]. The involvement of IL-17 in tissue remodeling is supported by increases in production of a range of matrix metalloproteinases such as MMP-1, MMP-3, MMP-9, MMP-13, and TIMP-1 in fibroblasts. This discovery may prove significant in diseases such as myocardial infarction, rheumatoid arthritis, and inflammatory bowel disease [91-95].

6.3. Endothelial Cells

IL-17F stimulates endothelial cells to produce TGF-beta, IL-2, MCP-1, and lymphotoxin-beta [46]. Interestingly, this cytokine has a conflicting effect on angiogenesis. IL-17 has a negative effect in human endothelial cells while it promotes angiogenesis and tumor growth *in vitro* with mouse fibroblasts, human umbilical venous cord endothelial cell (HUVECs), tumor cell lines, and human dermal microvascular endothelial cell (HMVECs). In order to promote angiogenesis, IL-17 up-regulates pro-angiogentic factors such as vascular endothelial growth factor (VEGF), PGE1, PGE2, keratinocyte-derived cytokines, and MIP-2 [96] found in fibroblasts and tumor cells. In human endothelial cell culture, IL-17 is found to enhance the effects of various growth factors, such as basic fibroblast growth factor (bFGF), human growth factor (HGF), and VEGF, to induce proliferation of vascular endothelial cells [97]. Additionally, the levels of IL-6, IL-10, IL-12, and PGE2 are induced [98]. IL-17 maintains the ability to induce IL-6 production through a STAT3 pathway in endothelial cells, $CD11c^+$ cells, $CD11b^+$ cells, melanoma cells, MEF, splenic, and tumor cells [99].

IL-25 (also known as IL-17E) is produced by brain capillary endothelial cells, and protects against pro-inflammatory cytokine-induced excessive blood-brain barrier collapse through a protein kinase C epsilon-dependent pathway [100]. Reversibly, endothelial cells also modulate Th17 differentiation. Playing crucial roles in immunity and homeostasis, the C-type lectin-like receptors (CLECs) recognize pathogens as well as self-antigens. Expressed at low levels by endothelial cells, CLEC-1 enhances the differentiation of Th17 cells while it suppresses the differentiation of Tregs [101]. The expression level of CLEC-1 can be positively modulated by Treg cells and immunosuppressive cytokines IL-10 or TGF-beta, and down-regulated by inflammatory stimuli. In addition, VEGF induced by lipopolysaccharides (LPS) plays a key role in the activation of naïve T cells and subsequent polarization to Th1 and Th17 cells in the airway [102].

6.4. Vascular Smooth Muscle Cells (VSMCs)

In cultured cells, IL-17 preferentially increases synthesis of pro-inflammatory mediators such as IL-6 and chemokines CXCL8 (IL-8) and CCL20 in VSMCs rather than endothelial cells [11]. IL-17 promotes C-reactive protein expression in hepatocytes and VSMCs via p38 Mitogen Activated Protein Kinase (MAPK), extracellular signal–regulated kinases 1 and 2 (ERK1/2)-dependent nuclear factor-kappaB (NF-kappaB) and CCAAT enhancer binding protein-beta (C/EBPbeta) activation [103]. In addition, IL-17 induces the migration of VSMCs in a MMP-9-dependent fashion and induces MMP-9 expression via similar pathways [104].

6.5. Monocytic Cells and Macrophages

While the effects of IL-17 and IL-17F have been heavily scrutinized, the characterization of the other cytokines in the IL-17 family and their effects have been limited. Despite this, IL-17B and IL-17C have been discovered to stimulate the release of TNF-alpha and IL-1beta from THP-1 cells (human acute monocytic leukemia cell line), a monocytic cell line, in a time and dose-dependent manner. Under the same conditions IL-17 was found to be ineffective. This difference in biological responses among IL-17 family members implies that they operate through diverse receptor-linked signal pathways. It has been demonstrated that IL-17B and IL-17C do not bind to the IL-17 receptor extracellular domain [48]. In addition, IL-17 stimulates the production and expression of pro-inflammatory cytokines IL-1beta and TNF-alpha by human macrophages, which is completely suppressed by IL-4 and IL-10 but is partially inhibited by IL-13 and TGF-beta [98].

6.6. Structural Lung Cells

IL-17 is a weak stimulus of IL-8 and IL-6 production, but has the ability to markedly enhance IL-8 and IL-6 responses to other stimuli. This point is seen in structural lung cell hyper-responsiveness to TNF-alpha [105]. The modular effect of IL-17 is contributed to a reduction in IL-8 and IL-6 mRNA degradation and not its associated effect on IL-8 and IL-6 gene transcription.

7. Signaling Pathways Initiated by Binding of IL-17 Cytokines and IL-17 Receptors

7.1. IL-17 Receptor Family

The IL-17 cytokine receptor family consists of five members: IL-17RA, IL-17RB, IL-17RC, IL-17RD, and IL-17RE (Figure 2) [47]. The original IL-17 receptor, IL-17RA (IL-17R), was first characterized as a type I transmembrane protein containing 866 amino acids. Structurally, the receptor is comprised of three domains including a 293 amino acid extracellular domain, a transmembrane domain of 21 amino acids, and a cytoplasmic tail of 525 amino acids [106]. Expressed in extensive tissues and cell types, the mRNA of IL-17RA is generated in fibroblast cells, epithelial cells, T lymphocytes, and bone marrow stromal cells as well as kidneys, liver, lungs, and spleen tissue [106, 107]. Furthermore, in humans, the IL-17RA protein is detected in peripheral blood T lymphocytes and in vascular endothelial cells [108].

The foremost studied receptors of the IL-17 receptor family are IL-17RA and IL-17RC binding IL-17A and IL-17F respectively [109]. Although IL-17RA binds preferentially to IL-17A, it also has affinity for IL-17F [106, 110]. IL-17RC also has been shown to have affinity for IL-17A in humans, whereas in mouse, IL-17RC is specific only for IL-17F [109]. IL-17RA is shown to form pre-assembled complexes in the plasma membrane utilizing fluorescence resonance energy transfer (FRET) microscopy [111]. Even though each cytokine preferentially binds to one receptor over another, the biological activities of IL-17, IL-17F, and IL-17A/F require both IL-17RA and IL-17RC. Moreover, the IL-17 receptors may exist as heterodimers or trimers forming an IL-17RA/RC functional receptor complex or other combinations of complexes although the exact stoichiometry is still unknown [112-114].

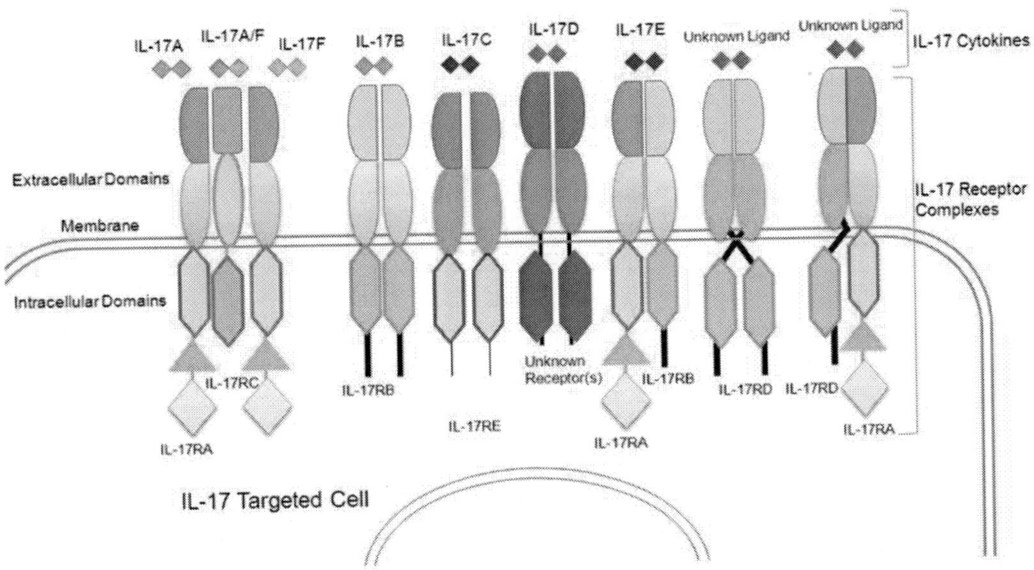

Figure 2. The Relationships between IL-17 Cytokines and the Receptors. The IL-17 cytokine family has six members, IL-17A, IL-17B, IL-17C, IL-17D, IL-17E and IL-17F. The IL-17 receptor family is made of five distinct receptors, IL-17RA, IL-17RB, IL-17RC, IL-17RD and IL-17RE. The IL-17 receptor complexes through which IL-17 cytokines induce signaling are shown. The exact stoichiometry of each IL-17 receptor complex has not been completely elucidated. Ligands for the IL-17RD or IL-17RA-IL-17RD complex are yet to be identified. The receptor through which IL-17D induces signaling needs to be verified

7.2. IL-17 Receptor Complex Components

Structurally, two fibronectin III-like domains are part of the extracellular region of IL-17RA (Figure 3). A common feature of type 1 cytokine receptor, the FN domain mediates protein-protein interactions and ligand binding [115, 116]. IL-17RA and IL-17RC are members of the "similar expression to fibroblast growth factor genes, IL-17 receptors and Toll-like receptors-IL-1R" (SEFIR) protein family defined by their cytoplasmic SEFIR domain. The structure of this protein family is homologous with the TLRs/IL-R (TIR) domain indicating that IL-17R may share Toll/IL-1 receptor signaling pathways. Despite activation of both the Toll/IL-1 receptor and IL-17 receptor pathways resulting in NF-kappaB initiation and induction of inflammatory cytokines [117], the pathways utilize separate mediators upstream of each receptor. Also, certain structural disparities exist among the SEFIR domains of Toll/IL-1 and IL-17 receptors. The structural elements found in the prototypical TIR, like the TIR box3 subdomain and the BB-loop domain, are not found within the SEFIR domain of IL-17R [115]. However, at the carboxyl-terminal side of the SEFIR domain, IL-17RA does contain a TIR-like loop (TILL) which is a region structurally similar to the BB-loops. Unique to IL-17RA, the functions and interactions of the TILL domain in IL-17R signaling are still yet to be understood [118, 119].

Another distinctive molecule in IL-17R signaling is ACT1 (NF-kappaB activator 1, also known as TRAF3 interacting protein 2). With a SEFIR domain located at its C terminus, ACT1 is also member of the SEFIR protein family like IL-17R [120, 121]. Even though ACT1 has been shown to be a negative regulator of B cell survival via its interaction with TRAF (TNF Receptor Associated Factor) 3, CD40, and B-cell activating factor (BAFF) signaling [122], it is an indispensable, positive component for IL-17-mediated biological activities with the IL-17 signaling pathway. Since IL-17 receptors and ACT1 both contain SEFIR domains, it is believed that the interaction of IL-17 receptor with ACT1 is mediated through this homologous domain. TRAF6 also plays a principle part in IL-17 signaling through interactions with ACT1 and its TRAF6 binding motif. This domain permits ACT1 binding to TRAF6, TRAF3, and TGF-beta-activated Kinase 1 (TAK1) thereby making TRAF family proteins critical in NF-kappaB activation [121].

IL-17A/F has a synergistic effect with TNF-alpha to increase CXCL1 expression in macrophages from mice. It has been established that the receptor for IL-17A/F uses TRAF6 for CXCL1 signaling. This was verified using a TRAF6 knock-out model showing diminished CXCL1 expression [51]. In addition to TRAF6, both TRAF3, and TAK1 are also involved in IL-17 signaling. This is possible since the TRAF6 binding motif of ACT1 can bind to TRAF3 as well as TAK1 [121]. It is probable that TAK1 may play a role in IL-17-induced NF-kappaB and c-Jun N-terminal kinase (JNK) activation since it is known to be involved in parallel activation of NF-kappaB and JNK by Toll/IL-1R [123]. Additionally, a domain called C/EBPbeta-activation domain (CBAD) has been identified in the C-terminus of IL-17RA. So far, this domain is only seen in IL-17RA, which is required for IL-17A-mediated glycogen synthase kinase 3beta (GSK3beta) phosphorylation of C/EBPbeta [118, 119].

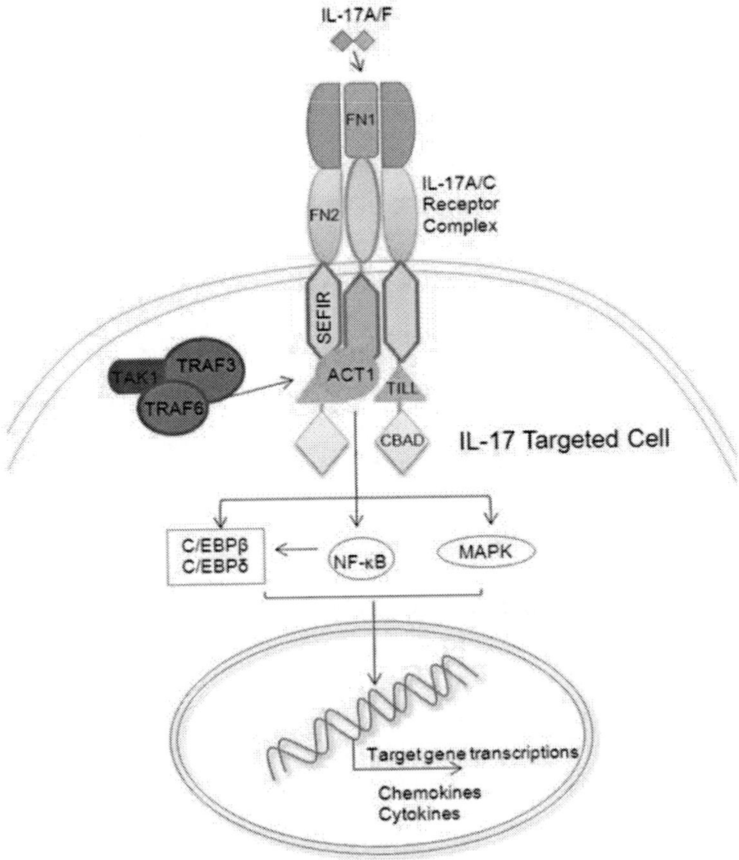

Figure 3. Major Structural characters of IL-17 Receptor and IL-17 Signaling Pathways. Extracellular domain of the IL-17 receptor has two fibronectin III-like domains (FN), which mediate ligand binding. Intracellular domain of IL-17 receptor contains a SEFIR domain, which defines IL-17 receptors as members of the "similar expression to fibroblast growth factor genes, IL-17 receptors and Toll-like receptors-IL-1R" (SEFIR) family. A TIR-like loop (TILL) and a C/EBPβ-activation domain (CBAD) are found in IL-17RA intracellular domain. ACT1 is a critical adaptor protein in IL-17 signaling pathways. ACT1 associates with TRAF6, TRAF3 and TAK1 in mediating signals downstream of IL-17 receptor. IL-17 receptor pathways activates NF-κB, MAPK, C/EBPβ and C/EBPδ, which lead to target gene transcriptions of pro-inflammatory chemokines and cytokines

7.3. IL-17 Receptor Signaling

The detailed mechanism of the IL-17 signaling pathway remains unresolved even though it has been established that IL-17 cytokines maintain the ability to induce a variety of cytokines and chemokines in several cell types via their receptor interactions (Figure 3). Published data reveal that IL-17 ultimately activates NF-kappaB and MAPK pathways to induce pro-inflammatory cytokines [93, 103, 124, 125]. Gel shift analyses show IL-17 activation of NF-kappaB via p50 and p65, which are the key components of the canonical NF-kappaB pathway [126]. The involvement of IL-17 in the non-canonical NF-kappaB has not been defined, but NF-kappaB-inducing kinase (NIK), a component of the non-canonical

NF-kappaB pathway, was demonstrated to be induced in response to IL-17A [124]. IL-17 has also been found to up-regulate the expression of CCAAT/enhancer binding proteins, C/EBP-delta and C/EBP-beta, which are important regulators in IL-6 transcription [126]. Furthermore, activation of the MAPK pathway by IL-17 signaling is important in controlling the stability of mRNA transcripts of pro-inflammatory cytokines. This is achievable via interactions with the AREs (AU-rich elements) of the 3' untranslated region of the transcripts [127]. The IL-17 activated MAPK pathway has additionally been implicated in triggering activator protein 1 (AP1) [128], but the exact mechanism by which IL-17 regulates mRNA stability needs to be clarified.

8. Th17/IL-17-Mediated Inflammations

8.1. Experimental Autoimmune Encephalomyelitis (EAE)/Multiple Sclerosis

Discovery of Tregs and Th17 subsets arose from closer examination of multiple sclerosis/EAE pathological conditions which cannot be rationalized by the Th1-Th2 paradigm. The EAE models exemplify the concept that Th17 cells are accountable for driving autoimmune inflammation. This is true since the disease is induced by passive transfer of IL-17-producing memory activated T cells [58]. Originally, Th1 cytokines, IL-12 and IFN-gamma were thought to induce EAE. However, it was demonstrated that these factors instead suppresses IL-17 production leading to protection from EAE development. This was verified in IFN-gamma-deficient or IL-12 mouse models where IL-17 activity occurs unchecked leading to exacerbated EAE disease conditions [58, 129, 130].

8.2. Inflammatory Skin Disease in Mice and Psoriasis in Humans

Several findings indicate that Th17 cells, but not Th1 cells, play an integral role in the pathogenesis of inflammatory skin disease. First, higher frequencies of Th17 cells, IL-17A^{+}, IL-17A/TNF-alpha^{+}, and IL-17A/IFN-gamma^{+} are found in psoriatic lesions while Th1 cell activity remains normal. Second, the effectiveness of anti-TNF-alpha therapy contributed to the suppression of Th17 cell expression. The final contributing factor is that IL-15 triggers IL-17 production from human T cell blasts which is significant given that IL-15 gene variants (single nucleotide polymorphisms, SNPs) are associated with psoriasis [1].

8.3. Inflammatory Bowel Disease (IBD) [1]

Th17 cells are more potent in transferring IBD than Th1 cells [131], while dextran sulfate sodium (DSS)-induced colitis is shown to be aggravated by IL-17. This finding is supported by the relative protection afforded to IL-17F knock-out. However, data gathered from IL-17A knock-out mice showed more severe disease development indicating that IL-17A is protective in DSS-induced colitis [132]. Furthermore, anti-IL-17A monoclonal antibody treatment, presumably neutralizing IL-17A, aggravates DSS-induced colitis [10]. These findings contribute to the complexity of the function of Th17 individual family members warranting future work to resolve this enigma.

8.4. Experimental Arthritis/Rheumatoid Arthritis [1]

Results suggest that IL-17/Th17 cells play a significant role in promoting the pathogenesis of arthritis. This claim is reinforced by the fact that IL-17 knock-out mice have reduced collagen-induced arthritis [133]. Mechanistically it is known that Th17 cells act on osteoblastic cells through the secretion of IL-17A. This association induces the expression of the receptor activator for nuclear factor-kappaB ligand (RANKL). The interaction between RANKL on osteoclast precursors results in the differentiation of osteoclasts and bone re-absorption [134]. In a clinical setting, patients with arthritis express higher concentrations of IL-17 in their joints than controls. Furthermore, explants from rheumatoid arthritis patients leads to up-regulation of MMP-1, -2, -3, -13 by IL-17 in primary synovial fibroblasts, whole synovial tissue, chondrocyte, and human cartilage cultures. Mild cartilage proteoglycan degradation is also witnessed upon IL-17 stimulation while this effect is exacerbated with TNF-alpha or Oncostatin M co-stimulation [135].

8.5. Atherosclerosis

Th17 cells are implicated in modulating autoimmune responses in atherosclerosis [136-138]. Generally, in disease states such as metabolic syndrome, hypertension, obesity, diabetes, and other cardiovascular disorders, perivascular tissue becomes dysfunctional. This result may be attributed to the simultaneous increase of detrimental adipocytokines such as leptin, resistin, IL-6, TNF-alpha, or IL-17 while the secretion of protective factors diminishes [139]. Experimentally, atherosclerosis can be induced in low density lipoprotein receptor deficient (LDLR-/-) mice fed a Western-type diet. In this model a significant 46% reduction in lesion size is seen in the aortic root when the subjects receive an IL-17R-/- bone marrow transplantation (BMT). Further examination of the atherosclerotic plaque composition revealed no significant changes in collagen content and neutrophil counts, but a reduction in mast cell numbers and an increase in macrophage numbers. Additionally, investigators observed a decrease in anti-oxLDL antibodies of the IgG class, a reduced IL-6 production, and an increased IL-10 production in IL-17R-/- BMT recipients. This finding indicates that signaling via the IL-17 receptor in bone marrow derived cells stimulates the atherosclerosis process [140].

ApoE-/- mice, another atherosclerosis mouse model [35], reveal similar results to that found in LDLR-/- mice. Again, there was a significant increase in the secretion of Th17 related cytokines, IL-17 and IL-6, and expression of RORgamma t. Also, suppression of Treg cell numbers, reduction of Treg related cytokine TGF-beta 1 secretion, and expression of Foxp3 was demonstrated when compared with age-matched wild-type C57BL/6J mice. Meanwhile, the expressions of Treg-related mediators are much lower in ApoE-/- mice than that in their age-matched wild-type littermates suggesting a potential role of Th17/Treg imbalance in the formation and progression of atherosclerosis [138]. Li and colleagues showed support for this hypothesis in IL-1 receptor associated kinase 1 deficient (IRAK-1-/-)

mice that have constitutively higher populations of Treg cells. However, attenuated STAT3 Ser727 phosphorylation and reduced expression of IL-17 and RORgamma t was seen in IRAK-1-/- CD4 Th cells when stimulated with T cell antigen receptor (TCR) agonists, IL-6, and TGF-beta when evaluated against wild-type cells. These findings reveal that IRAK-1 deletion results in decreased IL-17 expression and dampened inflammatory responses in acute and chronic inflammatory mice including atherosclerosis models [141].

Controversially, Mallat and colleagues published results implying novel SOCS3-controlled IL-17 regulatory pathways in atherosclerosis [137]. They discovered that the loss of suppressor of cytokine signaling 3 (SOCS3) in T cells induces both IL-17 and IL-10 generation, produces an anti-inflammatory macrophage phenotype, and leads to unexpected IL-17-dependent reduction in lesion development and vascular inflammation. Also, a hindrance of atherosclerotic lesion development, a reduction in endothelial vascular cell adhesion molecule-1 expression, and a decrease in vascular T cell infiltration is shown when IL-17 is administered *in vivo*. These results may provide valuable implications for understanding the increased susceptibility to vascular inflammation in patients with dominant-negative STAT3 mutations and defective Th17 cell differentiation [137]. However, the enigma of how pro-inflammatory cytokine IL-17 plays a suppressive role in inflammatory autoimmune atherosclerosis in SOCS3-/- mice remains unsolved. Interestingly, adult SOCS3-null mice on a leukemia-inhibitory factor (LIF)-null background succumb to a spontaneous fatal inflammatory disease characterized by neutrophilia and inflammatory-cell tissue infiltrates, suggesting that genetic reduction of embryonic LIF production rescues placentation in SOCS3-/- embryos but does not prevent inflammatory disease [142, 143]. Obviously, future work is needed to solve this controversy.

8.6. Diabetes

IL-17/Th17 cells play a key role in promoting diabetes. The dendritic cell (DC)-dependent loss in Tregs causes an increase in the number of T cells secreting inflammatory cytokines, such as IFN-gamma (Th1 cells) and IL-17 (Th17 cells). A loss of DCs leads to a loss of Tregs, and a decrease in Foxp3 expression on remaining Tregs, while DC expansion cause an increase in Treg expression which is sufficient to prevent type 1 autoimmune diabetes [144]. Even though treatment with either neutralizing anti-IL-17 or recombinant IL-25 (IL-17E) has any effect on diabetes progression in young (<5 weeks) non-obese diabetic (NOD) mice, a model of spontaneous autoimmune diabetes, intervention averts diabetes when treatment is started at 10 weeks of age ($P < 0.001$). Insulitis scoring and immunofluorescence staining show that both anti-IL-17 and IL-25 considerably reduce peri-islet T cell infiltration and decrease GAD65 (Glutamate decarboxylase) autoantibody levels. In addition, Berberine, an alkaloid derivative from *Berberis vulgaris L.*, inhibits Th17 differentiation by activating ERK1/2 and inhibits Th1 differentiation by inhibiting MAPK and JNK activation. Berberine suppresses the activity of STAT1 and STAT4 through down-regulation of p38 MAPK and JNK activation and inhibits the progression of type 1 diabetes in NOD mice by the ubiquitin-proteasome pathway [145]. Similarly, innocuous IFN-gamma induced by adjuvant-free antigen suppresses diabetes and restores normoglycemia in NOD mice through inhibition of IL-17 secretion [146]. In addition to the results gathered in mouse models of diabetes, the production of pro-inflammatory cytokines IL-17, TNF-alpha, and peripheral blood lymphocytes secrete IL-6 in patients with type 1 diabetes mellitus (T1D) is reduced when the autoantigen-specific T cell clones are curbed [147]. Another completed experiment suggests that there is a significant, although modest, increase in the frequency of IL-17-producing cells in lymphocytes from long-term patients with T1D compared with healthy controls [148]. These studies indicate that IL-25 treatment results in a T-cell-mediated dominant protective effect against autoimmunity and that Th17 cells participate in the pathogenesis of autoimmune diabetes [13].

9. Protective Roles of IL-17 in Infections

Th17 cells are associated with various experimental and human autoimmune diseases as a result of their aptitude to act as powerful tissue inflammation initiators. Functionally, the ability of Th17 cells to elicit tissue inflammation plays a crucial part in host defense. This is especially true in the removal of invaded pathogens, which are not sufficiently handled by Th1 or Th2 cells. This principle has been supported in experimentation involving host defense against *Klebsiella pneumoniae* in IL-23p19 knock-out (gene deficient, KO) mice and IL-17R KO models. IL-17KO mice are highly susceptible to *K. pneumoniae* infection due to the recruitment of neutrophils to the *K. pneumoniae* infection site by IL-17 [149, 150]. Additionally IL-17 and IL-17F anti-infection neutrophil infiltration was demonstrated in acute respiratory tract infection mouse models where neutrophils migrate into the lung in the clearance of mycoplasma pneumonia [151]. In another knock-out model, IL-17RA KO mice were revealed to be more prone to *Candida albicans* infection in comparison to wild type mice while protection from lethal doses of this pathogen was seen when wild type mice were given IL-17 injection [152]. IL-17 or IL-17, in conjunction with IL-22, also up-regulates the expression of antimicrobial molecules such as defensins, antibiotics found in the skin, lungs, and gastrointestinal tract [52, 153] in keratinocytes. IL-17 also enhances the production of mucin in the airways, which traps pathogens in the lung mucosa [82].

10. Interplay between Th17 Cells and Tregs

10.1. Recent Progress in Characterizing Tregs

Over the past 5 years, publications within the field have illustrated that a lack of Foxp3 expression directly results in the absence of Tregs. More significant is the established link between this deficiency and the development of both human and mouse autoimmune diseases. These findings have led to rapidly expanded knowledge in Tregs development. In the previous sections, 3.2, 5, 8.5, 8.6, we have discussed some recent progress in characterizing Tregs. In this section, we will focus on additional advances in the field.

(1) Human blood, isolated from an out bred population in a pathogenic environment, contains up to 30% $CD4^{+}CD25^{+}$ cells. However, it is important to note that out of the 30% only 2–4% of the cells with the highest CD25 expression could be considered regulatory. The best-accepted alternative in determining Tregs other than

CD25 expression is the lack of cell surface CD127 (IL-7 receptor). Furthermore, HLA-DR expression identifies a functionally distinct population of what appears to be terminally differentiated human Tregs [154].

(2) Tregs have been attributed with the ability to recognize a broad range of antigenic specificities with greater reactivity to self antigens. The ability to recognize a highly diverse repertoire is accredited to a distinct set of T-cell receptors (TCRs) and partial overlapping with the repertoire of conventional T cells [155].

(3) Data has established an alternative, extrathymic pathway for Treg differentiation [156]. This claim is supported by data that shows TGF-beta signaling is not necessary for thymic expression of Foxp3, nor is TGF-beta signaling or responsiveness required for *in vitro* suppressive activity of $Foxp3^+$ Tregs. In addition, TGF-beta treatment is capable of inducing both Foxp3 and RORgamma t expression, which leads exclusively to Treg differentiation.

(4) $Foxp3^+CD4^+CD25^+$ Treg subsets that maintain immunologic homeostasis have been considered to be a homogeneous population of naturally occurring, thymus-derived $CD4^+CD25^+$ T cells (nTregs). However, similar $Foxp3^+$ Tregs can be derived from $CD25^-$ precursors *in vivo*, and *ex vivo* with IL-2 and TGF-beta (adaptive Tregs, inducible Tregs, or iTregs). These two subsets differ in their principal antigen specificities, T-cell receptor signal strength, and co-stimulatory requirements needed for their generation. Although IL-6 can convert nTregs to Th17 cells, iTregs induced by IL-2 and TGF-beta are resistant to IL-6 and thereby may retain suppressive function at inflammatory sites. Thus, nTregs and iTregs may have different roles in the adaptive immune response [157].

(5) As a cell group, Tregs are defined as lymphocytes with particular phenotypic markers that share the ability to suppress, by various mechanisms, inflammatory responses. The group can be further divided into subsets such as IL-10 secreting type I Tregs and type III Tregs that produce TGF-beta, as well as iTregs and nTregs [158].

(6) Evidence exists which indicates APC-Treg interactions are key in preserving immune tolerance [159]. In autoimmunity, chronically activated Antigen Presenting Cells (APCs) under the influence of intracellular signaling pathways, such as phosphatidylinositol-3 kinase (PI3K), JAK-STAT, MAPK, and NF-kappaB pathways, can escape surveillance by Tregs. This leads to the activation of T cells that are obstinate to Treg regulation. Also, Tregs can be rendered defective by APCs and the inflammatory cytokines secreted by APCs such as TNF-alpha, IL-6, IL-1beta, and IL-23. Moreover, this effect can perpetuate itself by reciprocally enhancing the activity of the IL-17-producing pathogenic Th17 T cell subset.

(7) It has also been discovered that treatment with nTregs resolves colitis (a $CD45RB^{high}CD4^+$ T cell transfer model of colitis), though only when iTregs are also present. Even though iTregs require Foxp3 for suppressive activity and phenotypic stability like nTregs, their gene expression profile is distinct from the established nTregs' "genetic signature". This implies that iTregs and nTregs have developmental and possibly mechanistic distinction. The functional role for iTregs *in vivo* and the ability of iTregs and nTregs to act in concert to maintain tolerance is revealed by the above results [160, 161].

(8) Epigenetics is defined by regulation of gene expression without altering nucleotide sequence in the genome. Several epigenetic markers, such as histone acetylation and methylation, and cytosine residue methylation in CpG dinucleotides, have been reported at the Foxp3 locus. In particular, CpG dinucleotides at the Foxp3 locus are methylated in naïve $CD4^{+}CD25^{-}$ T cells, activated $CD4^{+}$ T cells, and TGF-beta-induced iTregs, whereas they are completely demethylated in nTregs. The DNA methyltransferases DNMT1 and DNMT3b are associated with the Foxp3 locus in $CD4^{+}$ T cells. Methylation of CpG residues represses Foxp3 expression, whereas complete demethylation is required for stable Foxp3 expression [162].

(9) The extracellular matrix composition provides contextual cues to leukocyte functionality in inflamed and healing tissues. An illustration of this is the function of low and high molecular weight hyaluronate (HA). Generated during active inflammation, low molecular weight hyaluronate (LMW-HA) is a TLR ligand and an endogenous "danger signal", whereas high molecular weight-HA (HMW-HA) acts inversely in healing or intact tissues to actively promote immune tolerance by augmenting Treg function. This is seen in a human iT(Reg) model where HMW-HA provides a co-stimulatory signal through cross-linking CD44. The co-stimulatory signal promotes Foxp3 expression; however, this effect is not seen with LMW-HA. Thus, HMW-HA contributes to the maintenance of immune homeostasis in uninjured tissue and effectively communicates an "all-clear" signal to down-regulate the adaptive immune system through Tregs after tissue matrix integrity has been restored [163].

(10) Tregs isolated from patients with an array of autoimmune diseases exhibit diminished regulatory function when evaluated against Tregs isolated from healthy controls using *in vitro* functional assays and phenotypic analysis [164]. Moreover, several investigators have shown that loss of Treg function results in autoimmune responses [165, 166]. Yang and colleagues have characterized a Treg-specific, IL-2-dependent apoptotic pathway and demonstrated that Treg apoptotic/survival pathways are therapeutic targets for Treg-based immunotherapy [19].

(11) Bluestone and colleagues recently reported that substantial percentages of cells have transient or unstable expression of the transcription factor Foxp3. These 'exFoxp3' T cells have an activated-memory T cell phenotype and secrete pro-inflammatory cytokines. Furthermore, exFoxp3 cell levels are elevated in inflamed tissues in autoimmune conditions and it has been shown that the adoptive transfer of auto reactive exFoxp3 cells induces rapid onset of diabetes. Finally, analysis of the T cell receptor repertoire suggests that exFoxp3 cells can be developed from both natural and adaptive Tregs. Therefore, it can be concluded that Foxp3 instability could lead to the generation of potentially auto reactive effector T cells resulting in major implications necessary for comprehending autoimmune disease pathogenesis [167].

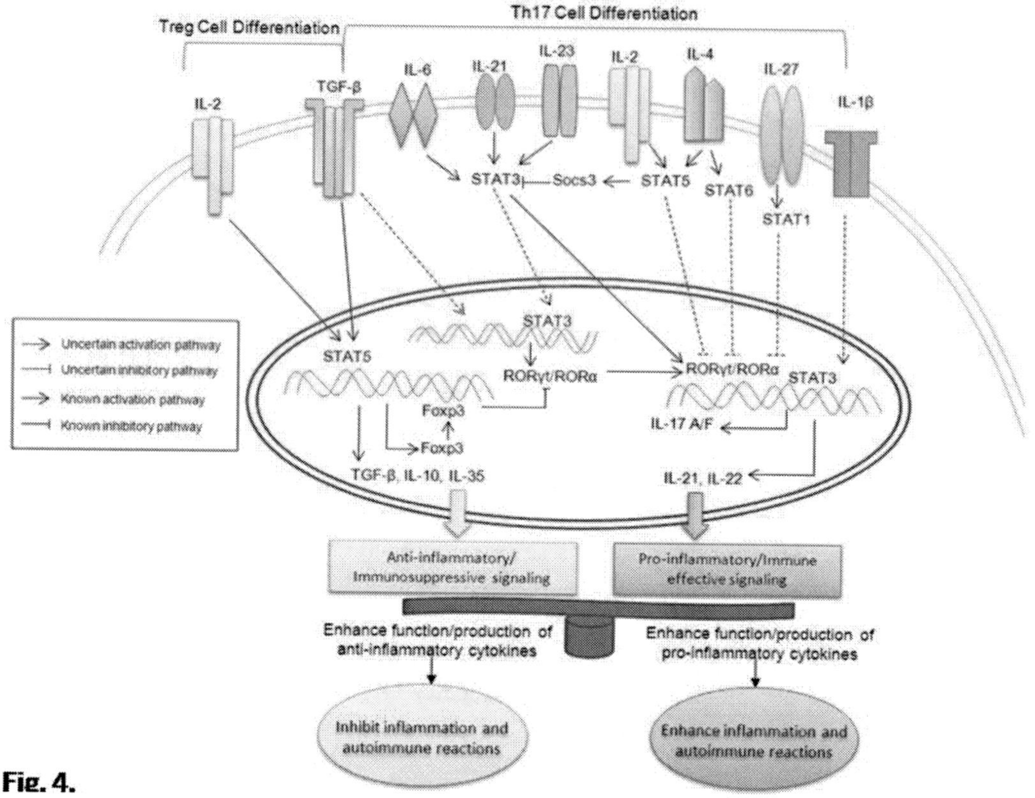

Fig. 4.

Figure 4. Tregs and Th17 cell Interplay. Differentiation of CD4+ T Cell is important for immune regulation and host defense. Signaling pathways in Tregs and Th17 differentiation show specific cytokines, receptors, and transcription factors involved in each lineage. TGF-β is a common factor in the differentiation of two lineages with opposing effects. Effector cytokines of Tregs enhance anti-inflammatory/immunosuppressive effects and thus inhibit inflammation and autoimmune reactions. Effector cytokines of Th17 are pro-inflammatory and enhance inflammation and autoimmune reactions

10.2. Reciprocally Interconnected Th17 and Tregs

Valuable in the prevention of autoimmune diseases and immune suppression, regulatory T cells are a subset of T helper cells [168]. IFN-gamma and IL-4, produced by Th1 and Th2 respectively, have been demonstrated to inhibit the differentiation of Th17 cells, while Tregs have been linked to the suppression of Th1 and Th2 cell immune responses [169]. *In vitro*, however, Tregs have not been shown to negatively impact IL-17 production or Th17 cell development [36, 170] (see next section for the new results). The developmental programs of Th17 cells and adaptive Tregs are reciprocally interconnected [1, 60, 171] (Figure 4). A naïve T cell can be driven to express Foxp3, upon TCR stimulation, and become a Treg cell in the presence of the immunosuppressive cytokine TGF-beta. Attributed with a broad inhibitory effect on the immune system, only TGF-beta induces the transcription factor Foxp3, which is necessary for Treg cell induction and maintenance. Meanwhile, the transcription factor RORgamma t is critical in Th17 development in naïve T cells. When T helper cells are differentiated into Tregs, RORgamma t expression is progressively diminished while the opposite is seen in Th17 development when IL-6 is present [1, 172, 173]. During an immune response IL-6 is secreted in response to activation by microbial antigens by APCs such as dendritic cells. The IL-6 produced then acts in concert with TGF-beta to inhibit adaptive Treg generation and to induce Th17 cell differentiation [60, 62, 64]. The reciprocal relationship between adaptive Treg and Th17 cells is demonstrated by the balance of TGF-beta and IL-6 in adaptive Treg and Th17 differentiation. In the absence of inflammatory stimuli, TGF-beta-induced Tregs keep the immune system in balance by suppressing autoimmunity. In the presence of inflammatory assault, adaptive Treg generation is suppressed and Th17-induced pro-inflammatory responses are activated. This point is illustrated in IL-6 KO (knock out) mice where generation of Th17 is retarded and adaptive Treg generation increases [60, 61]. It can be deduced that IL-6 and IL-21 act as switch factors controlling the Foxp3/RORgamma t balance [173]. Both Th17-specific transcription factors RORgamma t and RORalpha physically interact with Treg-specific transcription factor Foxp3 to antagonize each other's functions [173, 174]. This fact is supported by results generated from IL-6 deficient mice which show a severe defect in the generation of Th17 cells and increased numbers of Tregs in peripheral T cell repertoire [57, 60]. Additionally, the growth factor IL-2, which promotes Treg generation and survival, inhibits Th17 generation. This phenomenon occurs in the periphery where IL-2 receptor signaling promotes the generation of Tregs while the production of Th17 cells is inhibited [175, 176]. IL-2 deficient mice confirm these findings [175]. IL-2 along with other common gamma-chain cytokines such as IL-4, IL-7, IL-9, IL-15, and IL-21, has the capability to activate STAT5. This finding supports the above stated IL-2 results since the absence of STAT5 abrogates Treg differentiation. Furthermore, both STAT5 deficient and IL-2 deficient mice have elevated serum levels of IL-17. It can be concluded that the action of STAT5 works through binding the Foxp3 gene and inhibiting Th17 differentiation [177].

10.3. Inhibition of Th17 by Tregs

The ability of Th17 cells to resist human Foxp3$^+$ Treg suppression has been implicated by research to date. This idea has been experimentally challenged utilizing CD39 expression, an ectonucleotidase, which hydrolyzes ATP on a subset of human natural Tregs. Despite the fact that both CD4$^+$CD25^{high}CD39$^+$ and CD4$^+$CD25^{high}CD39$^-$ T cells suppress responder T cell IFN-gamma production and proliferation, only the predominantly Foxp3$^+$ CD4$^+$CD25^{high}CD39$^+$ suppress IL-17 production. CD4$^+$CD25^{high}CD39$^-$ T cells, however, are capable of producing IL-17. These results indicate a critical function of CD4$^+$CD25^{high}Foxp3$^+$CD39$^+$ in constraining pathogenic Th17 cells in patients with autoimmune inflammation [178].

Conclusion

The additional characterization of Th9, Tfh, Tregs, and Th17 CD4$^+$ T helper cell subsets has significantly improved on the old Th1-Th2 paradigm. The new multisubset model substantially expands our understanding roles of T helper cell subsets and their role in the processes of innate and adaptive immunity. Despite significant advances in the understanding of Th17 and Tregs T-helper subsets, the interplay between these two groups and their effect in various inflammatory and autoimmune disease models needs to be further elucidated. Future investigation to further characterize Th17 and Treg interactions in inflammation and autoimmune diseases would provide further insight and novel approaches to the advancement of therapeutics for diseases such as atherosclerosis, diabetes, and arthritis.

References

[1] Korn, T; Bettelli, E; Oukka, M; Kuchroo, VK. IL-17 and Th17 Cells. *Annu Rev Immunol.,* 2009, 27, 485-517.

[2] Hue, S; Ahern, P; Buonocore, S; et al. Interleukin-23 drives innate and T cell-mediated intestinal inflammation. *J Exp Med.,* 2006, 203, 2473-2483.

[3] Fujino, S; Andoh, A; Bamba, S; et al. Increased expression of interleukin 17 in inflammatory bowel disease. *Gut*, 2003, 52, 65-70.

[4] Csiszar, A; Ungvari, Z; Koller, A; Edwards, JG; Kaley, G. Aging-induced proinflammatory shift in cytokine expression profile in coronary arteries. *Faseb J.,* 2003, 17, 1183-1185.

[5] de Boer, OJ; van der Meer, JJ; Teeling, P; et al. Differential expression of interleukin-17 family cytokines in intact and complicated human atherosclerotic plaques. *J pathol.,* 2009.

[6] Kotake, S; Kamatani, N. The Role of IL-17 in Joint Destruction. *Drug News Perspect.,* 2002, 15, 17-23.

[7] Paradowska, A; Masliniski, W; Grzybowska-Kowalczyk, A; Lacki, J. The function of interleukin 17 in the pathogenesis of rheumatoid arthritis. *Arch Immunol Ther Exp.,* (Warsz) 2007, 55, 329-334.

[8] Komiyama, Y; Nakae, S; Matsuki, T; et al. IL-17 plays an important role in the development of experimental autoimmune encephalomyelitis. *J Immunol.,* 2006, 177, 566-573.

[9] Zhang, Z; Zheng, M; Bindas, J; Schwarzenberger, P; Kolls, JK. Critical role of IL-17 receptor signaling in acute TNBS-induced colitis. *Inflamm Bowel Dis.,* 2006, 12, 382-388.

[10] Ogawa, A; Andoh, A; Araki, Y; Bamba, T; Fujiyama, Y. Neutralization of interleukin-17 aggravates dextran sulfate sodium-induced colitis in mice. *Clin Immunol.,* 2004, 110, 55-62.

[11] Rao, DA; Eid, RE; Qin, L; et al. Interleukin (IL)-1 promotes allogeneic T cell intimal infiltration and IL-17 production in a model of human artery rejection. *J Exp Med.,* 2008, 205, 3145-3158.

[12] Amadi-Obi, A; Yu, CR; Liu, X; et al. TH17 cells contribute to uveitis and scleritis and are expanded by IL-2 and inhibited by IL-27/STAT1. *Nat Med.,* 2007, 13, 711-718.

[13] Emamaullee, JA; Davis, J; Merani, S; et al. Inhibition of Th17 cells regulates autoimmune diabetes in NOD mice. *Diabetes*, 2009, 58, 1302-1311.

[14] Bending, D; De La Pena, H; Veldhoen, M; et al. Highly purified Th17 cells from BDC2.5NOD mice convert into Th1-like cells in NOD/SCID recipient mice. *J Clin Invest.,* 2009.

[15] Mosmann, TR; Coffman, RL. TH1 and TH2 cells: different patterns of lymphokine secretion lead to different functional properties. *Annu Rev Immunol.,* 1989, 7, 145-173.

[16] Veldhoen, M; Uyttenhove, C; van Snick, J; et al. Transforming growth factor-beta 'reprograms' the differentiation of T helper 2 cells and promotes an interleukin 9-producing subset. *Nat Immunol.,* 2008, 9, 1341-1346.

[17] Soroosh, P; Doherty, TA. Th9 and allergic disease. *Immunology,* 2009, 127, 450-458.

[18] Dardalhon, V; Awasthi, A; Kwon, H; et al. IL-4 inhibits TGF-beta-induced Foxp3+ T cells and; together with TGF-beta; generates IL-9+ IL-10+ Foxp3(-) effector T cells. *Nat Immunol.,* 2008, 9, 1347-1355.

[19] Yang, XF. Factors regulating apoptosis and homeostasis of CD4+CD25highFOXP3+ regulatory T cells are new therapeutic targets. *Front Biosci.,* 2008, 13, 1472-1499.

[20] Ke, XY; Wang, J; Li, L; Chen, IH; Wang, H; Yang, XF. Roles of CD4+CD25high FOXP3+ Tregs in lymphomas and tumors are complex. *Front Biosci.,* 2008, 13, 3986-4001.

[21] King, C. New insights into the differentiation and function of T follicular helper cells. *Nat Rev Immunol.,* 2009, 9, 757-766.

[22] Fazilleau, N; Mark, L; McHeyzer-Williams, LJ; McHeyzer-Williams, MG. Follicular helper T cells: lineage and location. *Immunity.,* 2009, 30, 324-335.

[23] Basso, AS; Cheroutre, H; Mucida, D. More stories on Th17 cells. *Cell Res.,* 2009, 19, 399-411.

[24] Sallusto, F; Lenig, D; Mackay, CR; Lanzavecchia, A. Flexible programs of chemokine receptor expression on human polarized T helper 1 and 2 lymphocytes. *J Exp Med.,* 1998, 187, 875-883.

[25] Bonecchi, R; Bianchi, G; Bordignon, PP; et al. Differential expression of chemokine receptors and chemotactic responsiveness of type 1 T helper cells (Th1s) and Th2s. *J Exp Med.,* 1998, 187, 129-134.

[26] Lohmann, T; Laue, S; Nietzschmann, U; et al. Reduced expression of Th1-associated chemokine receptors on peripheral blood lymphocytes at diagnosis of type 1 diabetes. *Diabetes*, 2002, 51, 2474-2480.

[27] Okazaki, H; Kakurai, M; Hirata, D; et al. Characterization of chemokine receptor expression and cytokine production in circulating CD4+ T cells from patients with atopic dermatitis: up-regulation of C-C chemokine receptor 4 in atopic dermatitis. *Clin Exp Allergy.,* 2002, 32, 1236-1242.

[28] Anthony, RM; Rutitzky, LI; Urban, JF; Jr; Stadecker, MJ; Gause, WC. Protective immune mechanisms in helminth infection. *Nat Rev Immunol.,* 2007, 7, 975-987.

[29] Krakowski, M; Owens, T. Interferon-gamma confers resistance to experimental allergic encephalomyelitis. Eur *J Immunol.,* 1996, 26, 1641-1646.

[30] Tran, EH; Prince, EN; Owens, T. IFN-gamma shapes immune invasion of the central nervous system via regulation of chemokines. *J Immunol.,* 2000, 164, 2759-2768.

[31] Oppmann, B; Lesley, R; Blom, B; et al. Novel p19 protein engages IL-12p40 to form a cytokine; IL-23, with biological activities similar as well as distinct from IL-12. *Immunity.,* 2000, 13, 715-725.

[32] Cua, DJ; Sherlock, J; Chen, Y; et al. Interleukin-23 rather than interleukin-12 is the critical cytokine for autoimmune inflammation of the brain. *Nature*, 2003, 421, 744-748.

[33] Sakaguchi, S; Powrie, F. Emerging challenges in regulatory T cell function and biology. *Science*, 2007, 317, 627-629.

[34] Shevach, EM; DiPaolo, RA; Andersson, J; Zhao, DM; Stephens, GL; Thornton, AM. The lifestyle of naturally occurring CD4+ CD25+ Foxp3+ regulatory T cells. *Immunol Rev*, 2006, 212, 60-73.

[35] Yang, XF; Yin, Y; Wang, H. VASCULAR INFLAMMATION AND ATHEROGENESIS ARE ACTIVATED VIA RECEPTORS FOR PAMPs AND SUPPRESSED BY REGULATORY T CELLS. *Drug Discov Today Ther Strateg*, 2008, 5, 125-142.

[36] Annunziato, F; Cosmi, L; Santarlasci, V; et al. Phenotypic and functional features of human Th17 cells. *J Exp Med.*, 2007, 204, 1849-1861.

[37] Cosmi, L; De Palma, R; Santarlasci, V; et al. Human interleukin 17-producing cells originate from a CD161+CD4+ T cell precursor. *J Exp Med.*, 2008, 205, 1903-1916.

[38] Acosta-Rodriguez, EV; Rivino, L; Geginat, J; et al. Surface phenotype and antigenic specificity of human interleukin 17-producing T helper memory cells. *Nat Immunol.*, 2007, 8, 639-646.

[39] Tsuji, M; Komatsu, N; Kawamoto, S; et al. Preferential generation of follicular B helper T cells from Foxp3+ T cells in gut Peyer's patches. *Science*, 2009, 323, 1488-1492.

[40] Nurieva, RI; Chung, Y; Hwang, D; et al. Generation of T follicular helper cells is mediated by interleukin-21 but independent of T helper 1, 2, or 17 cell lineages. *Immunity.*, 2008, 29, 138-149.

[41] Eddahri, F; Denanglaire, S; Bureau, F; et al. Interleukin-6/STAT3 signaling regulates the ability of naive T cells to acquire B-cell help capacities. *Blood*, 2009, 113, 2426-2433.

[42] Chen, J; Liu, XS. Development and function of IL-10 IFN-gamma-secreting CD4(+) T cells. *J Leukoc Biol.*, 2009, 86, 1305-1310.

[43] Dumitriu, IE; Araguas, ET; Baboonian, C; Kaski, JC. CD4+ CD28 null T cells in coronary artery disease: when helpers become killers. *Cardiovasc Res*, 2009, 81, 11-19.

[44] Rouvier, E; Luciani, MF; Mattei, MG; Denizot, F; Golstein, P. CTLA-8, cloned from an activated T cell; bearing AU-rich messenger RNA instability sequences; and homologous to a herpesvirus saimiri gene. *J Immunol.*, 1993, 150, 5445-5456.

[45] Yao, Z; Painter, SL; Fanslow, WC; et al. Human IL-17, a novel cytokine derived from T cells. *J Immunol.*, 1995, 155, 5483-5486.

[46] Starnes, T; Robertson, MJ; Sledge, G; et al. Cutting edge: IL-17F; a novel cytokine selectively expressed in activated T cells and monocytes; regulates angiogenesis and endothelial cell cytokine production. *J Immunol.*, 2001, 167, 4137-4140.

[47] Hymowitz, SG; Filvaroff, EH; Yin, JP; et al. IL-17s adopt a cystine knot fold: structure and activity of a novel cytokine; IL-17F; and implications for receptor binding. *Embo J*, 2001, 20, 5332-5341.

[48] Allison, J; Thomas, H; Beck, D; et al. Transgenic overexpression of human Bcl-2 in islet beta cells inhibits apoptosis but does not prevent autoimmune destruction. *Int Immunol*, 2000, 12, 9-17.

[49] Lee, J; Ho, WH; Maruoka, M; et al. IL-17E; a novel proinflammatory ligand for the IL-17 receptor homolog IL-17Rh1. *J Biol Chem*, 2001, 276, 1660-1664.

[50] Wright, JF; Guo, Y; Quazi, A; et al. Identification of an interleukin 17F/17A heterodimer in activated human CD4+ T cells. *J Biol Chem*, 2007, 282, 13447-13455.

[51] Chang, SH; Dong, C. A novel heterodimeric cytokine consisting of IL-17 and IL-17F regulates inflammatory responses. *Cell Res.*, 2007, 17, 435-440.

[52] Liang, SC; Long, AJ; Bennett, F; et al. An IL-17F/A heterodimer protein is produced by mouse Th17 cells and induces airway neutrophil recruitment. *J Immunol.*, 2007, 179, 7791-7799.

[53] Ferretti, S; Bonneau, O; Dubois, GR; Jones, CE; Trifilieff, A. IL-17, produced by lymphocytes and neutrophils; is necessary for lipopolysaccharide-induced airway neutrophilia: IL-15 as a possible trigger. *J Immunol.*, 2003, 170, 2106-2112.

[54] Lockhart, E; Green, AM; Flynn, JL. IL-17 production is dominated by gammadelta T cells rather than CD4 T cells during Mycobacterium tuberculosis infection. *J Immunol.*, 2006, 177, 4662-4669.

[55] Harrington, LE; Hatton, RD; Mangan, PR; et al. Interleukin 17-producing CD4+ effector T cells develop via a lineage distinct from the T helper type 1 and 2 lineages. *Nat Immunol.*, 2005, 6, 1123-1132.

[56] Infante-Duarte, C; Horton, HF; Byrne, MC; Kamradt, T. Microbial lipopeptides induce the production of IL-17 in Th cells. *J Immunol.*, 2000, 165, 6107-6115.

[57] Korn, T; Bettelli, E; Gao, W; et al. IL-21 initiates an alternative pathway to induce proinflammatory T(H)17 cells. *Nature*, 2007, 448, 484-487.

[58] Langrish, CL; Chen, Y; Blumenschein, WM; et al. IL-23 drives a pathogenic T cell population that induces autoimmune inflammation. *J Exp Med.*, 2005, 201, 233-240.

[59] Liang, SC; Tan, XY; Luxenberg, DP; et al. Interleukin (IL)-22 and IL-17 are coexpressed by Th17 cells and cooperatively enhance expression of antimicrobial peptides. *J Exp Med.*, 2006, 203, 2271-2279.

[60] Bettelli, E; Carrier, Y; Gao, W; et al. Reciprocal developmental pathways for the generation of pathogenic effector TH17 and regulatory T cells. *Nature*, 2006, 441, 235-238.

[61] Korn, T; Oukka, M; Kuchroo, V; Bettelli, E. Th17 cells: effector T cells with inflammatory properties. *Semin Immunol*, 2007, 19, 362-371.

[62] Mangan, PR; Harrington, LE; O'Quinn, DB; et al. Transforming growth factor-beta induces development of the T(H)17 lineage. *Nature*, 2006, 441, 231-234.

[63] Nurieva, R; Yang, XO; Martinez, G; et al. Essential autocrine regulation by IL-21 in the generation of inflammatory T cells. *Nature*, 2007, 448, 480-483.

[64] Veldhoen, M; Hocking, RJ; Atkins, CJ; Locksley, RM; Stockinger, B. TGFbeta in the context of an inflammatory cytokine milieu supports de novo differentiation of IL-17-producing T cells. *Immunity.*, 2006, 24, 179-189.

[65] Zhou, L; Ivanov; II; Spolski, R; et al. IL-6 programs T(H)-17 cell differentiation by promoting sequential engagement of the IL-21 and IL-23 pathways. *Nat Immunol.*, 2007, 8, 967-974.

[66] Szabo, SJ; Sullivan, BM; Peng, SL; Glimcher, LH. Molecular mechanisms regulating Th1 immune responses. *Annu Rev Immunol.*, 2003, 21, 713-758.

[67] Glimcher, LH; Murphy, KM. Lineage commitment in the immune system: the T helper lymphocyte grows up. *Genes Dev*, 2000, 14, 1693-1711.

[68] Park, H; Li, Z; Yang, XO; et al. A distinct lineage of CD4 T cells regulates tissue inflammation by producing interleukin 17. *Nat Immunol.,* 2005, 6, 1133-1141.

[69] Chen, Y; Langrish, CL; McKenzie, B; et al. Anti-IL-23 therapy inhibits multiple inflammatory pathways and ameliorates autoimmune encephalomyelitis. *J Clin Invest.,* 2006, 116, 1317-1326.

[70] McGeachy, MJ; Bak-Jensen, KS; Chen, Y; et al. TGF-beta and IL-6 drive the production of IL-17 and IL-10 by T cells and restrain T(H)-17 cell-mediated pathology. *Nat Immunol.,* 2007, 8, 1390-1397.

[71] Brustle, A; Heink, S; Huber, M; et al. The development of inflammatory T(H)-17 cells requires interferon-regulatory factor 4. *Nat Immunol.,* 2007, 8, 958-966.

[72] Schraml, BU; Hildner, K; Ise, W; et al. The AP-1 transcription factor Batf controls T(H)17 differentiation. *Nature*, 2009, 460, 405-409.

[73] Harris, TJ; Grosso, JF; Yen, HR; et al. Cutting edge: An in vivo requirement for STAT3 signaling in TH17 development and TH17-dependent autoimmunity. *J Immunol.,* 2007, 179, 4313-4317.

[74] Ivanov; II; McKenzie, BS; Zhou, L; et al. The orphan nuclear receptor RORgammat directs the differentiation program of proinflammatory IL-17+ T helper cells. *Cell*, 2006, 126, 1121-1133.

[75] Wei, L; Laurence, A; Elias, KM; O'Shea, JJ. IL-21 is produced by Th17 cells and drives IL-17 production in a STAT3-dependent manner. *J Biol Chem*, 2007, 282, 34605-34610.

[76] Li, MO; Wan, YY; Flavell, RA. T cell-produced transforming growth factor-beta1 controls T cell tolerance and regulates Th1- and Th17-cell differentiation. *Immunity.,* 2007, 26, 579-591.

[77] Volpe, E; Servant, N; Zollinger, R; et al. A critical function for transforming growth factor-beta; interleukin 23 and proinflammatory cytokines in driving and modulating human T(H)-17 responses. *Nat Immunol.,* 2008, 9, 650-657.

[78] Yang, XO; Nurieva, R; Martinez, GJ; et al. Molecular antagonism and plasticity of regulatory and inflammatory T cell programs. *Immunity.,* 2008, 29, 44-56.

[79] Acosta-Rodriguez, EV; Napolitani, G; Lanzavecchia, A; Sallusto, F. Interleukins 1beta and 6 but not transforming growth factor-beta are essential for the differentiation of interleukin 17-producing human T helper cells. *Nat Immunol.,* 2007, 8, 942-949.

[80] Chen, Z; Laurence, A; O'Shea, JJ. Signal transduction pathways and transcriptional regulation in the control of Th17 differentiation. *Semin Immunol*, 2007, 19, 400-408.

[81] Wilson, NJ; Boniface, K; Chan, JR; et al. Development; cytokine profile and function of human interleukin 17-producing helper T cells. *Nat Immunol.,* 2007, 8, 950-957.

[82] Chen, Y; Thai, P; Zhao, YH; Ho, YS; DeSouza, MM; Wu, R. Stimulation of airway mucin gene expression by interleukin (IL)-17 through IL-6 paracrine/autocrine loop. *J Biol Chem*, 2003, 278, 17036-17043.

[83] McGeachy, MJ; Chen, Y; Tato, CM; et al. The interleukin 23 receptor is essential for the terminal differentiation of interleukin 17-producing effector T helper cells in vivo. *Nat Immunol.,* 2009, 10, 314-324.

[84] Manel, N; Unutmaz, D; Littman, DR. The differentiation of human T(H)-17 cells requires transforming growth factor-beta and induction of the nuclear receptor RORgammat. *Nat Immunol.,* 2008, 9, 641-649.

[85] Sundrud, MS; Koralov, SB; Feuerer, M; et al. Halofuginone inhibits TH17 cell differentiation by activating the amino acid starvation response. *Science*, 2009, 324, 1334-1338.

[86] Lee, WW; Kang, SW; Choi, J; et al. Regulating human Th17 cells via differential expression of IL-1 receptor. *Blood*, *115*, 530-540.

[87] Elyaman, W; Bradshaw, EM; Uyttenhove, C; et al. IL-9 induces differentiation of TH17 cells and enhances function of FoxP3+ natural regulatory T cells. *Proc Natl Acad Sci*, U S A 2009, 106, 12885-12890.

[88] Demoulin, JB; Van Roost, E; Stevens, M; Groner, B; Renauld, JC. Distinct roles for STAT1, STAT3, and STAT5 in differentiation gene induction and apoptosis inhibition by interleukin-9. *J Biol Chem*, 1999, 274, 25855-25861.

[89] Fossiez, F; Djossou, O; Chomarat, P; et al. T cell interleukin-17 induces stromal cells to produce proinflammatory and hematopoietic cytokines. *J Exp Med.,* 1996, 183, 2593-2603.

[90] Qiu, Z; Dillen, C; Hu, J; et al. Interleukin-17 regulates chemokine and gelatinase B expression in fibroblasts to recruit both neutrophils and monocytes. *Immunobiology*, 2009, 214, 835-842.

[91] Bamba, S; Andoh, A; Yasui, H; Araki, Y; Bamba, T; Fujiyama, Y. Matrix metalloproteinase-3 secretion from human colonic subepithelial myofibroblasts: role of interleukin-17. *J Gastroenterol*, 2003, 38, 548-554.

[92] Chabaud, M; Garnero, P; Dayer, JM; Guerne, PA; Fossiez, F; Miossec, P. Contribution of interleukin 17 to synovium matrix destruction in rheumatoid arthritis. *Cytokine*, 2000, 12, 1092-1099.

[93] Cortez, DM; Feldman, MD; Mummidi, S; et al. IL-17 stimulates MMP-1 expression in primary human cardiac fibroblasts via p38 MAPK- and ERK1/2-dependent C/EBP-beta ; NF-kappaB; and AP-1 activation. *Am J Physiol Heart Circ Physiol*, 2007, 293, H3356-3365.

[94] Jovanovic, DV; Martel-Pelletier, J; Di Battista, JA; et al. Stimulation of 92-kd gelatinase (matrix metalloproteinase 9) production by interleukin-17 in human monocyte/macrophages: a possible role in rheumatoid arthritis. *Arthritis Rheum*, 2000, 43, 1134-1144.

[95] Rifas, L; Arackal, S. T cells regulate the expression of matrix metalloproteinase in human osteoblasts via a dual mitogen-activated protein kinase mechanism. *Arthritis Rheum*, 2003, 48, 993-1001.

[96] Numasaki, M; Fukushi, J; Ono, M; et al. Interleukin-17 promotes angiogenesis and tumor growth. *Blood*, 2003, 101, 2620-2627.

[97] Takahashi, H; Numasaki, M; Lotze, MT; Sasaki, H. Interleukin-17 enhances bFGF-; HGF- and VEGF-induced growth of vascular endothelial cells. *Immunol Lett*, 2005, 98, 189-193.

[98] Jovanovic, DV; Di Battista, JA; Martel-Pelletier, J; et al. IL-17 stimulates the production and expression of proinflammatory cytokines; IL-beta and TNF-alpha; by human macrophages. *J Immunol.,* 1998, 160, 3513-3521.

[99] Wang, L; Yi, T; Kortylewski, M; Pardoll, DM; Zeng, D; Yu, H. IL-17 can promote tumor growth through an IL-6-Stat3 signaling pathway. *J Exp Med.,* 2009, 206, 1457-1464.

[100] Sonobe, Y; Takeuchi, H; Kataoka, K; et al. Interleukin-25 expressed by brain capillary endothelial cells maintains blood-brain barrier function in a protein kinase Cepsilon-dependent manner. *J Biol Chem*, 2009, 284, 31834-31842.
[101] Thebault, P; Lhermite, N; Tilly, G; et al. The C-type lectin-like receptor CLEC-1, expressed by myeloid cells and endothelial cells; is up-regulated by immunoregulatory mediators and moderates T cell activation. *J Immunol.,* 2009, 183, 3099-3108.
[102] Kim, YS; Hong, SW; Choi, JP; et al. Vascular endothelial growth factor is a key mediator in the development of T cell priming and its polarization to type 1 and type 17 T helper cells in the airways. *J Immunol.,* 2009, 183, 5113-5120.
[103] Patel, DN; King, CA; Bailey, SR; et al. Interleukin-17 stimulates C-reactive protein expression in hepatocytes and smooth muscle cells via p38 MAPK and ERK1/2-dependent NF-kappaB and C/EBPbeta activation. *J Biol Chem*, 2007, 282, 27229-27238.
[104] Cheng, G; Wei, L; Xiurong, W; Xiangzhen, L; Shiguang, Z; Songbin, F. IL-17 Stimulates Migration of Carotid Artery Vascular Smooth Muscle Cells in an MMP-9 Dependent Manner via p38 MAPK and ERK1/2-Dependent NF-kappaB and AP-1 Activation. *Cell Mol Neurobiol*, 2009.
[105] van den Berg, A; Kuiper, M; Snoek, M; et al. Interleukin-17 induces hyperresponsive interleukin-8 and interleukin-6 production to tumor necrosis factor-alpha in structural lung cells. *Am J Respir Cell Mol Biol*, 2005, 33, 97-104.
[106] Yao, Z; Spriggs, MK; Derry, JM; et al. Molecular characterization of the human interleukin (IL)-17 receptor. *Cytokine*, 1997, 9, 794-800.
[107] Silva, WA; Jr.; Covas, DT; Panepucci, RA; et al. The profile of gene expression of human marrow mesenchymal stem cells. *Stem Cells*, 2003, 21, 661-669.
[108] Moseley, TA; Haudenschild, DR; Rose, L; Reddi, AH. Interleukin-17 family and IL-17 receptors. *Cytokine Growth Factor Rev*, 2003, 14, 155-174.
[109] Kuestner, RE; Taft, DW; Haran, A; et al. Identification of the IL-17 receptor related molecule IL-17RC as the receptor for IL-17F. *J Immunol.,* 2007, 179, 5462-5473.
[110] Ye, P; Rodriguez, FH; Kanaly, S; et al. Requirement of interleukin 17 receptor signaling for lung CXC chemokine and granulocyte colony-stimulating factor expression; neutrophil recruitment; and host defense. *J Exp Med.,* 2001, 194, 519-527.
[111] Kramer, JM; Yi, L; Shen, F; et al. Evidence for ligand-independent multimerization of the IL-17 receptor. *J Immunol.,* 2006, 176, 711-715.
[112] Toy, D; Kugler, D; Wolfson, M; et al. Cutting edge: interleukin 17 signals through a heteromeric receptor complex. *J Immunol.,* 2006, 177, 36-39.
[113] Wright, JF; Bennett, F; Li, B; et al. The human IL-17F/IL-17A heterodimeric cytokine signals through the IL-17RA/IL-17RC receptor complex. *J Immunol.,* 2008, 181, 2799-2805.
[114] Claudio, E; Sonder, SU; Saret, S; et al. The adaptor protein CIKS/Act1 is essential for IL-25-mediated allergic airway inflammation. *J Immunol.,* 2009, 182, 1617-1630.
[115] Novatchkova, M; Leibbrandt, A; Werzowa, J; Neubuser, A; Eisenhaber, F. The STIR-domain superfamily in signal transduction; development and immunity. *Trends Biochem Sci*, 2003, 28, 226-229.
[116] Murphy, JM; Young, IG. IL-3, IL-5, and GM-CSF signaling: crystal structure of the human beta-common receptor. *Vitam Horm*, 2006, 74, 1-30.

[117] Miggin, SM; Palsson-McDermott, E; Dunne, A; et al. NF-kappaB activation by the Toll-IL-1 receptor domain protein MyD88 adapter-like is regulated by caspase-1. *Proc Natl Acad Sci*, U S A 2007, 104, 3372-3377.

[118] Maitra, A; Shen, F; Hanel, W; et al. Distinct functional motifs within the IL-17 receptor regulate signal transduction and target gene expression. *Proc Natl Acad Sci*, U S A 2007, 104, 7506-7511.

[119] Shen, F; Li, N; Gade, P; et al. IL-17 receptor signaling inhibits C/EBPbeta by sequential phosphorylation of the regulatory 2 domain. *Sci Signal*, 2009, 2, ra8.

[120] Chang, SH; Park, H; Dong, C. Act1 adaptor protein is an immediate and essential signaling component of interleukin-17 receptor. *J Biol Chem*, 2006, 281, 35603-35607.

[121] Qian, Y; Liu, C; Hartupee, J; et al. The adaptor Act1 is required for interleukin 17-dependent signaling associated with autoimmune and inflammatory disease. *Nat Immunol.,* 2007, 8, 247-256.

[122] Qian, Y; Qin, J; Cui, G; et al. Act1, a negative regulator in CD40- and BAFF-mediated B cell survival. *Immunity.,* 2004, 21, 575-587.

[123] Sato, S; Sanjo, H; Takeda, K; et al. Essential function for the kinase TAK1 in innate and adaptive immune responses. *Nat Immunol.,* 2005, 6, 1087-1095.

[124] Awane, M; Andres, PG; Li, DJ; Reinecker, HC. NF-kappa B-inducing kinase is a common mediator of IL-17-; TNF-alpha-; and IL-1 beta-induced chemokine promoter activation in intestinal epithelial cells. *J Immunol.,* 1999, 162, 5337-5344.

[125] Hata, K; Andoh, A; Shimada, M; et al. IL-17 stimulates inflammatory responses via NF-kappaB and MAP kinase pathways in human colonic myofibroblasts. *Am J Physiol Gastrointest Liver Physiol*, 2002, 282, G1035-1044.

[126] Ruddy, MJ; Wong, GC; Liu, XK; et al. Functional cooperation between interleukin-17 and tumor necrosis factor-alpha is mediated by CCAAT/enhancer-binding protein family members. *J Biol Chem*, 2004, 279, 2559-2567.

[127] Anderson, P. Post-transcriptional control of cytokine production. *Nat Immunol.,* 2008, 9, 353-359.

[128] Shen, MM. Identification of differentially expressed genes in mouse development using differential display and in situ hybridization. *Methods*, 2001, 24, 15-27.

[129] Gran, B; Zhang, GX; Yu, S; et al. IL-12p35-deficient mice are susceptible to experimental autoimmune encephalomyelitis: evidence for redundancy in the IL-12 system in the induction of central nervous system autoimmune demyelination. *J Immunol.,* 2002, 169, 7104-7110.

[130] Zhang, GX; Gran, B; Yu, S; et al. Induction of experimental autoimmune encephalomyelitis in IL-12 receptor-beta 2-deficient mice: IL-12 responsiveness is not required in the pathogenesis of inflammatory demyelination in the central nervous system. *J Immunol.,* 2003, 170, 2153-2160.

[131] Elson, CO; Cong, Y; Weaver, CT; et al. Monoclonal anti-interleukin 23 reverses active colitis in a T cell-mediated model in mice. *Gastroenterology*, 2007, 132, 2359-2370.

[132] Yang, XO; Chang, SH; Park, H; et al. Regulation of inflammatory responses by IL-17F. *J Exp Med.,* 2008, 205, 1063-1075.

[133] Nakae, S; Nambu, A; Sudo, K; Iwakura, Y. Suppression of immune induction of collagen-induced arthritis in IL-17-deficient mice. *J Immunol.,* 2003, 171, 6173-6177.

[134] Sato, K; Suematsu, A; Okamoto, K; et al. Th17 functions as an osteoclastogenic helper T cell subset that links T cell activation and bone destruction. *J Exp Med.,* 2006, 203, 2673-2682.

[135] Moran, EM; Mullan, R; McCormick, J; et al. Human rheumatoid arthritis tissue production of IL-17A drives matrix and cartilage degradation: synergy with tumour necrosis factor-alpha; Oncostatin M and response to biologic therapies. *Arthritis Res Ther*, 2009, 11, R113.

[136] Eid, RE; Rao, DA; Zhou, J; et al. Interleukin-17 and interferon-gamma are produced concomitantly by human coronary artery-infiltrating T cells and act synergistically on vascular smooth muscle cells. *Circulation*, 2009, 119, 1424-1432.

[137] Taleb, S; Romain, M; Ramkhelawon, B; et al. Loss of SOCS3 expression in T cells reveals a regulatory role for interleukin-17 in atherosclerosis. *J Exp Med.,* 2009, 206, 2067-2077.

[138] Xie, JJ; Wang, J; Tang, TT; et al. The Th17/Treg functional imbalance during atherogenesis in ApoE(-/-) mice Th17/Treg imbalance in atherogenesis. *Cytokine*, 2009.

[139] Guzik, TJ; Marvar, PJ; Czesnikiewicz-Guzik, M; Korbut, R. Perivascular adipose tissue as a messenger of the brain-vessel axis: role in vascular inflammation and dysfunction. *J Physiol Pharmacol*, 2007, 58, 591-610.

[140] van Es, T; van Puijvelde, GH; Ramos, OH; et al. Attenuated atherosclerosis upon IL-17R signaling disruption in LDLr deficient mice. *Biochem Biophys Res Commun*, 2009, 388, 261-265.

[141] Maitra, U; Davis, S; Reilly, CM; Li, L. Differential regulation of Foxp3 and IL-17 expression in CD4 T helper cells by IRAK-1. *J Immunol.,* 2009, 182, 5763-5769.

[142] Robb, L; Boyle, K; Rakar, S; et al. Genetic reduction of embryonic leukemia-inhibitory factor production rescues placentation in SOCS3-null embryos but does not prevent inflammatory disease. *Proc Natl Acad Sci*, U S A 2005, 102, 16333-16338.

[143] Chen, Z; Laurence, A; Kanno, Y; et al. Selective regulatory function of Socs3 in the formation of IL-17-secreting T cells. *Proc Natl Acad Sci*, U S A 2006, 103, 8137-8142.

[144] Darrasse-Jeze, G; Deroubaix, S; Mouquet, H; et al. Feedback control of regulatory T cell homeostasis by dendritic cells in vivo. *J Exp Med.,* 2009, 206, 1853-1862.

[145] Cui, G; Qin, X; Zhang, Y; Gong, Z; Ge, B; Zang, YQ. Berberine differentially modulates the activities of ERK; p38 MAPK; and JNK to suppress Th17 and Th1 T cell differentiation in type 1 diabetic mice. *J Biol Chem*, 2009, 284, 28420-28429.

[146] Jain, R; Tartar, DM; Gregg, RK; et al. Innocuous IFNgamma induced by adjuvant-free antigen restores normoglycemia in NOD mice through inhibition of IL-17 production. *J Exp Med.,* 2008, 205, 207-218.

[147] Boehm, BO; Rosinger, S; Sauer, G; et al. Protease-resistant human GAD-derived altered peptide ligands decrease TNF-alpha and IL-17 production in peripheral blood cells from patients with type 1 diabetes mellitus. *Mol Immunol*, 2009, 46, 2576-2584.

[148] Bradshaw, EM; Raddassi, K; Elyaman, W; et al. Monocytes from patients with type 1 diabetes spontaneously secrete proinflammatory cytokines inducing Th17 cells. *J Immunol.,* 2009, 183, 4432-4439.

[149] Aujla, SJ; Chan, YR; Zheng, M; et al. IL-22 mediates mucosal host defense against Gram-negative bacterial pneumonia. *Nat Med.,* 2008, 14, 275-281.

[150] Happel, KI; Dubin, PJ; Zheng, M; et al. Divergent roles of IL-23 and IL-12 in host defense against Klebsiella pneumoniae. *J Exp Med.,* 2005, 202, 761-769.

[151] Wu, Q; Martin, RJ; Rino, JG; Breed, R; Torres, RM; Chu, HW. IL-23-dependent IL-17 production is essential in neutrophil recruitment and activity in mouse lung defense against respiratory Mycoplasma pneumoniae infection. *Microbes Infect*, 2007, 9, 78-86.

[152] Huang, W; Na, L; Fidel, PL; Schwarzenberger, P. Requirement of interleukin-17A for systemic anti-Candida albicans host defense in mice. *J Infect Dis*, 2004, 190, 624-631.

[153] Oppenheim, JJ; Biragyn, A; Kwak, LW; Yang, D. Roles of antimicrobial peptides such as defensins in innate and adaptive immunity. *Ann Rheum Dis.,* 2003, 62 Suppl 2, ii17-21.

[154] Costantino, CM; Baecher-Allan, CM; Hafler, DA. Human regulatory T cells and autoimmunity. Eur *J Immunol.,* 2008, 38, 921-924.

[155] Pacholczyk, R; Kern, J. The T-cell receptor repertoire of regulatory T cells. *Immunology*, 2008, 125, 450-458.

[156] Buckner, JH; Ziegler, SF. Functional analysis of FOXP3. *Ann N Y Acad Sci*, 2008, 1143, 151-169.

[157] Horwitz, DA; Zheng, SG; Gray, JD. Natural and TGF-beta-induced Foxp3(+)CD4(+) CD25(+) regulatory T cells are not mirror images of each other. *Trends Immunol.,* 2008, 29, 429-435.

[158] George, J. Mechanisms of disease: the evolving role of regulatory T cells in atherosclerosis. *Nat Clin Pract Cardiovasc Med.,* 2008, 5, 531-540.

[159] Andre, S; Tough, DF; Lacroix-Desmazes, S; Kaveri, SV; Bayry, J. Surveillance of antigen-presenting cells by CD4+ CD25+ regulatory T cells in autoimmunity: immunopathogenesis and therapeutic implications. *Am J Pathol.,* 2009, 174, 1575-1587.

[160] Haribhai, D; Lin, W; Edwards, B; et al. A central role for induced regulatory T cells in tolerance induction in experimental colitis. *J Immunol.,* 2009, 182, 3461-3468.

[161] Chatila, TA. Regulatory, T. cells: key players in tolerance and autoimmunity. *Endocrinol Metab Clin North Am,* 2009, 38, 265-272, vii.

[162] Lal, G; Bromberg, JS. Epigenetic mechanisms of regulation of Foxp3 expression. *Blood*, 2009, 114, 3727-3735.

[163] Bollyky, PL; Falk, BA; Wu, RP; Buckner, JH; Wight, TN; Nepom, GT. Intact extracellular matrix and the maintenance of immune tolerance: high molecular weight hyaluronan promotes persistence of induced CD4+CD25+ regulatory T cells. *J Leukoc Biol.,* 2009, 86, 567-572.

[164] Valencia, X; Lipsky, PE. CD4+CD25+FoxP3+ regulatory T cells in autoimmune diseases. *Nat Clin Pract Rheumatol*, 2007, 3, 619-626.

[165] von Herrath, MG; Harrison, LC. Antigen-induced regulatory T cells in autoimmunity. *Nat Rev Immunol.,* 2003, 3, 223-232.

[166] Yan, Y; Chen, Y; Yang, F; et al. HLA-A2.1-restricted T cells react to SEREX-defined tumor antigen CML66L and are suppressed by CD4+CD25+ regulatory T cells. *Int J Immunopathol Pharmacol*, 2007, 20, 75-89.

[167] Zhou, X; Bailey-Bucktrout, SL; Jeker, LT; et al. Instability of the transcription factor Foxp3 leads to the generation of pathogenic memory T cells in vivo. *Nat Immunol.,* 2009, 10, 1000-1007.

[168] Sakaguchi, S. Naturally arising CD4+ regulatory t cells for immunologic self-tolerance and negative control of immune responses. *Annu Rev Immunol.,* 2004, 22, 531-562.

[169] Kim, JM; Rasmussen, JP; Rudensky, AY; Regulatory, T. cells prevent catastrophic autoimmunity throughout the lifespan of mice. *Nat Immunol.,* 2007, 8, 191-197.

[170] O'Connor, RA; Malpass, KH; Anderton, SM. The inflamed central nervous system drives the activation and rapid proliferation of Foxp3+ regulatory T cells. *J Immunol.,* 2007, 179, 958-966.

[171] Bettelli, E; Korn, T; Oukka, M; Kuchroo, VK. Induction and effector functions of T(H)17 cells. *Nature*, 2008, 453, 1051-1057.

[172] Yang, L; Anderson, DE; Baecher-Allan, C; et al. IL-21 and TGF-beta are required for differentiation of human T(H)17 cells. *Nature*, 2008, 454, 350-352.

[173] Zhou, L; Lopes, JE; Chong, MM; et al. TGF-beta-induced Foxp3 inhibits T(H)17 cell differentiation by antagonizing RORgammat function. *Nature*, 2008, 453, 236-240.

[174] Du, J; Huang, C; Zhou, B; Ziegler, SF. Isoform-specific inhibition of ROR alpha-mediated transcriptional activation by human FOXP3. *J Immunol.,* 2008, 180, 4785-4792.

[175] Laurence, A; Tato, CM; Davidson, TS; et al. Interleukin-2 signaling via STAT5 constrains T helper 17 cell generation. *Immunity.,* 2007, 26, 371-381.

[176] Mucida, D; Park, Y; Kim, G; et al. Reciprocal TH17 and regulatory T cell differentiation mediated by retinoic acid. *Science*, 2007, 317, 256-260.

[177] Adamson, AS; Collins, K; Laurence, A; O'Shea, JJ. The Current STATus of lymphocyte signaling: new roles for old players. *Curr Opin Immunol*, 2009, 21, 161-166.

[178] Fletcher, JM; Lonergan, R; Costelloe, L; et al. CD39+Foxp3+ regulatory T Cells suppress pathogenic Th17 cells and are impaired in multiple sclerosis. *J Immunol.,* 2009, 183, 7602-7610.

In: Regulatory T Cells
Editor: Ren S. Hayashi, pp. 33-56

ISBN: 978-1-61761-554-2

Chapter 2

Human CD8 Tregs as Contender in Chronic Persisting Infections

Simone A. Joosten and Tom H. M. Ottenhoff *
Department of Infectious Diseases, Leiden University Medical Center, Leiden, The Netherlands

Abstract

Regulatory T-cells (Tregs) are key regulators of the immune system, however their role in infectious diseases remains rather controversial. Chronic persistent infections in humans such as parasites, viruses and (myco)bacteria all can result in the induction of both $CD4^+$ and $CD8^+$ Tregs. In these chronic persistent infections, Tregs may dampen inflammation to limit tissue damage, but they may also inhibit ensuing effector immunity, thereby impairing pathogen clearance. All of these may depend on timing and location of Treg responses, early during infections Tregs may be essential balancing the trafficking of effector cells between the sites of immune induction and infection whereas in later stages Tregs may hamper pathogen clearance. A remarkably increased frequency of Tregs has been observed at the site of infection, suggesting active local involvement. The vast majority of studies has focused on the role of (various subsets of) $CD4^+$ Treg cells. However, $CD8^+$ Treg cells play a prominent role as well, but surprisingly little is known about these cells, particularly in humans. The antigen specificity of such $CD8^+$ Tregs has been demonstrated for several pathogens and the genetic restriction of the T-cell responses can comprise classical as well as non-classical HLA class I family members. Here we review and discuss human $CD8^+$ Tregs and $CD8^+$ Treg subsets, with particular emphasis on their properties, effector mechanisms and their role in chronic, persistent infectious diseases.

* Corresponding author: Department of Infectious Diseases, Leiden University Medical Center, Albinusdreef 2, 2333 ZA Leiden, The Netherlands, Phone +31 71 526 2620, Fax +31 71 526 6758, t.h.m.ottenhoff@lumc.nl

Introduction

Upon infection with a pathogen, the immune system in the ideal scenario would rapidly control the infection, clear the pathogen and avoid inflammation associated damage to host tissues. As a first step, pathogen-derived danger signals evoke innate immune activation through activation of pathogen recognition receptors (PRRs). In a subsequent phase, adaptive immunity is induced: the pathogen is phagocytosed, degraded and its peptidic fragments presented by antigen presenting cells (APC) to helper T-cells (Th), cytotoxic T-cells (CTL), and also regulatory T-cells (Treg). Th cells in collaboration with CTL will mediate effector immunity, launching an attack on the infected cells. To prevent excessive tissue damage and limit the extent and duration of inflammation, regulatory cells are present in both the innate (macrophages, dendritic cells) and the adaptive immune system. Among the latter, NKT-cells and Tregs are probably the best studied. Tregs can be subdivided in either naturally occurring ($CD4^+$) Tregs or inducible ($CD4^+$ or $CD8^+$) Tregs, the latter of which can be provoked by specific pathogen derived antigens. A number of studies have shown that Tregs play a critical role in the outcome of infection and inflammation [1].

Tregs are endowed with unique functional properties, although they share many features with classical T-cells: they express T-cell lineage markers, including CD3 and CD4 or CD8, CD45 isoforms and a functional T-cell receptor. Once activated through their T-cell receptor, Tregs can suppress proliferation of any T-cell in its vicinity, irrespective of the presence of antigen or the specificity of that T-cell, a phenomenon called "bystander suppression" [2]. Bystander suppression pertains mostly to cytokine secreting Tregs, in contrast to Tregs that interact with target cells in an antigen specific manner and use cell-cell contact to mediate suppression. In addition, some Tregs can alter the environment to promote outgrowth of new Tregs, a phenomenon called "infectious tolerance" [2]. Thus, Tregs have a central role as guardians of the immune system through multiple mechanisms.

Induction of Tregs during infectious diseases may have both beneficial and harmful effects for the host. The host can benefit from pathogen mediated Treg induction particularly during chronic infections, as Tregs limit inflammatory responses and thereby reduce host tissue damage [3]. However, the downside may be that the pathogen may not be fully eradicated, allowing the establishment of a persistent infection. This enhances the risk of disease reactivation later in life, when facing (temporal) immune dysfunction, e.g. during co-infections, co-morbidity or other forms of immune suppression (figure 1).

In addition to these relatively direct effects, Treg activation and function may also favor immunological memory, since persistent infection at a low pathogen load may permanently boost memory cells [3]. However, this might also lead to boosting of Tregs since recent evidence suggests a link between memory T-cells and $CD8^+$ Tregs [4]. The pathogen itself has direct benefits from the induction of regulatory immunity because dampening effector immunity allows the pathogen to survive for an extended period of time [3]. Secondly, induction of regulation not only promotes extended pathogen but also host survival as a consequence of the decreased inflammation and damage, an advantage for pathogens that need their host for their transmission and propagation [3]. Thus, the net effects of Tregs in the context of infectious disease are complex and multi-faceted, largely depending on the type of infection and pathogen involved, as well as on the kinetics, order and magnitude of T-cell subsets are induced (figure 1).

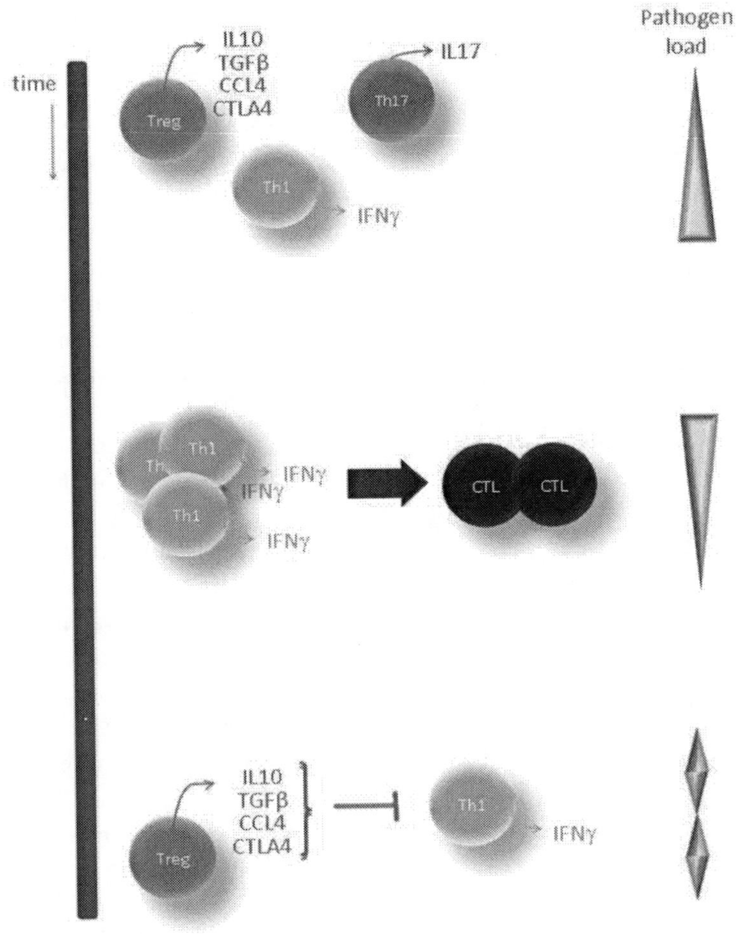

Figure 1. kinetics of T-cell populations over time upon infection. Early during infection pathogen loads increase due to the absence of specific effector immunity. Once proper, specific effector immunity is induced the pathogen burden will diminish, finally establishing a dynamic equilibrium between host effector immunity and the pathogen resulting in a chronic infection

In this review, we will focus on newly discovered human $CD8^+$ Tregs and their potential significance in various infectious diseases, with special attention to mycobacterial infection and tuberculosis (TB) vaccination.

Treg Subsets, Markers and Plasticity

Treg subsets have been well characterized in mice, where phenotypical analysis identified many subtypes, including thymus-derived naturally occurring $CD4^+CD25^+FoxP3^+$ Tregs and several subsets of induced or adaptive Tregs [1]. Induced Tregs develop in the periphery in the presence of self or foreign (pathogen derived) antigens and can be subdivided into Tr1 and Th3 cells, both expressing $CD4^+CD25^+$. Tr1 cells suppress via interleukin-10 (IL-10), whereas Th3 utilize transforming growth factor-β (TGF-β) [1]. In addition to $CD4^+$ regulatory T-cell subsets, also $CD8^+$ T-cells have been described that can inhibit T-cell proliferation and cytokine production [1]. However, these $CD8^+$ T-cells have been characterized to a much lesser extent compared to their $CD4^+$ counterparts, and their phenotype and effector molecules are largely unknown.

In contrast to murine Treg subsets, human Tregs -with the notable exception of $CD4^+CD25^{HI}$ T-cells- have not been studied and described in great detail, particularly human $CD8^+$ Tregs. The array of human Tregs seems broader and less strictly compartmentalized compared to murine subsets. Human Tregs are typically identified by functional analysis since their phenotypical identification and quantification is more difficult due to the lack of specific and unique markers. The IL-2 receptor chain CD25 e.g. is abundantly expressed also on activated T-cells and some $CD8^+$ Treg subsets completely lack CD25 expression [5]. Other surface markers frequently expressed by Tregs, including CTLA4 (Cytotoxic T-Lymphocyte Associated Antigen-4, CD152), GITR (Glucocorticoid-Induced TNF Receptor) and LAG-3 (Lymphocyte Activation Gene-3, CD223) can also be (transiently) expressed by activated non-regulatory T-cells. Expression of the forkhead transcription factor FoxP3 is frequently seen as a prerequisite for Treg function, since mutations in the FoxP3 gene result in autoimmune diseases due to dysfunctional Treg cells [6]. Moreover, ectopic expression of FoxP3 converted murine $CD4^+$ T-cells into Tregs [7]. However, in humans not all $FoxP3^+$ cells are Tregs [8] and FoxP3 is transiently upregulated also on activated non-regulatory T-cells [9], implying that FoxP3 is not a very specific marker for human Tregs. Others have suggested that circulating $FoxP3^+$ non-Treg cells might be a reservoir of silent Tregs that regain their function upon activation [10]. Thus, in humans FoxP3 is clearly expressed on naturally occurring Tregs, but it is not an exclusive marker for Tregs, necessitating functional evaluation to assess Treg activity.

The origin of human Tregs is largely unknown; some subsets are thymus derived whereas others originate in the periphery. $CD4^+CD25^{HI}FoxP3^+$ Tregs in the human periphery had closely related T-cell receptor sequences as compared with memory $CD4^+$ T-cells suggesting that Tregs can differentiate from memory cells in the periphery [11]. In our previous studies we had also observed that $CD8^+$ Tregs activated by BCG stimulation were only detectable in donors that were previously sensitized with mycobacteria, suggesting that memory immune responses were critically involved in activation of these $CD8^+$ Tregs, and may have been their source of origin [4].

Since the discovery of different T-cell polarities, the distinct phenotypical and functional properties of these T-cells have been considered to be relatively stable characteristics of these cells. Especially Tregs have mostly been considered to be terminally differentiated, "end-stage" cells. However a number of recent studies have now demonstrated that T-cells subsets, including Tregs have high level of plasticity. Human $CD4^+Foxp3^+$ Tregs can differentiate into IL-17 producing cells when stimulated with allogeneic monocytes in the presence of IL-2 and IL-15 [12]. These IL-17 producing cells also expressed the transcription factor RORγt, and the differentiation towards IL-17 production was dependent on histone/protein deacetylase activity. Intermediate stages were identified in which cells can have dual activity, e.g. produce IL-17 whilst expressing FoxP3, suggesting that there is a functional gradient with polar phenotypes at the extremes after stimulation [12]. Moreover, these IL-17 producing FoxP3+ cells retained their suppressive capacity and depending on their activation status either produced IL-17 or suppressed proliferation, both of which functions were completely reversible [13]. Co-expression of FoxP3 and RORγt was detected in human tonsils at an increased frequency compared to peripheral blood [14]. On the other hand, memory T-cells have been identified that dependent on the context will secrete IL-10 and/ or IL-17 [15]. This indicates that FoxP3 expressing Tregs are not end-stage differentiated cells but are still plastic components of the immune system that can be re-programmed under specific conditions to activate the immune system. Although this has thus far only been demonstrated for $CD4^+$ Tregs we speculate that similar plasticity may occur in $CD8^+$ Tregs, resulting for example in effector T-cell activity. In our previous experiments we have indeed observed that $CD8^+$ Treg populations also had CTL activity and in preliminary experiments we have shown that clonal populations can both suppress proliferation and lyse target cells [4,16].

CD8 Tregs in Human Disease

Although early studies in patients with disseminating or lepromatous leprosy identified human $CD8^+$ suppressor T-cell clones [17,18], surprisingly human $CD8^+$ Tregs have not been characterized in detail yet, in contrast to human $CD4^+$ Tregs. Distinct subsets and/or lineage markers have not been identified, and descriptions of human $CD8^+$ Tregs are scarce. CD28 expression seems to discriminate at least 2 different subsets of $CD8^+$ Tregs, but more recent literature describes multiple phenotypical and functional distinctive subsets even within CD28 positive or negative populations.

In the thymus of healthy individuals, $CD8^+$ Tregs exist that express CD25, FoxP3, GITR and CTLA4 and suppress in a contact dependent manner, thus seemingly representing the equivalent of classical $CD4^+CD25^+$ Tregs [19]. In the circulation of healthy individuals $CD8^+$ Tregs have been described that express the integrin CD103 [20]. In addition, markers including LAG-3, CD27, CD38 and HLA-DR have been described on human Tregs in different disease conditions [21].

Human $CD8^+$ Tregs have been implicated in various inflammatory disorders including inflammatory bowel disease (IBD) [22], ankylosing spondylitis (AS) [23], type 1 diabetes [24] and multiple sclerosis (MS) [25]. In ankylosing spondylitis Tregs were not capable to clear or limit the inflammation, but seemed to be a consequence of the inflammation since they were increased during active inflammation [23]. This seems to be exceptional since mostly it is the lack of Tregs rather than its excess that is associated with chronic inflammation; for example, in IBD patients the absence of Tregs in the lamina propria was associated with a lack of mucosal tolerance and the presence of persistent inflammation [22]. Similarly, an increase in Tregs upon immune modulation in both diabetes and MS patients was associated with reduced relapse rates and reduced inflammation, indicating amelioration of disease activity concomitant with increased Treg function [24,25].

Tolerance induction is required to induce long-term graft survival in recipients of solid organ transplants, and Tregs can be the executers of functional tolerance: in cardiac allograft recipients $CD8^+CD28^-$ Tregs have been described that modulate APCs towards a more tolerogenic phenotype that could induce alloantigen-unresponsiveness of $CD4^+$ T-cells [26]. Furthermore, $CD8^+CD28^-CD27^+$ cells were expanded in cardiac allograft recipients and had the highest frequency in recipients that did not experience acute rejection episodes, suggesting a role in maintenance of tolerance [21]. Induction of $CD8^+$ T-cell mediated tolerance did not imply loss of immunological memory for donor allo-antigens, indicating that Tregs most probably need to be permanently active to maintain graft survival [27]. A subset of $CD8^{+}LAP^{+}CD103^{+}PD\text{-}1^{+}PD\text{-}L1^{+}CTLA4^{+}$ Tregs is critically involved in remission from lupus after hematopoietic stem cell transplantation [28]. These cells were auto-antigen specific and suppressed predominantly through TGF-β, in a cell contact independent manner.

Therapeutic tolerance induction in autoimmune diseases and solid organ transplantation aims at activation and expansion of Tregs to dampen pathogenic inflammatory processes. In a special case of allo-antigen sensitization, namely pregnancy, the human body applies tolerance induction via Tregs to prevent inflammation and pathology upon recognition of foreign antigens. During pregnancy the mother develops tolerance to the allogeneic fetus; mediated by amongst others increased frequencies of $CD8^{+}$ Tregs. These cells are present at the site of feto-maternal contact, the placenta, but not in peripheral blood of the pregnant women [29,30]. Tolerance during pregnancy does not only involve placental T-cells, but also Tregs specific for minor histocompatibility antigens expressed by fetal cells can be detected in the maternal circulation. These $CD8^{+}$ Tregs were detectable in the mothers decades after pregnancy, suggesting a long term differentiation and persistence of these cells. In many mothers these cells suppressed in a trans-vivo DTH assay, through a mechanism which was largely dependent on CTLA4 [31].

Clearly, tolerance is not always beneficial to the host, for example during chronic persistent infections and during tumor development. Tumors may use Tregs as mechanism of immune escape; interestingly the frequency of $CD8^{+}CD28^{-}$ Tregs was increased among tumor-infiltrating-lymphocytes compared to PBMCs, still the frequency of circulating $CD8^{+}CD28^{-}$ Tregs was increased in patients compared to healthy controls [32]. Moreover, the number of $CD8^{+}CD28^{-}$ Tregs in the tumor-infiltrating-lymphocytes negatively correlated with patient survival [32]. In infectious diseases, $CD8^{+}$ Tregs have been associated mainly with chronic, persistent infections like leprosy [18], HIV [33], EBV [34], hepatitis C [35] and tuberculosis [4].

Classically $CD8^{+}$ T-cells are known for their long-term memory capacity, as well as their long-term maintenance of effector functions. Although $CD8^{+}$ Tregs may represent a distinct subset of $CD8^{+}$ T-cells, they also might be related to, or even evolve from one or several different other $CD8^{+}$ T-cell subsets. If $CD8^{+}$ Tregs can evolve from existing (memory) $CD8^{+}$ cells, therapeutic interventions that aim at $CD8^{+}$ T-cell memory induction may in fact also promote $CD8^{+}$ Treg formation. Classical $CD8^{+}$ central memory T-cells can differentiate into effector cells to generate a swift response in case the antigen (pathogen) is re-encountered. Similarly, it might be beneficial not only to have a pool of rapidly responding pre-existing effector cells, but also a pool of similarly rapid responsive regulatory cells that can dampen uncontrolled inflammatory responses towards specific pathogens. Preliminary evidence for this concept is supported by our own observations, in which only donors that were primed *in vivo* with mycobacteria had detectable Treg responses following stimulation with live mycobacteria [4] (ref.). This contrasted to donors that were mycobacterially naïve, suggesting that memory is necessary for activation and expansion of $CD8^{+}$ Tregs [4]. However, the requirements for $CD8^{+}$ Treg induction, activation and expansion remain incompletely understood.

Mechanism of Suppression by CD8⁺ Tregs

The mechanisms utilized by human CD8$^+$ Tregs to exert suppression have not been studied in great detail. In some studies on human CD8$^+$ Tregs IL-10 was critically involved in inhibition of cellular proliferation, either alone [32] or in combination with TGF-β and CTLA4 [27]. However, most papers on human Tregs describe cell-cell contact as requirement for suppression, and such pathways may involve CTLA4 [19,23,27,31], but in most cases remain undefined. The finding that Treg/T-cell contact is frequently required for inhibition of T-cell proliferation is somewhat paradoxical since many studies have described IL-10 production by CD8$^+$ Tregs [5,20,30,34], a factor well known to inhibit T-cell proliferation. Since cell contact is required for suppression, despite the production of IL-10, one could speculate that IL-10 is not sufficient to inhibit proliferation or that close cell contact is necessary for (perhaps membrane bound) IL-10 to exert its effects. This contrasts with especially the murine data on IL-10 and its role in suppression [36]. Also in the case of human CD4$^+$CD25$^+$ Tregs IL-10 has been demonstrated as suppression factor. The role of IL-10 in CD8$^+$ Treg effector activity thus needs further investigation; it remains possible that the effects of IL-10 are indirect, and that IL-10 may play a role via induction or inhibition of other cytokines or surface molecules that are the final mediators of suppression.

A novel suppression factor produced by CD8$^+$ Tregs and capable of directly inhibiting T-cell proliferation, which we have recently identified, is the chemokine CCL4 (MIP-1β). These results extend the function of this chemokine, which is otherwise well-known for its role in chemotaxis [4]. This finding was recently confirmed by an independent study using anti-CD3 antibodies for activation of CD8$^+$ Tregs [37]. The role of CCL4 and its receptor (CCR5) will be discussed separately below.

Finally, suppression by CD8$^+$ Tregs may also be mediated by TGF-β, either in soluble or in membrane-bound form [16,28,38-40]. Initially, membrane bound TGF-β was described on CD4$^+$ Tregs [38,39]; TGF-β in these studies was functionally involved in inhibition of target cell proliferation but no secreted TGF-β was detectable and cell-cell contact was required for this effect. TGF-β was detected at the cell surface of these cells and its activity was blocked by addition of its latency associated peptides (LAP)[39]. More recently, we and others have identified membrane bound TGF-β (mTGF-β) to play a similar role on CD8$^+$ Tregs [16,28]; other studies have described a role for soluble secreted TGF-β from CD8$^+$ Tregs in mediating suppression [40]. Taken together, these new studies add both soluble, secreted as well as mTGF-β to the array of suppressor effector molecules used by CD8$^+$ Tregs.

An alternative mechanism of suppression by CD8$^+$ Tregs is via modulation of APCs. CD8$^+$CD28$^-$ Tregs can upregulate ILT3 and ILT4 (ILT= immunoglobulin like transcript) on professional APCs as well as endothelial cells, thereby modulating these APCs towards a tolerogenic phenotype. CD4$^+$ T-cells that interact with these tolerogenic APCs become anergic and may obtain regulatory functions [26,41]. Thus a CD8$^+$ Treg could modulate the APC, which in turn promotes induction of CD4$^+$ Tregs that inhibit effector T-cells. Such a model shares many features with the old concepts of CD8$^+$ T-suppressor inducer and CD8$^+$ T-suppressor effector cells, which has been looked upon not too favorably the past two decades [42].

CD4$^+$ and CD8$^+$ Tregs not only regulate CD4$^+$ Th-cells (figure 1B), but also potently inhibit CD8$^+$ cytotoxic T-cell mediated effector immunity. For example, circulating

$CD4^+CD25^+FoxP3^+$ Tregs from HIV patients inhibited HIV-gag-specific cytolysis and $CD8^+$ effector function [43]. $CD8^+CD25^-FoxP3^-IL\text{-}10^+$ Tregs derived from HIV patients were capable of inhibiting HIV-gag induced degranulation as measured by CD107a/b expression as surrogate for cytotoxicity [5]. The effects of Tregs on cytotoxicity were not only detectable *in vitro* but also *in vivo*, where imaging recently demonstrated inhibition of killing of peptide loaded B cells in the presence of $CD4^+CD25^+$ Tregs [44] (figure 1B).

In addition, $CD4^+CD25^+$ and $CD8^+$ Treg seem to modulate antigen presenting cells towards an anti-inflammatory phenotype. These anti-inflammatory APCs promoted outgrowth of additional ($CD4^+$) Tregs, thus augmenting the regulatory pool [26,41,45] (figure 1B). A new addition to the family of Treg mediators are mast cells and basophils [46,47]. Mast cells may counteract Treg function through IL-6 and OX40/OX40L and skew differentiation towards Th17 cells [47]. On the other hand Tregs may directly inhibit the activity of mast cells, basophils and eosinophils [48]. However degranulation of mast cells may also transiently inhibit Tregs through downregulation of Treg effector molecules [49]. Thus Tregs have an active interplay with previously unknown, new innate immune cells.

The CCL4-CCR5 Axis as a New Player in CD8 Treg Mediated Suppression

We have recently identified human $CD8^+$ Tregs that suppressed proliferation by production of CCL4 (MIP-1β) [4]; as mentioned, these results were recently confirmed by others in a different setting [37]. The only known receptor for CCL4 is CCR5. CCR5 has been studied in detail in the context of HIV infection, since a deletion mutant (CCR5Δ32) prevents HIV entry. The CCR5Δ32 deletion prevents surface expression of CCR5 and as a consequence impairs functional signaling through CCR5. The role of CCR5 and CCL4 has been studied to a much more limited extent in other (infectious) diseases. We found $CD8^+$ Tregs suppressing via CCL4 only in donors that had been primed with mycobacterial antigens but not in naïve donors [4], suggesting that these cells arise from memory cells, and may play a role in anti-mycobacterial immunity.

Interestingly, CCL4 has been identified as a TB susceptibility gene in humans [50]. CCR5 knockout mice clear *Mycobacterium tuberculosis* (Mtb) infection better compared to wild-type mice, during the early stages of infection these mice have increased numbers of lymphocytes in the lung and augmented inflammation [51]. Moreover, AIDS patients that are heterozygous for the CCR5Δ32 mutation seem to be protected against opportunistic infections with several pathogens including *Mycobacterium avium*, compatible with the notion that reduced immune regulation via CCL4/CCR5 can improve control of infection at least in (myco)bacterial infections [52]. CCR5Δ32 mutations have also been studied in the context of chronic *hepatitis C* infection (HCV) and indeed the frequency of homozygous mutants was increased amongst patients with HCV infection [53]. In addition, the number of $CD8^+$ T-cells was increased in the circulation of mutants but in contrast to our expectations viral loads were higher, suggesting that this mutation did not improve control of viral infection directly [53]. Similarly, CCR5Δ32 mutant individuals are more likely to develop disease upon *West Nile Virus* infection, but this seemed the result of decreased migration to the brain rather than decreased regulation, thereby increasing local viral loads [54]. Taken together, the role of the

CCR5 pathway in infectious diseases is poorly identified, but in mycobacterial infections seems to have a clear immunoregulatory role, both in humans and in mice, whereas in viral infections the aberrant migration of lymphocytes seems dominant. Future research should clarify the potential dual role of CCR5 in infectious diseases.

Of interest, treatment with a monoclonal antibody against CD3 was reported to induce $CD8^+$ Tregs that suppressed partially through CCL4 [37]. These humanized monoclonal antibodies are used as experimental therapeutics in type 1 diabetes. Thus far, however, these $CD8^+$ Tregs have not yet been studied in these patients.

Evidence for Involvement of Tregs in Chronic Human Infectious Diseases

Human $CD4^+$ Tregs in Infectious Diseases

Mycobacterial infections

Infections with *M. tuberculosis* complex pathogens typically result in life-long infection. It has long been thought that the chronicity of infection is promoted by immunoregulatory host factors. A first observation suggesting this was when the suppressive capacity of T-cells was demonstrated in the 1970s in an animal model of BCG (Bacillus Calmette Guerin) vaccination [55]. Cells from BCG vaccinated animals were purified and transferred into naive animals, which were subsequently infected with BCG. Animals that had received cells from BCG vaccinated animals had a reduced immune response upon infection compared to untreated animals [55]. This was one of the first indications that a priming immune response might not be primarily and only directed towards controlling subsequent infection, and established the concept of "suppressor cells" in mycobacterial infections. More than a decade later, our laboratory as well as others were able to isolate and clone $CD4^+$ and $CD8^+$ Tregs from human lepromatous leprosy patients [17,18]. Tregs were cloned from both peripheral blood and skin lesions and suppressed T-cell proliferation in an antigen-dependent manner [17,18]. A specific *M. leprae* epitope was later identified that exclusively induced $CD4^+$ Tregs in the context of particular HLA-class II molecules, whereas the same epitope activated Th cell responses when presented by other HLA-class II molecules. These results thus showed both genetic restriction and genetic inter-individual variability in Treg induction in response to this human pathogen [56].

At the site of tuberculin skin test (PPD) injection the number of $CD4^+$ memory and $CD4^+$ Tregs have recently been found to increase in parallel in the human skin. The Tregs in these skin sites expressed CD25, CD27 and CD39 in the absence of CD127 and are regulated in the same manner as $CD4^+$ memory T-cells [57]. PBMCs from skin test positive individuals stimulated with mycobacterial antigens expressed increased levels of FoxP3 and CTLA4 mRNA [58].

Recently, many different groups have analyzed the presence of Treg sub-populations in patients with (active) tuberculosis. Initially, IL-10 secreting cells were identified in anergic tuberculosis patients, *i.e.* patients with culture proven tuberculosis but a negative tuberculin skin test. Decreased responses to Mtb but not other pathogen-derived antigens were observed *in vitro* [59,60], indicating again antigen-specificity of suppression. Furthermore, patients with active tuberculosis disease had increased frequencies of classical $CD4^+$ Tregs ($CD25^+$ or ^{HI} and/or $FoxP3^+$), which normalized upon treatment. The frequency of these CD4 Tregs was increased most at the site of infection when compared to the circulation [61-65]. Some studies have also suggested a role for Tregs in containment or control of local infection, due to their increased frequency in granulomas [4,61]. *In vitro* depletion of Tregs increased the response (IFNγ production) against mycobacterial antigens [61-63]. The expression of FoxP3 in whole blood correlated well with phenotypical Treg induction in time, and with active tuberculosis [66,67]. FoxP3 mRNA expression could be used to differentiate latent tuberculosis from active disease when combined with IL-8 and IL-12 mRNA levels [68]. T-cells isolated from BAL predominantly produced IL-10 and were capable of suppressing T-cell proliferation [65].

The collective data from human studies in TB indicate that Tregs are mostly associated with active disease, particularly at the site of infection, where they downregulate immune responses to Mtb antigens, and thus probably can impair efficient effector immunity and concomitant eradication of Mtb. Recent mouse experiments have indeed confirmed this notion [64,69-71]: depletion of Tregs prior to infection resulted in reduced Mtb outgrowth, strongly supporting the idea that Tregs are "bad guys" in tuberculosis when it comes to control of bacterial growth, especially early during infection.

Parasitic infections

One of the most elegant demonstrations that Tregs are significant players in infectious disease and can impair pathogen clearance is the mouse model of leishmaniasis. Establishment of chronic infection with *Leishmania major* was completely dependent on the activity of $CD4^+CD25^+$ Tregs, which were antigen specific and present at the site of infection [72,73]. Adoptive transfer of $CD4^+CD25^+$ Tregs abrogated pathogen clearance by $CD4^+CD25^-$ resulted in increased parasite numbers and increased lesion size [73]. A low number of persistent parasites was required to maintain permanent effector immunity [73]. These studies have been reviewed in detail in reference [3]. Interestingly, $CD4^+CD25^+$ Tregs could also be isolated from skin lesions of patients with cutaneous leishmaniasis, and suppressed T-cell proliferation *in vitro* [74], demonstrating that functional Tregs indeed are present to the site of infection in human lesions.

In human malaria infection, $CD4^+CD25^+FoxP3^+$ Tregs were associated with higher parasite growth *in vivo* and decreased antigen specific immunity *in vitro* [75]. In vitro stimulation with malarial antigens increased the percentage of $CD25^+$ and $FoxP3^+$ T-cells, suggesting antigen specificity of the Treg populations [76]. Increased frequencies of $CD4^+CD25^+$ cells have also been described in *Schistosoma mansoni* [46,77], *Paracoccidioides brasiliensis* [78] and filarial infections [79], but their effect on parasite clearance was not established.

Other bacterial infections

In mucosal bacterial infections Tregs also have been implicated. In line with findings in inflammatory bowel disease, which was associated with a deficiency of Tregs [22], Treg frequencies were inversely associated with pathology in *H. pylori* infection: infected children had less tissue damage than infected adults, despite similar bacterial loads, and had more abundant levels of Tregs and Treg derived cytokines IL-10 and TGF-β [80], although other factors might have contributed to this difference as well. In addition, adults with gastric adenocarcinoma as a consequence of *H. pylori* infection had increased frequencies of Tregs in the tumor compared to normal gastric tissues [81]. This increase in Tregs could be a consequence of the infection, with subsequent downregulation of effector immunity that might have permitted tumor outgrowth, or alternatively, have been a consequence of the tumor. The precise cause-effect relationship remains to be resolved, given its important therapeutic implications.

Viral infections

Increased numbers of circulating $CD4^{+}CD25^{HI}$ cells at the time of infection have been described in *hepatitis B* [82,83] and *human papillomavirus* (HPV) [84-86] infections. In lymph nodes and tumors from patients with HPV-induced malignancies, $CD4^{+}CD25^{+}$ Tregs are present that are specific for HPV antigens [86]. These Tregs require antigen specific activation before they can inhibit proliferation and cytokine secretion by effector T-cells.

Furthermore, circulating $CD4^{+}CD25^{+}FoxP3^{+}$ Treg frequencies were increased in patients with acute *Dengue fever* [87]. The ratio of Tregs over effector T-cells was particularly increased in patients with milder disease, suggesting an association with protection from more severe infections which are characterized by hyper inflammation. Upon West Nile Virus infection, symptomatic donors had significantly lower Treg frequencies compared to asymptomatic infected individuals. Mice lacking Tregs developed lethal West Nile Virus infections, suggesting that Tregs protect against lethality of severe West Nile Virus disease [88].

The frequency of Tregs might represent a possible biomarker that discriminates between responders and non-responders early after initiation of anti-viral treatment, such as exemplified by *hepatitis B* infection [89]: non-responders maintained or further increased high frequencies of Tregs, while responders did not, supporting a role for Tregs in abrogation of pathogen clearance. Indeed, the increased frequency of $CD4^{+}CD25^{+}$ Tregs was significantly increased in chronic *hepatitis B* infection, and their frequency was highest in those with the highest viral loads in serum [90]. This suggested that Tregs directly inhibited cellular immunity and thereby viral clearance, resulting in chronic persistent *hepatitis B* infection [90].

Human $CD8^{+}$ Tregs in Infectious Diseases

In most studies, including those on infectious diseases in humans, Treg research has mainly focused on the role of various $CD4^{+}$ Treg subsets, partly due to parallel studies in mice, the emerging role of these cells in immune homeostasis, and the availability of markers to detect these $CD4^{+}$ Tregs. However, this has led to a scarcity of attention for the contribution of $CD8^{+}$ Tregs, which is ironical since CD8/OKT8 T-cells were originally described to be the main source of human T suppressor cells. The possible role of human $CD8^{+}$ Tregs has been described in only few infectious diseases, including particularly leprosy [17,18], tuberculosis [4], HIV [33] and HCV [35].

Mycobacterial infections

As mentioned above, $CD8^+$ Tregs were initially described in peripheral blood and lesions of leprosy patients with lepromatous disease. Such patients typically display an antigen specific defect in T-cell responsiveness to *M. leprae* [17,18]. Recently we identified two new subsets of $CD8^+$ Tregs in individuals that had been primed with mycobacterial antigens *in vivo* as evident from their response to PPD [4,16]. Together, our previous and recent studies suggest a significant role for $CD8^+$ Tregs in mycobacterial infections. The mechanisms of their induction and activation, as well as their effects on effector cells need to be assessed in more detail.

Viral infections

In HIV infection, $CD8^+$ Tregs can suppress HIV specific and unrelated immune responses via TGF-β1 [33]. These TGF-β producing $CD8^+$ Tregs had peptide specificities different from that of IFNγ secreting $CD8^+$ T-cells, suggesting that certain HIV epitopes can specifically induce Tregs, in analogy to our own findings for CD4 Tregs in leprosy. Moreover, $CD8^+CD25^-FoxP3^-IL\text{-}10^+$ Tregs from HIV patients were able to inhibit cytolysis and cytokine production by other $CD8^+$ T-cells, but these cells suppressed through a cell-cell contact dependent mechanism rather than through TGF-β [5]. In macaques peripheral $CD8^+CD25^+FoxP3^+$ Tregs were induced a few days after peak viral load during simian immunodeficiency virus infection, and correlated with viral load and inversely with CD4 counts [91,92]. Moreover, in these animals the presence and number of $CD8^+$ Tregs was associated with poor prognosis [91]. The expansion of $CD8^+$ Tregs was most pronounced in lymphoid and colorectal mucosal tissues, being preferential sites for viral replication [92]. Expansion of $CD8^+$ Tregs correlated directly with acute phase viremia and inversely with the magnitude of antiviral T-cell responses [92]. HIV infected humans also have increased numbers of $CD8^+CD25^+FoxP3^+$ T-cells [92]. Taken together these data suggest that HIV/SIV infections can directly induce $CD8^+$ Tregs that inhibit effector immune responses and efficient viral clearance.

Transplant recipients under chronic immunosuppression have a poor control of EBV (*Epstein Barr Virus*) infection, which correlated with the induction of a population of EBV specific $CD8^+FoxP3^+$ T-cells that suppressed $CD4^+$ T-cell proliferation [34]. However, these patients received immunosuppression to prevent graft rejection, including corticosteroids that are known to skew towards tolerogenic APCs that can favor Treg outgrowth [93]. Therefore, Treg induction might not be a direct effect of viral infection, although EBV specific, but rather a combined result of infection and immunosuppressive drug treatment.

Antigen Specificity & HLA Restriction of Tregs

Although in many infectious diseases Treg appearance and kinetics are associated with the timing and evolution of the infection, antigen specificity has only been addressed in a limited number of infections. As discussed before, specific induction of Tregs by selected pathogen derived antigens occurs in HIV and leprosy [33,56]. In HCV, antigen specific stimulation *in vitro* followed by tetramer staining allowed analysis of antigen specific Tregs [35]. In patients, circulating $CD8^+$ tetramer positive HCV specific cells expressed FoxP3,

CTLA4 and GITR and had a reduced proliferative capacity, while suppressing proliferation of responder T-cells in a contact-dependent manner [35]. Moreover, intrahepatic tetramer positive $CD8^+IL\text{-}10^+$ T-cells were mostly present in areas with reduced hepatic inflammation and tissue damage [94]. Functional analysis of circulating $CD3^+CD8^+CD25^-$ Tregs demonstrated that HCV specific immunity was suppressed in a TGF-β dependent manner, similar to the above findings in HIV infection [95]. A significant role for Tregs as inhibitors of clearance of HCV was indicated by longitudinal follow up studies that showed chronic HCV infection only in patients with persistent Treg responses, while loss of $CD4^+CD25^{HI}$ Treg activity was associated with clearance of HCV [96]. In conclusion, both $CD4^+$ and $CD8^+$ Tregs are found in several chronic, persistent infections and are associated with reduced pathogen clearance.

However, in addition to infectious diseases antigen specific $CD8^+$ Tregs to minor histocompatibility antigens have been identified that utilize CTLA4 for suppression [31]. These Tregs were identified in mothers that had been pregnant from sons with a minor histocompatibility difference (HY or HA-1), suggesting that even under physiological circumstances, in the absence of danger signals, antigen specific $CD8^+$ Tregs can be induced.

The genetic restriction of $CD8^+$ Tregs so far was limited to the analysis of HLA class Ia molecules, however more recently also Tregs restricted to non-classical HLA class Ib molecules have been identified. Mycobacteria reside in phagosomes of antigen presenting cells, especially macrophages, and these phagosomes are enriched in HLA-E, suggesting a relatively large presentation of mycobacterial peptides on human HLA-E [97]. Indeed, human $CD8^+$ T-cells were recently found to recognize Mtb peptides presented by human HLA-E, and to have functional properties of both Tregs and CTLs [16]. Interestingly, all T-cells that recognized peptides in the context of HLA-E had a more or less suppressive activity, suggesting that peptide presentation through HLA-E may skew towards Tregs. Multiple Mtb derived peptides bound HLA-E and were preferentially recognized by individuals with memory responses towards mycobacteria.

Location of Tregs Suggests Significant Role in Containment of Pathogens, but Timing of Treg Activity is Crucial

Tregs mediate there suppressive effect either through local secretion of suppressive cytokines or in a cell-cell contact dependent fashion, indicating that close proximity to the effector cells is required for optimal suppression. This would mean that Tregs need to be present at the site of disease rather than in the circulation. During pregnancy Tregs are specifically increased in the placenta and in cancer Tregs are highly prevalent among tumor infiltrating lymphocytes [29,30,32]. Also in infections Tregs can be isolated from lesions for example from leishmaniasis, fungal infections and leprosy skin lesions [18,74,78].

Liver biopsies from patients with chronic HCV infection revealed the presence of high numbers of $FoxP3^+$, mostly $CD4^+$, T-cells, in the portal tract infiltrates even though there was no association with HCV disease stage [98]. This was associated with HCV infection rather than general liver damage, since patients with primary biliary cirrhosis did not have increased numbers of $FoxP3^+$ cells in liver biopsies [98]. HCV antigen-specificity of liver $CD8^+IL\text{-}10^+$

T-cells was demonstrated using tetramers on tissue samples, and tetramer positive cells were indeed mostly located in areas with limited tissue damage and weak inflammation [94].

In mycobacterium induced granulomas we found 20 fold increased numbers of $CD8^+LAG\text{-}3^+CCL4^+$ Tregs compared to normal distant lymph node tissue [4]. Likewise, $FoxP3^+$ cells were increased in the vicinity of tuberculoid granulomas [61] and $CD4^+CD25^+$ Tregs were increased at the site of (extra)pulmonary tuberculosis compared to the circulation in tuberculosis patients [61-63,99].

The preferential location of these Treg subsets (both $CD4^+$ and $CD8^+$) at the site of infection in various diseases suggests an active role for Tregs at the site of disease, presumably involved in controlling local effector immunity and inflammation. Especially the localization close to the pathogen may suggest that Tregs can directly inhibit effector mechanisms involved in pathogen killing, thus promoting the development of persistent infection. Alternatively, Tregs may be induced at the site of infection, as was recently demonstrated for memory T-cells. Mice lacking secondary lymphoid organs developed granulomas, controlled Mtb growth and had antigen specific T-cells, including memory T-cells [100]. Even transferred T-cells were primed locally in the lungs, in a granuloma dependent manner [100]. The bacterial load in mice lacking the secondary lymphoid organs is similar compared to WT, suggesting that not only effector and memory T-cells are induced, but also Tregs, potentially also within the granuloma.

Tregs are the first cells to become activated after contact with antigen, showing STAT5 phosphorylation within hours after antigen triggering [101]. The rapid activation of Tregs may be to minimize the risk of autoimmunity/inflammatory tissue damage during the immune response to foreign antigen. Indeed, in viral infections in mice Tregs seem critical in the earliest stages after *herpes simplex* virus infection [102]. Depletion of Tregs before infection dramatically accelerated fatal infection with increased viral loads [102]. It was suggested that Tregs facilitate early protective responses to local infection by allowing a timely entry of immune cells into the infected tissue (figure 2). However, during infection with a lymphotropic virus, *Friend* virus, depletion of Tregs decreased viral levels in the spleen due to increased virus-specific $CD8^+$ T-cells [103]. Thus the effect of Tregs on viral infections may depend on the viral target cells and not comprise a general mechanism in the defense against viral infections.

However in mycobacterial infections, the presence of Tregs hampers pathogen clearance, since depletion of Tregs prior to infection reduced bacterial burden. This indicates that Tregs favor pathogen outgrowth, even in the early stages after infection (figure 2) [64,69-71]. This discrepancy between viral and (myco) bacterial infections remains unexplained, but may be intrinsic to the nature of the pathogen.

During malaria infections Tregs may play different roles depending on the stage of infection [104]. During the initial phase of infection they may hamper the inflammatory response required for clearance of infected cells, whereas later in infection, once parasitemia is under control, they may prevent the onset of severe immune-mediated pathology [104].

This emphasizes once more that the role of Tregs in infectious diseases is complex, multifactorial and may differ depending on time after infection but also on the type of pathogen (figure 2).

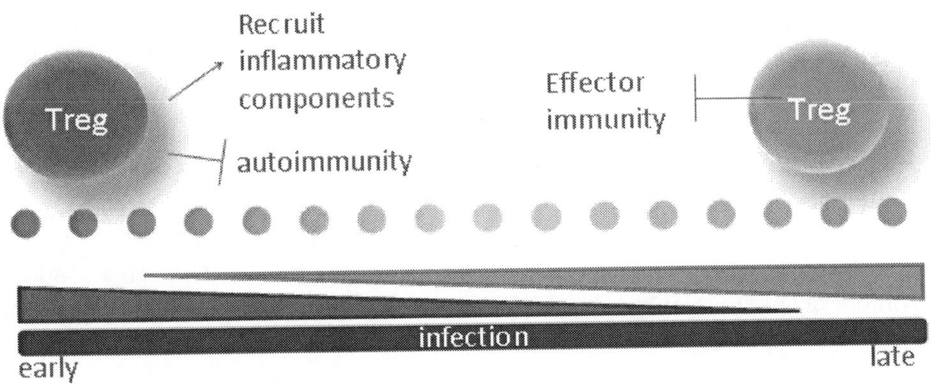

Figure 2. differential role of Tregs depending on stage of infection. Early after infection Tregs may be beneficial since they recruit inflammatory mediators and dampen autoimmunity (blue). However in later stages Tregs may prevent full effector immunity to become active and thus promote persistent infections (orange)

Tregs in TB Vaccination May Inhibit TB Protection: Lessons for Rational Vaccine Design

Protection induced by BCG vaccination against pulmonary TB in adults is poor and highly variable, even though the same vaccine affords good protection against severe TB in infants. There are many possible explanations for the lack of sufficient and consistent protection by BCG in adults (such as the masking and the blocking hypothesis) [105], but one which we want to discuss here is the recent discovery that mycobacteria, including BCG, can induce Tregs with high efficiency. BCG vaccination in animals results in induction of Tregs that hamper efficient immunity upon re-vaccination [55]. Our previous work showed the presence of human Tregs in leprosy [17], and in healthy mycobacterium responsive individuals [4]. $CD4^+$ and $CD8^+$ Treg activity was detectable in donors that had memory responses to PPD *in vitro,* most likely as a consequence of priming by environmental mycobacteria (NTM). We hypothesize that priming with NTM, as well as vaccination with BCG can both induce strong Treg activation *in vivo* (our unpublished observations)[106]. This implies that during subsequent infection with *M. tuberculosis* these Tregs will further expand, which could regulate and impair the development of optimal effector immunity. Indeed, when Tregs are depleted after BCG immunization but before MTb challenge, the lung mycobacterial load was significantly reduced, due to fine tuning of Th1 responses [71]. However, this effect may wane rapidly because the levels of Th1 responses were similar already a month after CD25 depletion [71]. These studies support the idea that vaccination with BCG can already induce Tregs, which subsequently influence effector immunity upon encounter of the pathogen.

At present, several novel vaccines against tuberculosis are entering clinical trials, most of them designed primarily to boost immunity induced by prior BCG vaccination [107]. It is known that BCG revaccination does not enhance BCG's protection to tuberculosis [108]. which might be a consequence, at least in part, of its efficient Treg induction: BCG vaccination could induce Tregs that subsequently dampen effector immunity against *M. tuberculosis*. Such BCG induced Tregs could then be enhanced by booster vaccines, and thereby further limit the efficacy of prime-boost vaccinations. Similarly, in tuberculosis endemic countries people are exposed to many environmental mycobacteria that share a high level of similarity with pathogenic mycobacteria; this cross-reactivity might be sufficient to induce similar Treg activation and expansion, and enhance susceptibility to tuberculosis. The effects of both BCG and environmental mycobacteria on Treg induction need to be analyzed in more detail since identification of the components involved in Treg induction may allow the design of vaccines that induce selectively protective effector cells. Also for other human diseases, more detailed understanding of the requirements for Treg induction, activation and expansion is urgently needed.

Conclusion

Tregs and chronic infections, which heavily influence and manipulate the host immune system, seem to be soul mates in many ways. Both the host and pathogen benefit from Tregs: the pathogen because Treg activity limits effector immunity, and the host because Treg activity limits inflammatory tissue damage. However, the host pays a price by persisting colonization by the pathogen, with the concurrent risk of disease reactivation later in life. Specific modulation of Tregs during infection may increase effector immunity and promote pathogen clearance, however this should be done with great care to prevent excessive pathogen induced tissue damage and disturbance of physiological homeostasis. Design of novel vaccines against major infectious diseases like tuberculosis should take into account current knowledge on Treg induction by both the vaccine and the pathogen itself.

Acknowledgments

The authors gratefully acknowledge all funding that made the work discussed in this review possible over the past few years. Especially we would like to acknowledge the European Commission for the 6th framework project TBVAC (contract LSHP-CT-2003-503367); the 7th framework project NEWTBVAC (contract HEALTH-F3-2009-241745); the Bill and Melinda Gates Foundation Grand Challenges in Global Health (GC 06-74 and GC12-82); the Netherlands Leprosy Relief foundation (grants ILEP 702.02.68 and 702.02.70); The Netherlands Organization for Scientific Research (VENI grant 916.86.115) and the "Gisela Thier foundation" of the Leiden University Medical Center.

References

[1] Mills, KH; Regulatory, T. cells: friend or foe in immunity to infection? *Nat Rev Immunol*, 2004, 4, 841-855.

[2] Tang, Q; Bluestone, JA. The Foxp3+ regulatory T cell: a jack of all trades; master of regulation. *Nat Immunol.,* 2008, 9, 239-244.

[3] Belkaid, Y; Regulatory, T. cells and infection: a dangerous necessity. *Nat Rev Immunol.,* 2007, 7, 875-888.

[4] Joosten, SA; van Meijgaarden, KE; Savage, ND; de Boer, T; Triebel, F; van der Wal, A; de Heer, E; Klein, MR; Geluk, A; Ottenhoff, TH. Identification of a human CD8+ regulatory T cell subset that mediates suppression through the chemokine CC chemokine ligand 4. *Proc Natl Acad Sci.,* U S A 2007, 104, 8029-8034.

[5] Elrefaei, M; Ventura, FL; Baker, CA; Clark, R; Bangsberg, DR; Cao, H. HIV-specific IL-10-positive CD8+ T cells suppress cytolysis and IL-2 production by CD8+ T cells. *J Immunol.,* 2007, 178, 3265-3271.

[6] Bacchetta, R; Passerini, L; Gambineri, E; Dai, M; Allan, SE; Perroni, L; gna-Bricarelli, F; Sartirana, C; Matthes-Martin, S; Lawitschka, A; Azzari, C; Ziegler, SF; Levings, MK; Roncarolo, MG. Defective regulatory and effector T cell functions in patients with FOXP3 mutations. *J Clin Invest.,* 2006, 116, 1713-1722.

[7] Fontenot, JD; Gavin, MA; Rudensky, AY. Foxp3 programs the development and function of CD4+CD25+ regulatory T cells. *Nat Immunol.,* 2003, 4, 330-336.

[8] Morgan, ME; van Bilsen, JH; Bakker, AM; Heemskerk, B; Schilham, MW; Hartgers, FC; Elferink, BG; van der Zanden, L; de Vries, RR; Huizinga, TW; Ottenhoff, TH; Toes, RE. Expression of FOXP3 mRNA is not confined to CD4+CD25+ T regulatory cells in humans. *Hum Immunol.,* 2005, 66, 13-20.

[9] Wang, J; Ioan-Facsinay, A; van der Voort, EI; Huizinga, TW; Toes, RE. Transient expression of FOXP3 in human activated nonregulatory CD4+ T cells. Eur *J Immunol.,* 2007, 37, 129-138.

[10] Zelenay, S; Lopes-Carvalho, T; Caramalho, I; Moraes-Fontes, MF; Rebelo, M; Demengeot, J. Foxp3+ CD25- CD4 T cells constitute a reservoir of committed regulatory cells that regain CD25 expression upon homeostatic expansion. *Proc Natl Acad Sci.,* U S A 2005, 102, 4091-4096.

[11] Vukmanovic-Stejic, M; Zhang, Y; Cook, JE; Fletcher, JM; McQuaid, A; Masters, JE; Rustin, MH; Taams, LS; Beverley, PC; Macallan, DC; Akbar, AN. Human CD4+ CD25hi Foxp3+ regulatory T cells are derived by rapid turnover of memory populations in vivo. *J Clin Invest.,* 2006, 116, 2423-2433.

[12] Koenen, HJ; Smeets, RL; Vink, PM; van, RE; Boots, AM; Joosten, I. Human CD25highFoxp3pos regulatory T cells differentiate into IL-17-producing cells. *Blood.,* 2008, 112, 2340-2352.

[13] Beriou, G; Costantino, CM; Ashley, CW; Yang, L; Kuchroo, VK; Baecher-Allan, C; Hafler DA. IL-17-producing human peripheral regulatory T cells retain suppressive function. *Blood.,* 2009, 113, 4240-4249.

[14] Voo, KS; Wang, YH; Santori, FR; Boggiano, C; Wang, YH; Arima, K; Bover, L; Hanabuchi, S; Khalili, J; Marinova, E; Zheng, B; Littman, DR; Liu, YJ. Identification of IL-17-producing FOXP3+ regulatory T cells in humans. *Proc Natl Acad Sci.,* U S A 2009, 106, 4793-4798.

[15] Rivino, L; Gruarin, P; Haringer, B; Steinfelder, S; Lozza, L; Steckel, B; Weick, A; Sugliano, E; Jarrossay, D; Kuhl, AA; Loddenkemper, C; Abrignani, S; Sallusto, F; Lanzavecchia, A; Geginat, J. CCR6 is expressed on an IL-10-producing; autoreactive

memory T cell population with context-dependent regulatory function. *J Exp Med.,* 2010, 207, 565-577.

[16] Joosten, SA; van Meijgaarden, KE; van Weeren, PC; Kazi, F; Geluk, A; Savage, ND; Drijfhout, JW; Flower, DR; Hanekom, WA; Klein, MR; Ottenhoff, TH. Mycobacterium tuberculosis peptides presented by HLA-E molecules are targets for human CD8 T-cells with cytotoxic as well as regulatory activity. *PLoS Pathog.,* 2010, 6, e1000782.

[17] Ottenhoff, TH; Elferink, DG; Klatser, PR; de Vries, RR. Cloned suppressor T cells from a lepromatous leprosy patient suppress *Mycobacterium leprae* reactive helper T cells. *Nature.,* 1986, 322, 462-464.

[18] Modlin, RL; Kato, H; Mehra, V; Nelson, EE; Fan, XD; Rea, TH; Pattengale, PK; Bloom, BR. Genetically restricted suppressor T-cell clones derived from lepromatous leprosy lesions. *Nature.,* 1986, 322, 459-461.

[19] Cosmi, L; Liotta, F; Lazzeri, E; Francalanci, M; Angeli, R; Mazzinghi, B; Santarlasci, V; Manetti, R; Vanini, V; Romagnani, P; Maggi, E; Romagnani, S; Annunziato, F. Human CD8+CD25+ thymocytes share phenotypic and functional features with CD4+CD25+ regulatory thymocytes. *Blood.,* 2003, 102, 4107-4114.

[20] Uss, E; Rowshani, AT; Hooibrink, B; Lardy, NM; van Lier, RA; ten Berge, IJ. CD103 is a marker for alloantigen-induced regulatory CD8+ T cells. *J Immunol.,* 2006, 177, 2775-2783.

[21] Colovai, AI; Mirza, M; Vlad, G; Wang, S; Ho, E; Cortesini, R; Suciu-Foca, N. Regulatory CD8+CD28- T cells in heart transplant recipients. *Hum Immunol.,* 2003, 64, 31-37.

[22] Brimnes, J; Allez, M; Dotan, I; Shao, L; Nakazawa, A; Mayer, L. Defects in CD8+ regulatory T cells in the lamina propria of patients with inflammatory bowel disease. *J Immunol.,* 2005, 174, 5814-5822.

[23] Jarvis, LB; Matyszak, MK; Duggleby, RC; Goodall, JC; Hall, FC; Gaston, JS. Autoreactive human peripheral blood CD8+ T cells with a regulatory phenotype and function. Eur *J Immunol.,* 2005, 35, 2896-2908.

[24] Bisikirska, B; Colgan, J; Luban, J; Bluestone, JA; Herold, KC. TCR stimulation with modified anti-CD3 mAb expands CD8+ T cell population and induces CD8+CD25+ Tregs. *J Clin Invest.,* 2005, 115, 2904-2913.

[25] Tennakoon, DK; Mehta, RS; Ortega, SB; Bhoj, V; Racke, MK; Karandikar, NJ. Therapeutic induction of regulatory; cytotoxic CD8+ T cells in multiple sclerosis. *J Immunol.,* 2006, 176, 7119-7129.

[26] Chang, CC; Ciubotariu, R; Manavalan, JS; Yuan, J; Colovai, AI; Piazza, F; Lederman, S; Colonna, M; Cortesini, R; Dalla-Favera, R; Suciu-Foca, N. Tolerization of dendritic cells by T(S) cells: the crucial role of inhibitory receptors ILT3 and ILT4. *Nat Immunol.,* 2002, 3, 237-243.

[27] Cai, J; Lee, J; Jankowska-Gan, E; Derks, R; Pool, J; Mutis, T; Goulmy, E; Burlingham, WJ; Minor, H. antigen HA-1-specific regulator and effector CD8+ T cells; and HA-1 microchimerism; in allograft tolerance. *J Exp Med.,* 2004, 199, 1017-1023.

[28] Zhang, L; Bertucci, AM; Ramsey-Goldman, R; Burt, RK; Datta, SK; Regulatory, T. cell (Treg) subsets return in patients with refractory lupus following stem cell transplantation; and TGF-beta-producing CD8+ Treg cells are associated with immunological remission of lupus. *J Immunol.,* 2009, 183, 6346-6358.

[29] Tilburgs, T; Roelen, DL; van der Mast, BJ; van Schip, JJ; Kleijburg, C; de Groot-Swings, GM; Kanhai, HH; Claas, FH; Scherjon, SA. Differential distribution of CD4(+)CD25(bright) and CD8(+)CD28(-) T-cells in decidua and maternal blood during human pregnancy. *Placenta.,* 2006, 27 Suppl A:S47-S53.

[30] Shao, L; Jacobs, AR; Johnson, VV; Mayer, L. Activation of CD8+ regulatory T cells by human placental trophoblasts. *J Immunol.,* 2005, 174, 7539-7547.

[31] van Halteren, AG; Jankowska-Gan, E; Joosten, A; Blokland, E; Pool, J; Brand, A; Burlingham, WJ; Goulmy, E. Naturally acquired tolerance and sensitization to minor histocompatibility antigens in healthy family members. *Blood.,* 2009, 114, 2263-2272.

[32] Filaci, G; Fenoglio, D; Fravega, M; Ansaldo, G; Borgonovo, G; Traverso, P; Villaggio, B; Ferrera, A; Kunkl, A; Rizzi, M; Ferrera, F; Balestra, P; Ghio, M; Contini, P; Setti, M; Olive, D; Azzarone, B; Carmignani, G; Ravetti, JL; Torre, G; Indiveri, F. CD8+ CD28- T regulatory lymphocytes inhibiting T cell proliferative and cytotoxic functions infiltrate human cancers. *J Immunol.,* 2007, 179, 4323-4334.

[33] Garba, ML; Pilcher, CD; Bingham, AL; Eron, J; Frelinger, JA. HIV antigens can induce TGF-beta(1)-producing immunoregulatory CD8+ T cells. *J Immunol.,* 2002, 168, 2247-2254.

[34] Popescu, I; Macedo, C; Abu-Elmagd, K; Shapiro, R; Hua, Y; Thomson, AW; Morelli, AE; Storkus, WJ; Metes, D. EBV-specific CD8+ T cell reactivation in transplant patients results in expansion of CD8+ type-1 regulatory T cells. *Am J Transplant*, 2007, 7, 1215-1223.

[35] Billerbeck, E; Blum, HE; Thimme, R. Parallel expansion of human virus-specific FoxP3- effector memory and de novo-generated FoxP3+ regulatory CD8+ T cells upon antigen recognition in vitro. *J Immunol.,* 2007, 179, 1039-1048.

[36] O'Garra, A; Vieira, P. T(H)1 cells control themselves by producing interleukin-10. *Nat Rev Immunol.,* 2007, 7, 425-428.

[37] Ablamunits, V; Bisikirska, BC; Herold, KC. Human regulatory CD8 T cells. *Ann N Y Acad Sci.,* 2008, 1150, 234-238.

[38] Nakamura, K; Kitani, A; Strober, W. Cell contact-dependent immunosuppression by CD4(+)CD25(+) regulatory T cells is mediated by cell surface-bound transforming growth factor beta. *J Exp Med.,* 2001, 194, 629-644.

[39] Savage, ND; de Boer, T; Walburg, KV; Joosten, SA; van Meijgaarden, K; Geluk, A; Ottenhoff, TH. Human anti-inflammatory macrophages induce Foxp3+ GITR+ CD25+ regulatory T cells; which suppress via membrane-bound TGFbeta-1. *J Immunol.,* 2008, 181, 2220-2226.

[40] Yamada, Y; Sugita, S; Horie, S; Yamagami, S; Mochizuki, M. Mechanisms of immune suppression for CD8+ T cells by human corneal endothelial cells via membrane-bound TGF{beta}. *Invest Ophthalmol Vis Sci*, 2010, 51, 2548-2557.

[41] Suciu-Foca, N; Cortesini, R. Central role of ILT3 in the T suppressor cell cascade. *Cell Immunol*, 2007, 248, 59-67.

[42] Moller, G. Do suppressor T cells exist? Scand *J Immunol.,* 1988, 27, 247-250.

[43] Kinter, AL; Horak, R; Sion, M; Riggin, L; McNally, J; Lin, Y; Jackson, R; O'shea, A; Roby, G; Kovacs, C; Connors, M; Migueles, SA; Fauci, AS. CD25+ regulatory T cells isolated from HIV-infected individuals suppress the cytolytic and nonlytic antiviral activity of HIV-specific CD8+ T cells in vitro. *AIDS Res Hum Retroviruses*, 2007, 23, 438-450.

[44] Mempel, TR; Pittet, MJ; Khazaie, K; Weninger, W; Weissleder, R; von, BH; von Andrian, UH; Regulatory, T. cells reversibly suppress cytotoxic T cell function independent of effector differentiation. *Immunity*, 2006, 25, 129-141.

[45] Tiemessen, MM; Jagger, AL; Evans, HG; van Herwijnen, MJ; John, S; Taams, LS. CD4+CD25+Foxp3+ regulatory T cells induce alternative activation of human monocytes/macrophages. *Proc Natl Acad Sci.,* U S A 2007, 104, 19446-19451.

[46] Maizels, RM; Pearce, EJ; Artis, D; Yazdanbakhsh, M; Wynn, TA. Regulation of pathogenesis and immunity in helminth infections. *J Exp Med.,* 2009, 206, 2059-2066.

[47] Piconese, S; Gri, G; Tripodo, C; Musio, S; Gorzanelli, A; Frossi, B; Pedotti, R; Pucillo, CE; Colombo, MP. Mast cells counteract regulatory T-cell suppression through interleukin-6 and OX40/OX40L axis toward Th17-cell differentiation. *Blood.,* 2009, 114, 2639-2648.

[48] Ozdemir, C; Akdis, M; Akdis, CA. T regulatory cells and their counterparts: masters of immune regulation. *Clin Exp Allergy*, 2009, 39, 626-639.

[49] de Vries, V; Wasiuk, A; Bennett, KA; Benson, MJ; Elgueta, R; Waldschmidt, TJ; Noelle, RJ. Mast cell degranulation breaks peripheral tolerance. *Am J Transplant.,* 2009, 9, 2270-2280.

[50] Jamieson, SE; Miller, EN; Black, GF; Peacock, CS; Cordell, HJ; Howson, JM; Shaw, MA; Burgner, D; Xu, W; Lins-Lainson, Z; Shaw, JJ; Ramos, F; Silveira, F; Blackwell, JM. Evidence for a cluster of genes on chromosome 17q11-q21 controlling susceptibility to tuberculosis and leprosy in Brazilians. *Genes Immun*, 2004, 5, 46-57.

[51] Algood, HM; Flynn, JL. CCR5-deficient mice control *Mycobacterium tuberculosis* infection despite increased pulmonary lymphocytic infiltration. *J Immunol.,* 2004, 173, 3287-3296.

[52] Ashton, LJ; Stewart, GJ; Biti, R; Law, M; Cooper, DA; Kaldor, JM. Heterozygosity for CCR5-Delta32 but not CCR2b-64I protects against certain intracellular pathogens. *HIV Med*, 2002, 3, 91-96.

[53] Ahlenstiel, G; Woitas, RP; Rockstroh, J; Spengler, U. CC-chemokine receptor 5 (CCR5) in hepatitis C--at the crossroads of the antiviral immune response? *J Antimicrob Chemother*, 2004, 53, 895-898.

[54] Lim, JK; Louie, CY; Glaser, C; Jean, C; Johnson, B; Johnson, H; McDermott, DH; Murphy, PM. Genetic deficiency of chemokine receptor CCR5 is a strong risk factor for symptomatic West Nile virus infection: a meta-analysis of 4 cohorts in the US epidemic. *J Infect Dis*, 2008, 197, 262-265.

[55] Ha, TY; Waksman, BH; Treffers, HP. The thymic suppressor cell. I. Separation of subpopulations with suppressor activity. *J Exp Med.,* 1974, 139, 13-23.

[56] Mutis, T; Cornelisse, YE; Datema, G; van den Elsen, PJ; Ottenhoff, TH; de Vries, RR. Definition of a human suppressor T-cell epitope. *Proc Natl Acad Sci.,* U S A 1994, 91, 9456-9460.

[57] Vukmanovic-Stejic, M; Agius, E; Booth, N; Dunne, PJ; Lacy, KE; Reed, JR; Sobande, TO; Kissane, S; Salmon, M; Rustin, MH; Akbar, AN. The kinetics of CD4+Foxp3+ T cell accumulation during a human cutaneous antigen-specific memory response in vivo. *J Clin Invest.,* 2008, 118, 3639-3650.

[58] Babu, S; Bhat, SQ; Kumar, NP; Kumaraswami, V; Nutman, TB. Regulatory, T. cells modulate Th17 responses in patients with positive tuberculin skin test results. *J Infect Dis*, 2010, 201, 20-31.

[59] Boussiotis, VA; Tsai, EY; Yunis, EJ; Thim, S; Delgado, JC; Dascher, CC; Berezovskaya, A; Rousset, D; Reynes, JM; Goldfeld, AE. IL-10-producing T cells suppress immune responses in anergic tuberculosis patients. *J Clin Invest.,* 2000, 105, 1317-1325.

[60] Delgado, JC; Tsai, EY; Thim, S; Baena, A; Boussiotis, VA; Reynes, JM; Sath, S; Grosjean, P; Yunis, EJ; Goldfeld, AE. Antigen-specific and persistent tuberculin anergy in a cohort of pulmonary tuberculosis patients from rural Cambodia. *Proc Natl Acad Sci.,* U S A 2002, 99, 7576-7581.

[61] Chen, X; Zhou, B; Li, M; Deng, Q; Wu, X; Le, X; Wu, C; Larmonier, N; Zhang, W; Zhang, H; Wang, H; Katsanis, E. CD4(+)CD25(+)FoxP3(+) regulatory T cells suppress Mycobacterium tuberculosis immunity in patients with active disease. *Clin Immunol,* 2007, 123, 50-59.

[62] Guyot-Revol, V; Innes, JA; Hackforth, S; Hinks, T; Lalvani, A. Regulatory, T. Cells are Expanded in *Blood.,* and Disease Sites in Tuberculosis Patients. *Am J Respir Crit Care Med.,* 2005, 173, 803-810.

[63] Ribeiro-Rodrigues, R; Resende, CT; Rojas, R; Toossi, Z; Dietze, R; Boom, WH; Maciel, E; Hirsch, CS. A role for CD4+CD25+ T cells in regulation of the immune response during human tuberculosis. *Clin Exp Immunol*, 2006, 144, 25-34.

[64] Scott-Browne, JP; Shafiani, S; Tucker-Heard, G; Ishida-Tsubota, K; Fontenot, JD; Rudensky, AY; Bevan, MJ; Urdahl, KB. Expansion and function of Foxp3-expressing T regulatory cells during tuberculosis. *J Exp Med.,* 2007, 204, 2159-2169.

[65] Sharma, PK; Saha, PK; Singh, A; Sharma, SK; Ghosh, B; Mitra, DK. FoxP3+ regulatory T cells suppress effector T-cell function at pathologic site in miliary tuberculosis. *Am J Respir Crit Care Med.,* 2009, 179, 1061-1070.

[66] Burl, S; Hill, PC; Jeffries, DJ; Holland, MJ; Fox, A; Lugos, MD; Adegbola, RA; Rook, GA; Zumla, A; McAdam, KP; Brookes, RH. FOXP3 gene expression in a tuberculosis case contact study. *Clin Exp Immunol*, 2007, 149, 117-122.

[67] Roberts, T; Beyers, N; Aguirre, A; Walzl, G. Immunosuppression during active tuberculosis is characterized by decreased interferon- gamma production and CD25 expression with elevated forkhead box P3, transforming growth factor- beta; and interleukin-4 mRNA levels. *J Infect Dis*, 2007, 195, 870-878.

[68] Wu, B; Huang, C; Kato-Maeda, M; Hopewell, PC; Daley, CL; Krensky, AM; Clayberger, C. Messenger RNA expression of IL-8, FOXP3, and IL-12beta differentiates latent tuberculosis infection from disease. *J Immunol.,* 2007, 178, 3688-3694.

[69] Kursar, M; Koch, M; Mittrucker, HW; Nouailles, G; Bonhagen, K; Kamradt, T; Kaufmann, SH. Cutting Edge: Regulatory T Cells Prevent Efficient Clearance of Mycobacterium tuberculosis. *J Immunol.,* 2007, 178, 2661-2665.

[70] Ozeki, Y; Sugawara, I; Udagawa, T; Aoki, T; Osada-Oka, M; Tateishi, Y; Hisaeda, H; Nishiuchi, Y; Harada, N; Kobayashi, K; Matsumoto, S. Transient role of CD4+CD25+ regulatory T cells in mycobacterial infection in mice. *Int Immunol*, 2010, 22, 179-189.

[71] Jaron, B; Maranghi, E; Leclerc, C; Majlessi, L. Effect of attenuation of Treg during BCG immunization on anti-mycobacterial Th1 responses and protection against Mycobacterium tuberculosis. *PLoS One*, 2008, 3, e2833.

[72] Suffia, IJ; Reckling, SK; Piccirillo, CA; Goldszmid, RS; Belkaid, Y. Infected site-restricted Foxp3+ natural regulatory T cells are specific for microbial antigens. *J Exp Med.,* 2006, 203, 777-788.

[73] Belkaid, Y; Piccirillo, CA; Mendez, S; Shevach, EM; Sacks, DL. CD4+CD25+ regulatory T cells control *Leishmania major* persistence and immunity. *Nature.,* 2002, 420, 502-507.

[74] Campanelli, AP; Roselino, AM; Cavassani, KA; Pereira, MS; Mortara, RA; Brodskyn, CI; Goncalves, HS; Belkaid, Y; Barral-Netto, M; Barral, A; Silva, JS. CD4+CD25+ T cells in skin lesions of patients with cutaneous leishmaniasis exhibit phenotypic and functional characteristics of natural regulatory T cells. *J Infect Dis*, 2006, 193, 1313-1322.

[75] Walther, M; Tongren, JE; Andrews, L; Korbel, D; King, E; Fletcher, H; Andersen, RF; Bejon, P; Thompson, F; Dunachie, SJ; Edele, F; de Souza, JB; Sinden, RE; Gilbert, SC; Riley, EM; Hill, AV. Upregulation of TGF-beta; FOXP3, and CD4+CD25+ regulatory T cells correlates with more rapid parasite growth in human malaria infection. *Immunity*, 2005, 23, 287-296.

[76] Finney, OC; Riley, EM; Walther, M. Phenotypic analysis of human peripheral blood regulatory T cells (CD4+FOXP3+CD127lo/-) ex vivo and after in vitro restimulation with malaria antigens. Eur *J Immunol.,* 2010, 40, 47-60.

[77] Watanabe, K; Mwinzi, PN; Black, CL; Muok, EM; Karanja, DM; Secor, WE; Colley, DG. T regulatory cell levels decrease in people infected with Schistosoma mansoni on effective treatment. *Am J Trop Med Hyg.,* 2007, 77, 676-682.

[78] Cavassani, KA; Campanelli, AP; Moreira, AP; Vancim, JO; Vitali, LH; Mamede, RC; Martinez, R; Silva, JS. Systemic and local characterization of regulatory T cells in a chronic fungal infection in humans. *J Immunol.,* 2006, 177, 5811-5818.

[79] Metenou, S; Dembele, B; Konate, S; Dolo, H; Coulibaly, SY; Coulibaly, YI; Diallo, AA; Soumaoro, L; Coulibaly, ME; Sanogo, D; Doumbia, SS; Traore, SF; Mahanty, S; Klion, A; Nutman, TB. At homeostasis filarial infections have expanded adaptive T regulatory but not classical Th2 cells. *J Immunol.,* 2010, 184, 5375-5382.

[80] Harris, PR; Wright, SW; Serrano, C; Riera, F; Duarte, I; Torres, J; Pena, A; Rollan, A; Viviani, P; Guiraldes, E; Schmitz, JM; Lorenz, RG; Novak, L; Smythies, LE; Smith, PD. Helicobacter pylori gastritis in children is associated with a regulatory T-cell response. *Gastroenterology*, 2008, 134, 491-499.

[81] Enarsson, K; Lundgren, A; Kindlund, B; Hermansson, M; Roncador, G; Banham, AH; Lundin, BS; Quiding-Jarbrink, M. Function and recruitment of mucosal regulatory T cells in human chronic Helicobacter pylori infection and gastric adenocarcinoma. *Clin Immunol*, 2006, 121, 358-368.

[82] Peng, G; Li, S; Wu, W; Sun, Z; Chen, Y; Chen, Z. Circulating CD4+ CD25+ regulatory T cells correlate with chronic hepatitis B infection. *Immunology*, 2008, 123, 57-65.

[83] Lin, CY; Tsai, MC; Huang, CT; Hsu, CW; Tseng, SC; Tsai, IF; Chen, YC; Yeh, CT; Sheen, IS; Chien, RN. Liver injury is associated with enhanced regulatory T-cell activity in patients with chronic hepatitis B. *J Viral Hepat*, 2007, 14, 503-511.

[84] Visser, J; Nijman, HW; Hoogenboom, BN; Jager, P; van, BD; Schuuring, E; Abdulahad, W; Miedema, F; van der Zee, AG; Daemen, T. Frequencies and role of regulatory T cells in patients with (pre)malignant cervical neoplasia. *Clin Exp Immunol*, 2007, 150, 199-209.

[85] Molling, JW; de Gruijl, TD; Glim, J; Moreno, M; Rozendaal, L; Meijer, CJ; van den Eertwegh, AJ; Scheper, RJ; von Blomberg, ME; Bontkes, HJ. CD4(+)CD25hi regulatory T-cell frequency correlates with persistence of human papillomavirus type 16 and T helper cell responses in patients with cervical intraepithelial neoplasia. *Int J Cancer*, 2007, 121, 1749-1755.

[86] van der Burg, SH; Piersma, SJ; de, JA; van der Hulst, JM; Kwappenberg, KM; van den, HM; Welters, MJ; Van Rood, JJ; Fleuren, GJ; Melief, CJ; Kenter, GG; Offringa, R. Association of cervical cancer with the presence of CD4+ regulatory T cells specific for human papillomavirus antigens. *Proc Natl Acad Sci.,* U S A 2007, 104, 12087-12092.

[87] Luhn, K; Simmons, CP; Moran, E; Dung, NT; Chau, TN; Quyen, NT; Thao le, TT; Van, NT; Dung, NM; Wills, B; Farrar, J; McMichael, AJ; Dong, T; Rowland-Jones, S. Increased frequencies of CD4+ CD25(high) regulatory T cells in acute dengue infection. *J Exp Med.,* 2007, 204, 979-985.

[88] Lanteri, MC; O'Brien, KM; Purtha, WE; Cameron, MJ; Lund, JM; Owen, RE; Heitman, JW; Custer, B; Hirschkorn, DF; Tobler, LH; Kiely, N; Prince, HE; Ndhlovu, LC; Nixon, DF; Kamel, HT; Kelvin, DJ; Busch, MP; Rudensky, AY; Diamond, MS; Norris, PJ. Tregs control the development of symptomatic West Nile virus infection in humans and mice. *J Clin Invest.,* 2009, 119, 3266-3277.

[89] Sprengers, D; Stoop, JN; Binda, RS; Kusters, JG; Haagmans, BL; Carotenuto, P; Artsen, A; van der Molen, RG; Janssen, HL. Induction of regulatory T-cells and interleukin-10-producing cells in non-responders to pegylated interferon-alpha therapy for chronic hepatitis B. *Antivir Ther,* 2007, 12, 1087-1096.

[90] Nan, XP; Zhang, Y; Yu, HT; Li, Y; Sun, RL; Wang, JP; Bai, XF. Circulating CD4+CD25high regulatory T cells and expression of PD-1 and BTLA on CD4+ T cells in patients with chronic hepatitis B virus infection. *Viral Immunol*, 2010, 23, 63-70.

[91] Karlsson, I; Malleret, B; Brochard, P; Delache, B; Calvo, J; Le, GR; Vaslin, B. FoxP3+ CD25+ CD8+ T-cell induction during primary simian immunodeficiency virus infection in cynomolgus macaques correlates with low CD4+ T-cell activation and high viral load. *J Virol*, 2007, 81, 13444-13455.

[92] Nigam, P; Velu, V; Kannanganat, S; Chennareddi, L; Kwa, S; Siddiqui, M; Amara, RR. Expansion of FOXP3+ CD8 T cells with suppressive potential in colorectal mucosa following a pathogenic simian immunodeficiency virus infection correlates with diminished antiviral T cell response and viral control. *J Immunol.,* 2010, 184, 1690-1701.

[93] de Jong, EC; Smits, HH; Kapsenberg, ML. Dendritic cell-mediated T cell polarization. *Springer Semin Immunopathol*, 2005, 26, 289-307.

[94] Abel, M; Sene, D; Pol, S; Bourliere, M; Poynard, T; Charlotte, F; Cacoub, P; Caillat-Zucman, S. Intrahepatic virus-specific IL-10-producing CD8 T cells prevent liver damage during chronic hepatitis C virus infection. *Hepatology*, 2006, 44, 1607-1616.

[95] Alatrakchi, N; Graham, CS; van, d; V; Sherman, KE; Exley, MA; Koziel, MJ. Hepatitis C virus (HCV)-specific CD8+ cells produce transforming growth factor beta that can suppress HCV-specific T-cell responses. *J Virol*, 2007, 81, 5882-5892.

[96] Smyk-Pearson, S; Golden-Mason, L; Klarquist, J; Burton, JR; Jr; Tester, IA; Wang, CC; Culbertson, N; Vandenbark, AA; Rosen, HR. Functional suppression by FoxP3+CD4+CD25(high) regulatory T cells during acute hepatitis C virus infection. *J Infect Dis*, 2008, 197, 46-57.

[97] Grotzke, JE; Harriff, MJ; Siler, AC; Nolt, D; Delepine, J; Lewinsohn, DA; Lewinsohn, DM. The Mycobacterium tuberculosis phagosome is a HLA-I processing competent organelle. *PLoS Pathog.,* 2009, 5, e1000374.

[98] Ward, SM; Fox, BC; Brown, PJ; Worthington, J; Fox, SB; Chapman, RW; Fleming, KA; Banham, AH; Klenerman, P. Quantification and localisation of FOXP3+ T lymphocytes and relation to hepatic inflammation during chronic HCV infection. *J Hepatol*, 2007, 47, 316-324.

[99] Hougardy, JM; Place, S; Hildebrand, M; Drowart, A; Debrie, AS; Locht, C; Mascart, F. Regulatory T cells depress immune responses to protective antigens in active tuberculosis. *Am J Respir Crit Care Med.,* 2007, 176, 409-416.

[100] Day, TA; Koch, M; Nouailles, G; Jacobsen, M; Kosmiadi, GA; Miekley, D; Kuhlmann, S; Jorg, S; Gamradt, P; Mollenkopf, HJ; Hurwitz, R; Reece, ST; Kaufmann, SH; Kursar, M. Secondary lymphoid organs are dispensable for the development of T-cell-mediated immunity during tuberculosis. Eur *J Immunol.,* 2010, 40, 1-11.

[101] O'Gorman, WE; Dooms, H; Thorne, SH; Kuswanto, WF; Simonds, EF; Krutzik, PO; Nolan, GP; Abbas, AK. The initial phase of an immune response functions to activate regulatory T cells. *J Immunol.,* 2009, 183, 332-339.

[102] Lund, JM; Hsing, L; Pham, TT; Rudensky, AY. Coordination of early protective immunity to viral infection by regulatory T cells. *Science*, 2008, 320, 1220-1224.

[103] Zelinskyy, G; Dietze, K; Sparwasser, T; Dittmer, U; Regulatory, T. cells suppress antiviral immune responses and increase viral loads during acute infection with a lymphotropic retrovirus. *PLoS Pathog.,* 2009, 5, e1000406.

[104] Finney, OC; Riley, EM; Walther, M; Regulatory, T. cells in malaria--friend or foe? *Trends Immunol*, 2010, 31, 63-70.

[105] Andersen, P; Doherty, TM. The success and failure of BCG - implications for a novel tuberculosis vaccine. *Nat Rev Microbiol*, 2005, 3, 656-662.

[106] Hanekom, WA. The immune response to BCG vaccination of newborns. *Ann N Y Acad Sci.,* 2005, 1062, 69-78.

[107] Kaufmann, SH. Envisioning future strategies for vaccination against tuberculosis. *Nat Rev Immunol.,* 2006, 6, 699-704.

[108] Rodrigues, LC; Pereira, SM; Cunha, SS; Genser, B; Ichihara, MY; de Brito, SC; Hijjar, MA; Dourado, I; Cruz, AA; Sant'Anna, C; Bierrenbach, AL; Barreto, ML. Effect of BCG revaccination on incidence of tuberculosis in school-aged children in Brazil: the BCG-REVAC cluster-randomised trial. *Lancet*, 2005, 366, 1290-1295.

In: Regulatory T Cells
Editor: Ren S. Hayashi, pp. 57-78

ISBN: 978-1-61761-554-2

Chapter 3

Role of Regulatory t Cells in the Treatment of Multiple Sclerosis

***K. Szklany*[1], *T. C. Koehler*[2*] *and W. van Eden*[3]**
[1]Department of Pharmaceutical Sciences, Faculty of Science, Utrecht University
[2]Pharmacology and Pathophysiology, Department of Pharmaceutical Sciences, Faculty of Science, Utrecht University
[3]Department of Infectious Diseases & Immunology, Faculty of Veterinary Medicine, Utrecht University

Abstract

In the last years knowledge of regulatory T cells has expanded. They play a role in autoimmune disease but it is still unclear how they suppress auto-reactive T cells exactly. The possibilities to use regulatory T cells in the treatment of MS are discussed in this chapter. The focus will be on the function of HLA-G and $CD4^+CD25^+$ T cells expressing FoxP3 and CTLA-4.

Two types of Tregs with different ways of suppression are known. Adaptive Tregs are induced in the periphery and mainly secrete TGF-β and IL-10. These cytokines mediate suppression of T cells through induction of FoxP3. Natural Tregs are generated in the thymus and suppress T cells through HLA-G, CTLA-4 and FoxP3. HLA-G interrupts the signalling in APCs and T cells resulting in suppression. FoxP3 blocks the expression of IL-2 and consequently inhibits proliferation of T cells. The expression of FoxP3 is induced by TGF-β. TGF-β also induces $CD4^+CD25^-$ T cells to convert into $CD4^+CD25^+$ T cells which express FoxP3 and hence suppress T cell proliferation. The suppression mechanism of CTLA-4 is still unknown. It might be a positive feedback with TGF-β resulting in increased expression of FoxP3 or a block of co-stimulatory molecules on APCs causing reduced secretion of cytokines following reduced activation of T cells.

* Corresponding author: Department of Pharmaceutical Sciences, Faculty of Science, Utrecht University, The Netherlands. Email: T.C.Koehler@uu.nl

T_H17 cells also have an impact on MS. T_H17 cells are induced by IL-6 and TGF-β. These cytokines start the differentiation and activate the transcription factor ROR-γt that makes a T_H17 population expand. The presence of IL-6 inhibits the expression of FoxP3. Presence of TGF-β and absence of IL-6 induces FoxP3 which inhibits the expression of ROR-γt leading to the production of pro-inflammatory cytokines. Hence, FoxP3 and ROR-γt are of great influence and a disbalance between these factors will lead to a skewed balance in favour of T_H17 cells.

Consequently, increased levels of TGF-β will induce expression of FoxP3 and could normalise the situation.

The current treatment of MS is mainly symptomatic. Glatiramer acetate can promote activation of FoxP3, hence enhancing the function of Tregs. IFN-β up-regulates the expression of HLA-G. Unfortunately, these drugs do not cure MS, since glatiramer acetate and IFN-β are not able to skew the dysbalance back to normal.

Because TGF-β and HLA-G are able to induce the function of Tregs; the conclusion is that administration of TGF-β and HLA-G appears to be a future treatment of MS.

1. Introduction

1.1. Introduction

Genetic and environmental factors can affect the immune system resulting in a defect. The neurological disease multiple sclerosis (MS) is an example of an autoimmune disease caused by a defect in the immune system. More than 2,000,000 people worldwide suffer from MS. [1] The pathophysiology of MS is not exactly known making it difficult to treat patients suffering from this disease.

Regulatory T cells (Tregs) are a group of cells that contribute to prevent the immune system to react against endogens which can cause an autoimmune reaction. These Tregs are able to suppress immunological T cells which are activated to attack endogens. [2] Through this suppression the effect of T cells is reduced, preventing the immune system to react against the endogen.

Different types of Tregs are identified; one type is generated in the thymus and is called natural Tregs. Another type is induced Tregs, they are induced in the lymph nodes in the periphery. [3; 4] How exactly Tregs mediate their suppressive function and if they can be used in the treatment of MS is still unknown. [3] But it is commonly known that Tregs play an important role in autoimmune disease and this suggests that these cells could be used in the treatment of MS.

The aim of this chapter was to answer the question if regulatory T cells can be used in the treatment of multiple sclerosis?

To answer this question the current knowledge of Tregs and MS is described. This chapter first describes the different subtypes of Tregs. The focus will be on the natural Tregs and the function of the transcription factor forkhead box protein 3 (FoxP3), the human leukocyte antigen G (HLA-G), a major histocompatibility complex (MHC) class I molecule, and the cytotoxic T lymphocyte-associated antigen-4 (CTLA-4), a co-stimulatory molecule. These molecules are thought to play an important regulating role in the immune system. Secondly, the pathophysiology of MS is described with a focus on T_H17 cells. Finally, the current treatment of MS and the mechanisms of actions are described.

In our concluding remarks we will connect our immunological insights on Tregs with the pathophysiology of MS, resulting in a number of possible treatment strategies.

2. Regulatory T cells

Regulatory T cells (Tregs) suppress the immune response to both foreign and self-antigens. [2] The Tregs are either of a natural or an adaptive type (Figure 1). The natural Tregs are identified as the $CD4^{+}CD25^{+}$ T cells and human leukocyte antigen G (HLA-G) regulatory cells. The adaptive regulatory T-cells are divided in subtypes being T regulatory 1 (T_R1) cells and T helper 3 (T_H3) cells. [4] The focus of this chapter is mainly on natural Tregs because it is thought that they play an important role in MS. [3] The adaptive Tregs are briefly described. Table 1 provides an overview of the characteristics of the four types of Tregs.

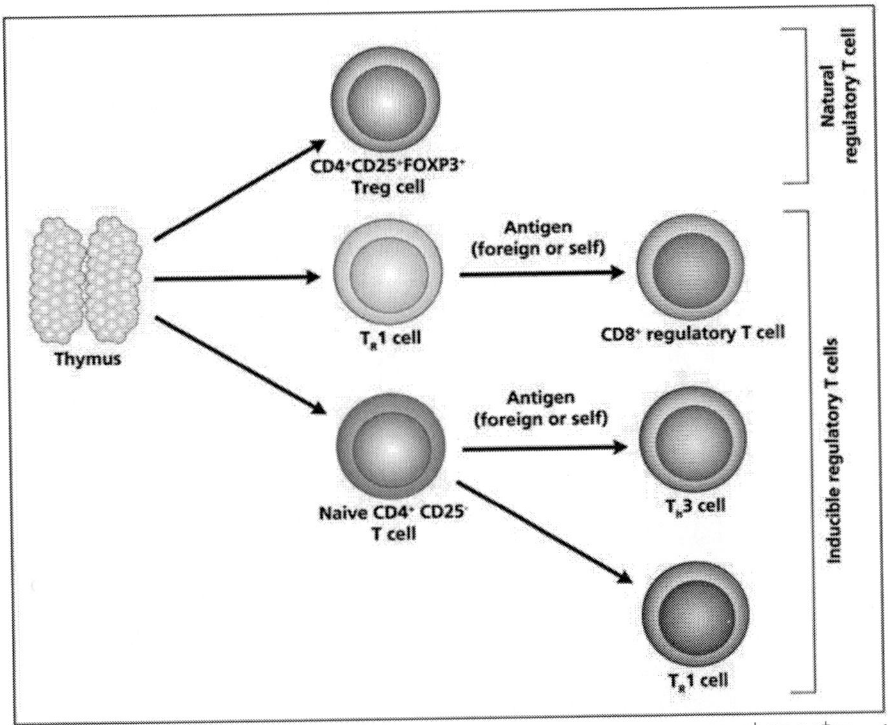

Figure 1. The different subtypes of regulatory T cells. The natural occurring $CD4^{+}CD25^{+}$ T cell generated in the thymus and the adaptive regulatory T cells induced in the periphery. (The $CD8^{+}$ subtype is not described in this chapter) Adapted from [2]

Table 1. Characteristic of the four types of Tregs. *Most secreted cytokine

	$CD4^{+}CD25^{+}$	HLA-G	T_R1	T_H3
Cell markers	CD25, CTLA-4, FoxP3, GITR	HLA-G	CD25, CTLA-4, FoxP3	CD25, CTLA-4, FoxP3
Cytokines secreted	IL-10, TGF-β	Soluble HLA-G	IL-10*, TGF-β, IL-4, IL-5	TGF-β*, IL-10, IL-4

2.1. Adaptive Regulatory T Cells

2.1.1. T_R1 cells

Characteristics/cell markers of T_R1 are as follows: CD25, cytotoxic T lymphocyte-associated antigen-4 (CTLA-4) and the transcription factor FoxP3. CD25 is an interleukin 2 (IL-2) receptor, necessary for the amplification of T cells. CTLA-4 is a co-stimulatory molecule that inhibits the signalling in the activation of T cells. FoxP3 interferes with the IL-2 promoter resulting in inhibition of the cell amplification. [5] T_R1 is induced in the periphery by an antigen presenting cell (APC) through a T-cell receptor (TCR) in an IL-10 rich environment. [3; 5] After activation T_R1 cells secrete mostly IL-10 but also transforming growth factor β (TGF-β), IL-4 and IL-5. These cytokines are able to suppress the function of APCs and effector T cells. [3]

When a foreign antigen gets through the physical barriers of the body it is phagocytised by an APC. Afterwards the APC presents a peptide of the phagocytised antigen to a T helper 1 (T_H1) cell and/or to a T helper 2 (T_H2) cell and these cells get activated. These activated T_H1 and T_H2 cells secrete mediators which induce inflammation to combat the damage made by foreign antigen. This response from T_H1 en T_H2 cells can be suppressed by T_R1 cells. When the concentration of an antigen increases in the body induction of T_R1 cells occurs. T_R1 cells possibly play a role in the treatment of allergy and/or autoimmune disease because the functions of T_H1 and T_H2 cells are down-regulated in the presence of T_R1 cells. [6]

2.1.2. T_H3 cells

T_H3 cells are present in mucosa and contain characteristics/cell markers like CD25, CTLA-4 and FoxP3. Compared to T_R1 cells the cell markers are the same but T_H3 secretes more TGF-β and less IL-10 than T_R1 cells. T_H3 cells also secrete IL-4 but no IL-5. [3; 4] T_H3 cells are activated in the mucosa by an APC and induced in the presence of IL-2 and TGF-β. [3] People who lack T_H3 cells have a higher risk of developing mucosa-related autoimmune diseases. [5]

2.2. Natural Regulatory T Cells

2.2.1. HLA-G

The HLA-G subtype of regulatory cells is able to suppress proliferation of T cells. [7; 8; 9; 10] HLA-G is a non-classical major histocompatibility complex (MHC) class I molecule that exists in different isoforms; Membrane bound HLA-G1, -G2, -G3, -G4 and soluble HLA-G5, -G6 and –G7. HLA-G1 and HLA-G5 are counterparts and are identical qua structure. [11; 12]

These HLA-G regulatory cells mediate suppression through a binding between soluble or membrane bound HLA-G and an inhibitory receptor on an APC and/or a T cell. These regulatory cells probably originate in the thymus. They express HLA-G1 and HLA-G5 but not FoxP3 and CD25 like other Tregs. [12] In contrast to T_R1 and T_H3 cells, HLA-G cells secrete no TGF-β [12] and IL-10 is only secreted when HLA-G cells are co-cultured with HLA-G negative cells. [8]

HLA-G expression is induced by inflammatory pathologies. [12; 13; 14; 15; 16] This occurs in the brain of MS patients too; the expression of HLA-G is increased compared to healthy controls. In contrast, peripheral expression of HLA-G is decreased in MS patients. [15; 17] The reason for this decrease could be an accumulation of HLA-G in the brain causing a peripheral decrease. [15] Wiendl et al. [15] compared the levels of HLA-G in MS patients with patients suffering from other neurological diseases and found a significant higher level of soluble HLA-G in MS patients. [15] HLA-G is expressed and secreted by macrophages and microglial cells. [12; 15]

The suppression by HLA-G cells occurs via three receptors being immunoglobulin-like transcript family (ILT) 2, ILT4 and killer cell immunoglobulin-like receptor (KIR). When HLA-G binds to i.e. an ILT2 receptor, four intracellular immunoreceptor tyrosine-based inhibitory motifs (ITIM) are phosphorylated. This phosphorylation recruits Src homology protein 1 and 2 and Src homology 2 domain-containing inositol phosphatases. When these phosphatases are activated they are able to suppress activating receptors resulting in inhibition of T cell proliferation. [7; 8; 9; 10; 18; 19; 20] The ILT2 receptor is expressed at a very low level in healthy tissue compared to inflammatory tissue. Since cells expressing this receptor are highly present in inflammatory tissues, this could be an explanation of the higher expression in inflammatory tissue. [15] It is known that IL-10 is very important to make the milieu suitable for HLA-G suppression. The exact role of IL-10 is still unknown. [8]

How exactly is HLA-G functioning suppressed? It is known that inhibitory receptors (ILT-2) play a role but is it direct or indirect inhibition via cytokines?

Wiendl et al. [21] and Feger et al. [12] suggested that suppression is mediated through a direct interaction between two APCs or between APC and T cell. Wiendl et al. [21] argue for a direct interaction because the ILT2 receptor is highly expressed on APCs. APCs are then able to reduce each other and themselves. [21] Feger et al. [12] present the argument that HLA-G cells make HLA-G negative cells secrete anti-inflammatory cytokines. The binding of HLA-G cells on the HLA-G negative cells occurs through the ILT2 receptor. The HLA-G cells then trigger the HLA-G negative cells to secrete anti-inflammatory cytokines. These cytokines then suppress the inflammatory cells. [12]

Bahri et al. [22] have the hypothesis that cell proliferation is inhibited through a dephosphorylation of the ζ-chain. HLA-G5 binds on a receptor (KIR, ILT2 or ILT4) resulting in activated phosphatases which remove phosphor groups from the ζ-chain. When the ζ-chain is no longer phosphorylated, TCR signalling decreases and the T cell is not able to enter cell proliferation. Unfortunately no decrease in the phosphorylation of the ζ-chain was found, [22] leading tot the assumption that the dephosporylation occurs later in the signalling pathway i.e. dephosphorylation of the proteins phosphorylated by ζ-chain-associated protein kinase 70 (ZAP70). After phosphorylation of the ζ-chain, ZAP70 phosphorylates proteins necessary for further signalling. A decrease in the phosphorylation of these proteins also results in a decreased TCR signalling.

2.2.2. $CD4^+CD25^+$

$CD4^+CD25^+$ regulatory T cells are the best characterised subtype so far. These cells are called natural Tregs because they are derived from the thymus. [3; 23] The cell markers of this subtype are mostly CD25 and CTLA-4 but FoxP3, glucocorticoid-induced tumor necrosis factor receptor (GITR) and CD39 are also present but in a lesser amount. [3; 5] The way of suppression is via cell-cell contact but cytokines may intervene as well. [5; 24] These cells only secrete a small amount of IL-10 and TGF-β but it might be enough to mediate suppression. [3; 5] $CD4^+CD25^+$ Tregs express the transcription factor FoxP3 which is important for development, function and maintenance of natural $CD4^+CD25^+FoxP3$ Tregs. [24; 25]

FoxP3 reduces the effect and function of effector T cells. FoxP3 interferes with the transcription factors activating protein 1 (AP-1) and nuclear factor of activating T cells (NFAT) at the IL-2 promoter. This interference prevents activation of the IL-2 gene. [5] When the IL-2 gene is blocked and can not be activated the synthesis of IL-2 is reduced. Activated T cells need IL-2 to proliferate and differentiate into effector T cells. In other words, FoxP3 blocks the synthesis of IL-2 followed by a decrease in proliferation of effector T cells. Finally, this results in a reduced effect of effector T cells. [5]

By secreting IL-10 $CD4^+CD25^+$ Tregs prevent DCs to secrete IL-12. The absence of IL-12 means no activation and differentiation of T_H1 cells and this can prevent infections. [5] $CD4^+CD25^+$ cells can also suppress T cells via cell-cell contact, this happens via CTLA-4 (described later) and GITR. [2; 26] $CD4^+CD25^+$ Tregs can be generated in the periphery too. In a TGF-β rich milieu $CD4^+CD25^-$ T cells can convert to $CD4^+CD25^+$ Tregs. These new generated $CD4^+CD25^+$ T cells induce the expression of FoxP3 and are then able to mediate suppression. [27; 28; 29; 30; 31; 32]

The importance of FoxP3 is shown in mice. If this gene is absent mice will develop an autoimmune-like disease where T cells keep on proliferating. [24] IPEX (Immune dysregulation, polyendocrinopathy, enteropathy, X-linked syndrome) is an autoimmune disease where mutations in the FoxP3 gene result in a reduced functioning of $CD4^+CD25^+$ Tregs. [5] These data conform the importance of FoxP3 in the regulatory mechanisms of the immune system.

2.3. 4 Mechanisms of Suppression

It is not exactly known how Tregs suppress other immune cells but so far four mechanisms of suppression are described:

1. by secretion of cytokines;
2. by metabolic disruption;
3. by targeting DCs;
4. by cytolysis (Figure 2). [24]

In this chapter the focus will be on cytokines and metabolic disruption. CTLA-4 is also discussed because it is thought to play an important role in suppression of APCs and T cells by Tregs.

2.3.1. Cytokines

TGF-β and IL-10 are cytokines that play a major role in the suppression mediated by Tregs. TGF-β converts $CD4^+CD25^+$ Tregs from $CD4^+CD25^-$ T cells making them able to express FoxP3, by blocking the IL-2 promoter the proliferation of T cells is blocked. TGF-β can also inhibit the growth of T_H1 cells through a direct binding on T_H1 cells. How this exactly occurs, remains to be elucidated. [5; 6; 28] IL-10 can bind DCs in such a way that maturation of the DC is inhibited. When DCs are not mature they are not able to activate T cells which results in a reduced functioning of the T cells. IL-10 is also able to inhibit binding of phosphatidylinositol 3-kinase (PI3K) to phosphorylated CD28 resulting in an interrupted signaltransduction of co-stimulatory molecules. This interruption inhibits the survival metabolism of T cells causing a reduced functioning of the T cells. [5; 6; 33]

2.3.2. Metabolic disruption

$CD4^+CD25^+FoxP3^+$ Tregs can suppress effector T cells through the adenosine receptor 2A (A_{2A} receptor) preventing expression of transcription factors (Figure 3). $CD4^+CD25^+FoxP3^+$ Tregs have a higher expression of CD39 and CD73 than other T cells. CD39 and CD73 are cell surface ectoenzymes; they convert adenosine triphosphate (ATP) and adenosine diphosphate (ADP) into adenosine. Adenosine can bind the A_{2A} receptor on an effector T cell or on a Treg. When binding on a Treg the expression of FoxP3 increases. Following this increase expression of CD39 enhances and more adenosine is formed. If adenosine binds an A_{2A} receptor on an effector T cell intracellular signalling in the T cell will prevent ZAP70 phosphorylation. Through this prevented phosporylation TCR signalling is inhibited and transcription factors like AP-1 will not be activated. Consequently, a reduced proliferation of effector T cells occurs because expression of IL-2 and CD25 (which are needed for proliferation) in/on the effector T cell is reduced. [24; 35] When the A_{2A} receptor on effector T cells is activated, production of pro-inflammatory cytokines IFN-γ, IL-4 and IL-6 is suppressed. This results in a reduced development of respectively T_H1, T_H2 and T_H17 cells. [24; 35] Through this suppression of T_H17 cells, expression of FoxP3 is even more increased since IL-6 (secreted by T_H17 cells) is absent and so is the inhibition of FoxP3. The final occurrence resulting from A_{2A} activation is up-regulation of co-stimulatory molecules CTLA-4 and programmed death 1 (PD1) and down-regulation of CD40. CTLA-4 and PD1 influence TCR signalling in a negative way but CD40 in a positive way, thus two ways of influencing TCR signalling co-exist. [35]

2.3.3. Targeting dendritic cells

Tregs can suppress the function of effector T cells by blocking co-stimulatory molecules on DCs so T cells will not become active. This probably occurs via CTLA-4 because it is highly expressed on the surface of Tregs. This is shown by blocking experiments using CTLA-4 antibodies. Blocking of CTLA-4 prevents suppression of DCs resulting in normally functioning effector T cells. [24] The suppressing function of CTLA-4 will be further discussed below.

Tregs can induce DCs to express the enzyme indoleamine 2,3-dioxygenase (IDO). This enzyme increases the production of pro-apoptotic metabolites generated from the catabolism of tryptophan. These pro-apoptotic metabolites mediate suppression of DCs in a way dependent on the binding between CTLA-4 and CD80/CD86. Ivars et al. [36] found that Tregs could down-regulate the expression of CD80/CD86 on DCs and in that way reduce the activation of T cells.

Lymphocyte activation gene (LAG3) could also play a role in the suppression of DCs. A possibly mechanism of action is the blocking of MHC-II molecules on immature DCs. The MHC-II molecule is necessary for maturation of DCs. Blocking the MHC-II molecule on DCs induces an immunoreceptor tyrosine-based activating motif (ITAM) inhibitory pathway via extracellular-signal-regulated kinase (ERK) and SRC-homology-2-domain-containing protein tyrosine phosphatase 1 (SHP1), finally resulting in suppression of maturation of DCs. [24]

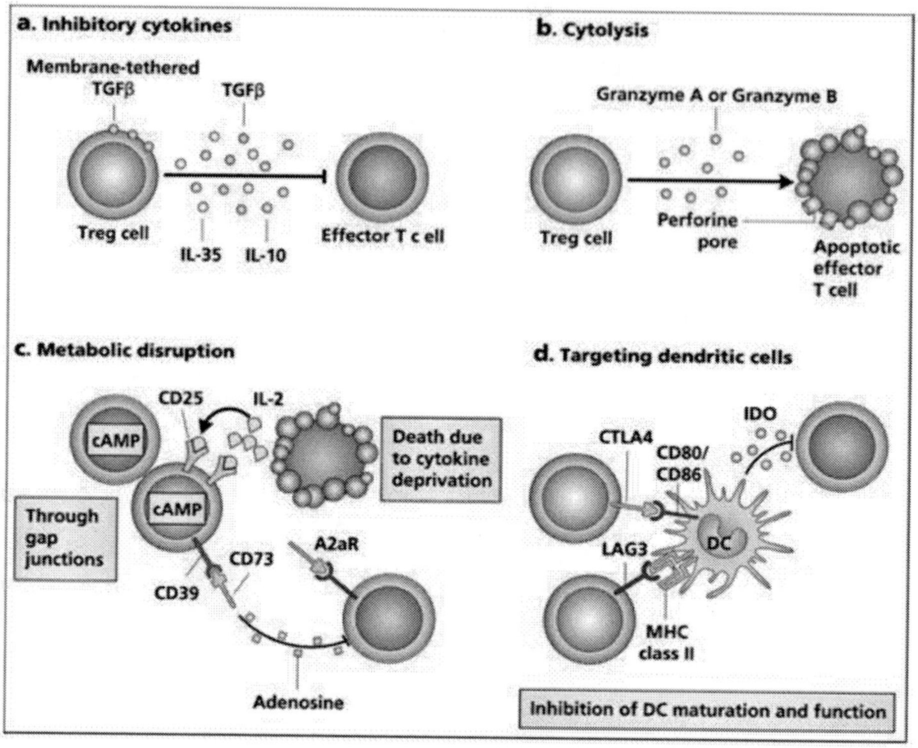

Figure 2. Four mechanisms of suppression by Tregs. a: Cytokines, the cytokines TGF-β, IL-10 and IL-35 suppress effector T cells. b: Cytolysis, granzyme A and B expressed by Tregs mediate apoptosis of effector T cells. c: Metabolic disruption, CD25 and CD39 convert ATP to adenosine. Adenosine binds the A_{2A} receptor on Tregs increasing the expression of FoxP3 and on effector T cells inhibiting the TCR signalling. d: Targeting dendritic cells, Tregs block the co-stimulatory molecules on dendritic cells preventing activation of effector T cells. Adapted from [24]

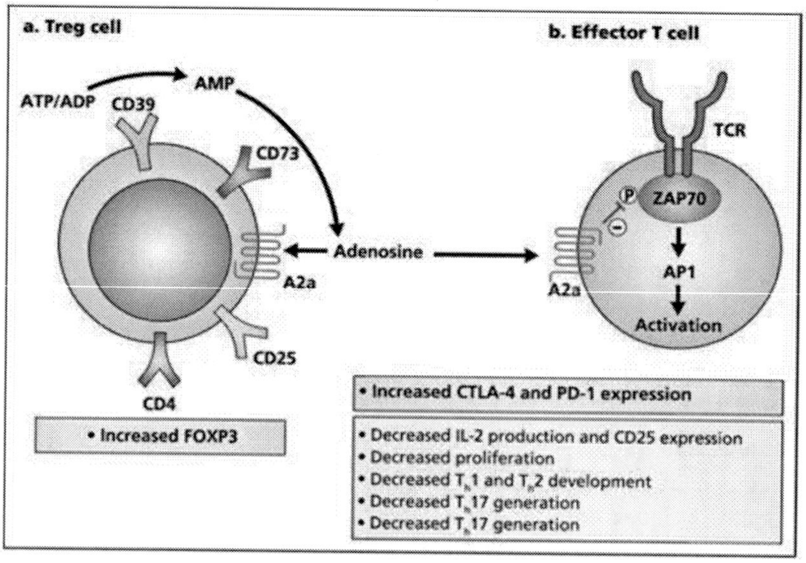

Figure 3. The conversion of ATP and ADP into adenosine and the effects when adenosine binds the A2A receptor on Tregs (a) and effector T cell (b). Adapted from [35]

2.3.4. Cytolysis

Tregs express granzyme A and granzyme B, these granzymes enter effector T cells via a perforin pore. The effector T cell undergo apoptosis mediated by Tregs. [24]

2.4. CTLA-4

That CTLA-4 inhibits the normal cell proliferation is common knowledge but the relation between CTLA-4 and regulatory T cells remains a mystery. Ten years of research have gone by since it was reported that Tregs constitutively express CTLA-4. [26; 37; 38] Because of this natural expression of CTLA-4 it is understandable that researchers attribute an important role to this molecule in the regulation of Tregs. We will discuss a number of hypotheses.

CTLA-4 is important in regulation of T cell proliferation. This molecule is expressed on naive T cells after activation. CTLA-4 inhibits the TCR signalling through binding with a B7-molecule on an APC. CTLA-4 competes with CD28 for binding on the B7-molecule. Normally, the binding between CD28 and B7 leads to increased survival and differentiation of T cells; it forms one of the necessary signals for T cell activation. The first signal in T cell activation is binding of the MHC-complex by the TCR and a CD4 co-receptor. This leads to activation of Lck and Fyn, a co-receptor associated kinase and receptor-associated kinase respectively. These kinases phosphorylate the ITAMs on the intracellular part of CD3 and on the ζ-chain (intracellular part of TCR) leading to activation of ZAP-70. The signalling pathway is hereby initiated, finally leading to cell proliferation and cell differentiation due to activation of transcription factors like nuclear factor κ-light-chain-enhancer of activated B cells (NF-κB), NFAT and AP-1. These transcription factors initiate expression of the IL-2 gene. An increased level of IL-2 makes the T cells proliferate. The second signal in T cell activation is co-stimulation through CD28 and B7. This increases the production of AP-1 and NF-κB which leads to a threefold increase in IL-2 expression. The second effect of this co-stimulation is stabilisation of IL-2 mRNA, increasing the IL-2 production 20-30 fold. The third and final signal in T cell activation is differentiation of the T cell into T_H1, T_H2 or T_H17 dependent on the cytokine profile. [5]

However, CTLA-4 has a higher affinity for B7 molecules than CD28. Thus, if CTLA-4 is present, it competes with CD28 for B7-molecules interrupting TCR signalling. This process is mediated by ITIM. This is a cytoplasmic tail of CTLA-4 which gets phosphorylated by the binding of B7. [5] Phosphorylated ITIM recruits phosphatases which remove phosphate groups from ITAMs phosphorylated during T cell activation. [5; 39] When the ITAMs are no longer phosphorylated the TCR signalling stops. This inhibitory signal reduces the generation of IL-2 and the T cell is less able to proliferate. [5]

Suppression through TGF-β and FoxP3

After activation it could be that natural Tregs secrete enough TGF-β to increase the expression of FoxP3. But there still is no agreement on natural Tregs secreting enough TGF-β or not. Takashi et al. and Thornton et al [40; 41] did not find an increased secretion of TGF-β following activation of Tregs. Chen et al. [42] on the other hand, have reported that Tregs activated through TCR and CTLA-4 secrete TGF-β. Maybe it is a positive feedback mechanism between CTLA-4 expression and TGF-β secretion. CTLA-4 is naturally expressed on Tregs. When binding with a B7 molecule occurs TGF-β is expressed on the surface of the Tregs. [42; 43; 44] TGF-β then up-regulates the expression of CTLA-4 that even more TGF-β is secreted. [31; 45] The increased level of TGF-β triggers the expression of FoxP3 and the Tregs fulfil their suppressing function. Stimulated CTLA-4 increases the TGF-β level resulting in an increased expression of FoxP3 that suppresses the proliferation of auto-reactive T cells reducing the disease activity in MS patients.

Physical interaction between Tregs and APCs

A physical interaction between $CD4^+CD25^+$ Tregs and APCs through CTLA-4 may limit the activation of auto-reactive T cells. When a B7 on an APC binds a CTLA-4 molecule on a Treg, the APC is not able to activate auto-reactive T cells and autoimmunity is prevented. [40; 41; 46; 47]

Another possibility is that CTLA-4 expression on Tregs modulates the expression of co-stimulatory molecules on APC i.e. down-regulation of the B7-molecule or adhesion molecules results in a reduced ability of T cells to adhere to the APC. [48] The consequence of this down-regulation is a decreased ability of the APC to activate $CD4^+$ T cells. [46]

Prolonged contact time through LFA-1

It is proven that CTLA-4 increases the expression of the adhesion molecule lymphocyte function-associated antigen 1 (LFA-1). Because of an increase of this molecule an easier and stronger binding between the Treg and APC or $CD4^+$ T cell might occur. A prolonged contact time makes it easier to activate the Tregs and to make a suppressive interaction between Tregs and APCs or $CD4^+$ T cells. [46; 49] In other words, increasing the expression of LFA-1 and the prolonged contact time between Tregs and APCs or $CD4^+$ T cells, might be a possible treatment of MS. Tregs with an impaired function would then be able to suppress APCs and $CD4^+$ T cells anyway because of the prolonged contact time.

3. Multiple Sclerosis

Multiple Sclerosis (MS) is an incurable autoimmune disease. [50] A chronically inflammation in the central nervous system (CNS) induced by a self-antigen causes damage of myelinated axons. This damage results in impaired signalling between brain and periphery leading to a reduced ability to move. A general consensus is that MS begins with an acute inflammation resulting in lesions. [50] These lesions are characterised by cell infiltrates around the damaged axons. [51] This inflammation causes an increase in the permeability of the blood brain barrier (BBB). Immune cells such as T lymphocytes, macrophages and DCs can pass this barrier and will diffuse to the place of inflammation. [5; 50] In healthy individuals these cells modulate the inflammation. However, in some cases (i.e. MS) it takes a different turn. In MS patients these newly arrived cells cause even more inflammation because they are auto-reactive and the self-antigen is never eliminated.

3.1. Pathophysiology

The exact pathophysiology of MS is unknown but T_H17 cells play an important role. [52; 53] The antigen causing MS is still not known; it probably is a self-antigen. The greatest focus is on myelin basic protein (MBP) because it is similar to the antigen causing MS in animal models. [50] When this unknown antigen meets an APC i.e. a dendritic cell, the antigen is taken up into the APC by phagocytosis. Following phagocytosis the APC presents a peptide of the antigen through the MHC-II molecule and can activate a naive T-cell. Dependent on the cytokine profile in the environment, T cells mature into a T_H1, T_H2 or T_H17 populations. In MS the role of T_H1 and T_H17 is important and the differentiation into those two T cells is mediated by IL-12 (T_H1) and IL-23, IL-6 and TGF-β (T_H17). [5] The knowledge of development and function of T_H17 cells is based on animal models and it is known that that the human situation could be quite different. [54; 55; 56; 57] During the start of inflammation T_H17 cells play a major role followed by influx of T_H1 cells. [5] The mature cells diffuse through the BBB and into the CNS. Here they will be reactivated by microglia cells, macrophages or DCs. These mature T cells are now able to secrete pro-inflammatory cytokines.

How the damage of the myelin exactly occurs is still unclear. Though, it is most likely that a number of mechanisms cause the damage. Below are some possible damaging mechanisms described. [58]

T_H17 cells secrete the pro-inflammatory cytokine IL-17 and this might cause demyelination.

T_H1 cells secrete interferon γ (IFN-γ) causing B-cells to get activated and differentiate into plasma cells. These plasma cells secrete antibodies against the myelin on the axon.

Astrocytes can damage the nerves in three different ways:

1. Damage of the dendrites on the neuron,
2. Damaging the oligodendrocytes
3. Damage through secreted nitric oxide (NO).

$CD8^+$-cells can also cause demyelation. They do this by releasing perforines in the vicinity of myelin cells, granzymes will enter the myelin cell via the perforins and the cell will undergo apoptosis.

3.1.1. T_H17 cells

T_H17 cells secrete IL-17 when activated and may cause damage of myelin axons when activated by a self-antigen. This damaging reaction can end up in an autoimmune reaction. The T_H17 cell is a recently discovered $CD4^+$ subtype. This subtype is similar to the T_H1 subtype but secretes IL-17 instead of IL-4 and IFN-γ. [5] IL-6, TGF-β, IL-21 and IL-23 are the most important cytokines in the differentiation and expansion of T_H17 cells. IL-6 and TGF-β are both necessary for naive $CD4^+$ T cells to differentiate into T_H17 cells (Figure 4). In an IL-6 and TGF-β rich environment the transcription factor retinoic acid-related orphan receptor-γt (ROR-γt) is activated and initiates the expression of the IL-23 receptor in T_H17 cells. When differentiated T_H17 cells secrete IL-21 it amplifies the T_H17 populations. [5; 32; 59] Following differentiation and amplification intracellular transcription factors are activated through binding of the IL-23 receptor and IL-23. This turns into an increased expression of IL-17 and IL-22. These cytokines will stabilise T_H17 cells and promote survival of these cells. [5; 59] IL-17 is a pro-inflammatory cytokine and induces expression of other pro-inflammatory cytokines, chemokines and matrix metalloproteases. [61] When a self-antigen causes inflammation it is possible that the resulting tissue damage can turn into autoimmunity. [61; 62] T_H17 cells also attract macrophages and neutrophils into the damaged tissue. In a normal inflammation these macrophages and neutrophils try to clean-up the inflammation. In the case of a self-antigen they promote even more damage because the antigen is never eliminated. [63]

Thus, by reducing the amount of IL-6, IL-21 and IL-23 and blocking the transcription factor ROR-γt the effect of T_H17 cells decreases.

3.1.2. IL-6

The necessity of IL-6 in the differentiation of T_H17 cells is proved in IL-6 deficient mice. In these mice T_H17 cells were absent and the FoxP3 level was increased. These mice did not develop experimental autoimmune encephalomyelitis (EAE) (a disease model for human MS). [32; 64; 65] In mice IL-6 alone is not able to initiate the differentiation of the T_H17 cells. TGF-β is necessary as co-stimulator but in a low concentration. Otherwise it will inhibit the transcription factor ROR-γt. [32; 60; 66]

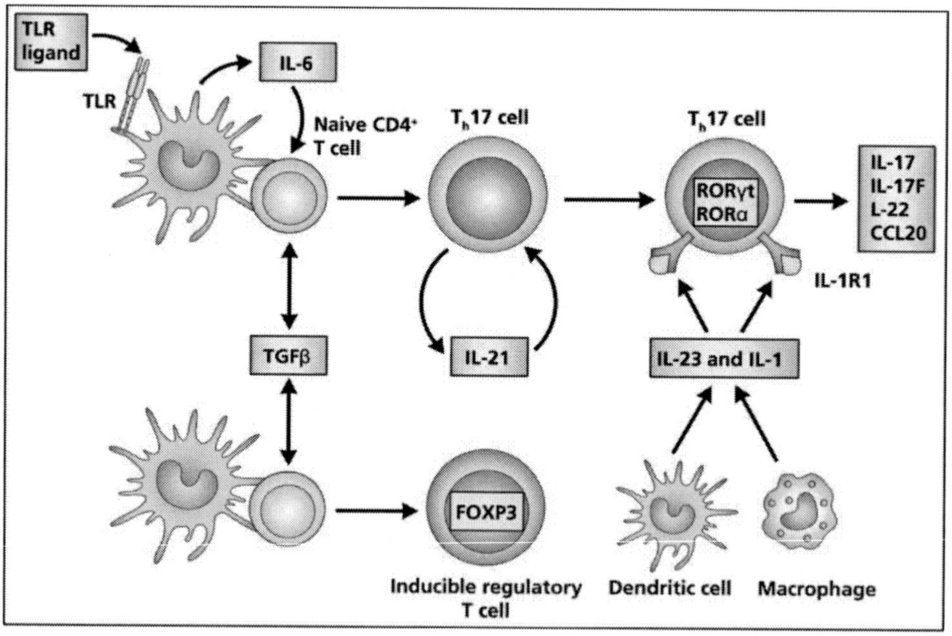

Figure 4. Differentiation and development of T_H17 cells. IL-6 and TGF-β initiate the differentiation. IL-21 expands the T_H17 cells and the ROR-γt is activated and starts to express IL-23R. IL-23 binds on the IL-23R increasing the expression of IL-17. Adapted from [63]

3.1.3. TGF-β

TGF-β can induce both Tregs and T_H17 cells. [60; 66] High levels of TGF-β induce $CD4^+CD25^-$ T cells to convert into FoxP3 Tregs. [32; 60] These FoxP3 Tregs suppress the development of effector T cells; this occurs in a non-inflammatory situation. [60; 66] In an inflammatory situation DCs secrete IL-6 and this cytokine inhibits the generation of Tregs. In this way IL-6 inhibits the suppressing function performed by Tregs. This reduced effect of Tregs is due to a decreased expression of the FoxP3 molecule. [32; 60; 66]

The presence of both IL-6 and TGF-β induces the generation of T_H17 cells. [32] From this Bettelli et al. [32] have suggested that Tregs and T_H17 cells arise from the same $CD4^+$ precursor T cells. The subtype they differentiate into depends on the cytokine profile. The presence of TGF-β alone induces Tregs but if both IL-6 and TGF-β are present the FoxP3 expression decreases and the expression of IL-17 increases. [32] This probably means that the balance between IL-6 and TGF-β decides if naive T cells differentiate into T_H17 or Treg cells. [61; 67] Thus, inhibiting the secretion of IL-6 and inducing the secretion of TGF-β increases the expression of FoxP3 and consequently reduces the differentiation of T_H17 cells. The secretion of IL-6 can be reduced by inhibiting DCs. This can be mediated by Tregs expressing HLA-G or CTLA-4 or by Tregs secreting IL-10. These molecules bind DCs and inhibit the signalling cascade resulting in a reduced secretion of IL-6 followed by a decreased differentiation into T_H17 cells.

3.1.4. ROR-γt

Previously as discussed, TGF-β and IL-6 increase the expression of ROR-γt. This makes $CD4^+$ cells differentiate into T_H17 cells. [60] FoxP3 inhibits the activity of ROR-γt with a decrease in the differentiation of T_H17 cells as result – this occurs in the presence of TGF-β and absence of IL-6. [60] The balance of the expression of ROR-γt and FoxP3 is important in the homeostasis of the immune system. If this balance is disturbed towards the ROR-γt it might cause an autoimmune reaction like MS because an unnecessary differentiation of T_H17 cells arises. [60] To normalise this disturbed balance the expression of FoxP3 needs to be increased, this occurs by an increase of TGF-β and a decrease of IL-6. The increase of TGF-β might be realised by administering TGF-β, which induces the expression of FoxP3 in Tregs. The decrease in IL-6 can be realised through suppression of DCs, like discussed above.

3.1.5. IL-23

IL-23 is very important for the survival of the T_H17 cells. Langrish et al. [67] proved this in an EAE model; animals with a deficiency in IL-23 had a defect in IL-17 expression. An IL-23 deficient mouse is resistant to EAE in contrast to an IL-17 deficient mouse that is susceptible to EAE. [62; 68; 69] IL-23 is not able to induce the differentiation of T_H17 cells but it increases the amount of secreted IL-17. Thus, IL-23 plays a role in the maintenance and stability of the T_H17 cells and not in the differentiation of T_H17 cells. [32; 70; 71] In other words, when the level of IL-23 is reduced the amount of IL-17 is reduced too, resulting in less damage of the myelin. IL-23 is secreted by DCs, thus by inhibiting DCs the secretion of IL-23 can be reduced and the effect of T_H17 cells as well. DCs can be inhibited by HLA-G, CTLA-4 and IL-10.

3.1.6. Balance T_H17 T cells and regulatory T cells

A dysregulation in the balance between transcription factors expressed by Tregs and T_H17 cells, FoxP3 and ROR-γt respectively, also forms an important part of the problem. [58; 60] Since the amount of $CD4^+CD25^+$ Tregs is the same in patients and in healthy controls, this dysregulation is not caused by the amount of Tregs. $CD4^+CD25^{high}$ Tregs isolated from MS patients do have a regulatory effect, but this effect is much less than the effect of

$CD4^+CD25^{high}$ Tregs from healthy controls. The difference between patient and healthy controls lies in the functioning of Tregs. [72; 73] This impaired function of Tregs is caused by a decreased expression of FoxP3 in patients with MS compared to healthy controls. [74; 75] When the expression of FoxP3 decreases, the balance will shift towards T_H17 cells and expression of the transcription factor ROR-γt will increase. This will end up in an increase of T_H17 cells and enhancement of the inflammation/autoimmune disease. [58] In other words, if the dysbalance between ROR-γt and FoxP3 is normalised, the disease activity in MS patients is reduced. To normalise the balance, it is necessary to induce the expression of FoxP3 and reduce the expression of ROR-γt. Through administering of TGF-β the expression of FoxP3 increases. The expression of ROR-γt decreases in the presence of FoxP3 but also through reducing the amount of IL-6.

3.2. MS Treatment

A cure for MS is at present not available. At present five products are thought to reduce disease activity and disease progression in MS patients. [76; 77] Two of these products are IFN-β and glatiramer acetate. These might induce HLA-G and FoxP3 respectively, normalising the dysbalance between Tregs and T cells that causes MS. [17; 78; 79] Thought provoking is the fact that treatment with IFN-β or glatiramer acetate is not curing MS. Thus, IFN-β and glatiramer acetate are not able to normalise the dysbalance. Natalizumab is another product; a monoclonal antibody that blocks vascular cell adhesion molecule (VCAM). This results in a decreased diffusion of immunological cells through the BBB which reduces disease activity. [58] The last products are mitoxantrone and methotrexate; they down-regulate all DNA causing a decreased proliferation of immunological cells but also give rise to a great number of adverse effects. [80; 81; 82]

Conclusion

Recent data suggest that regulatory T cells can be used in the treatment of MS. Especially through induction of transcription factor FoxP3, membrane bound or soluble HLA-G and CTLA-4. More research groups have found the role of FoxP3 and HLA-G to be important in regulating the balance between T_H17 cells and Tregs. [32; 60; 73] Because CTLA-4 is constitutively expressed on Tregs it is likely that it hs a connection with MS but the exact role is still unknown. [26; 37; 38] Induction of FoxP3 results in an increased amount of T cells able to suppress other T cells through blocking IL-2 and ROR-γt genes causing a reduced T cell proliferation. HLA-G suppresses both APCs and T cells through soluble and membrane bound HLA-G which interrupts signalling in APCs and T cells following a binding to the ILT-2 receptor. [26; 37; 38] CTLA-4 likely induces the secretion of TGF-β by Tregs and suppresses T_H17 cells as a consequence of inhibited cytokine secretion by APCs. [37; 41]

FoxP3 interferes with transcription factors promoting expression of IL-2 and ROR-γt resulting in a decrease in the development of T_H1 cells and T_H17 cells. Expression of FoxP3 is induced by TGF-β. Increased TGF-β levels also inhibit expression of ROR-γt. Thus, administered TGF-β increases the expression of FoxP3 and decreases expression of ROR-γt.

This might result in a normalised balance between FoxP3 and ROR-γt and a decreased transcription of T cell proliferation promoting factors.

HLA-G also plays an essential role in suppressing T cells. When HLA-G binds the ILT-2 receptor the signalling pathway in APCs and T cells is interrupted. ILT-2 is highly expressed on APC as well as HLA-G. This suggests that APCs are able to regulate themselves. A binding of HLA-G on an APC results in T cell suppression because of a reduced ability of the APC to activate T cells. Ligation of the ILT-2 receptor probably inhibits secretion of IL-6 and IL-23 by APCs and/or the peptide presentation through MHC-II molecules. A reduced secretion of IL-6 results in reduced differentiation of T_H17 cells and increased FoxP3 expression. Concluding: administering HLA-G or inducing HLA-G in the body inhibits the IL-6 and IL-23 secreting APCs, resulting in a reduced differentiation and stability of T_H17 cells. Thus, administering or inducing HLA-G could be a treatment for MS since the reduced differentiation of T_H17 cells provides an increase of FoxP3 expression and an induction of the function of Tregs.

Since CTLA-4 is always expressed on Tregs it is likely that it plays an important role. This role is still not elucidated but theories suggest that expression of CTLA-4 and TGF-β serves as a positive feedback resulting in expression of FoxP3. Another theory is the occurrence of a negative signal activated through a binding between CTLA-4 on a Treg and B7 on an APC resulting in reduced activation of T cells. It is most likely that both theories play a role in the manner Tregs suppress the effect of T cells. Thus, stimulating CTLA-4 with a ligand can activate Tregs to secrete TGF-β, which induces expression of FoxP3 resulting in increased suppression of T_H17 cells. Furthermore, when CTLA-4 binds on APCs signalling is inhibited and differentiation of T_H17 cells reduces because of a decreased amount of secreted IL-6 and IL-23. In other words, CTLA-4 appears to suppress T_H17 cells in two ways, suggesting an important role in the treatment of MS.

Research still needs to be done on the administration route and the possible occurring adverse effects. Will direct transfer of $CD4^+CD25^+FoxP3^+$ T cells into the blood or brain be the administration route of choice or will a larger effect take place when the Tregs are stimulated and induced in the body by administered TGF-β or HLA-G?

However, it is our firm believe that it will be possible to treat MS with Tregs through stimulating CTLA-4 and/or induction or transfer of FoxP3 Tregs and HLA-G cells.

Abbreviations

A_{2A} receptor	Adenosine receptor 2A. A receptor on Tregs and effector T cells. It can increase the FoxP3 expression and inhibit the signalling in effector T cells.
ADP	Adenosine di-phosphate. Converted from ATP and is converted to adenosine which binds the A_{2A} receptor.
AP-1	Activating protein 1. Transcription factor promoting the expression of IL-2.
APC	Antigen-presenting cell. Dendritic cell or macrophage presenting peptides to T helper cells via MHC-II.
ATP	Adenosine tri-phosphate. Is converted to adenosine which bind the A_{2A} receptor.
BBB	Blood brain barrier.

cAMP	cyclic adenosine mono-phosphate.
CD25	IL-2 receptor.
CNS	Central nervous system.
CTLA-4	Cytotoxic T lymphocyte antigen 4. Always expressed on Tregs and inhibits the signalling in APCs.
DC	Dendritic cell.
EAE	Experimental autoimmune encephalomyelitis. Animal model of MS.
ERK	Extracellular-signal-regulated kinase. Play a role in suppression of APCs by LAG3.
FoxP3	Forkhead box protein 3. Transcription factor expressed by Tregs which inhibits the expression of IL-2.
GITR	Glucocorticoid-induced tumor necrosis factor receptor. Co-stimulatory molecule on Tregs which maybe play a role in the suppression.
GR	Glucocorticoid receptor.
HLA-G	Human leukocyte antigen G. MHC-I molecule which plays an important role in suppression by Tregs.
IDO	Indoleamine 2,3-dioxygenase. Enzyme that increases the production of pro-apoptotic metabolites generated from tryptophan.
IFN-β	Interferon β.
IFN-γ	Interferon γ.
IL-10	Interleukine 10.
ILT-2	Immunoglobulin-like transcript family. Receptor that HLA-G binds to mediate suppression.
IPEX	Immune dysregulation, polyendocrinopathy, enteropathy, X-linked syndrome. Autoimmune disease caused by a mutation in FoxP3.
ITAM	Immunoreceptor tyrosine-based activating motif.
ITIM	Immunoreceptor tyrosine-based inhibiting motif.
KIR	Killer cell immunoglobulin-like receptor. HLA-G receptor.
LAG3	Lymphocyte activation gene. Blocking the MHC-II molecule on APCs.
LFA-1	Lymphocyte function-associated antigen 1. Adhesion molecule on Tregs.
MBP	Myelin basic protein. Possible self-antigen causing MS.
MHC-I	Major histocompatibility complex claas I.
MS	Multiple Sclerosis.
NF-κB	Nuclear factor κ-light-chain-enhancer of activated B cells. Transcription factor promoting cell proliferation.
NFAT	Nuclear factor of activating T cells. Transcription factor promoting transcription of IL-2.
NK-cells	Natural killer cells.
PD1	Programmed death 1. Co-stimulatory molecule inhibiting the TCR signalling.
PI3K	Phosphatidylinositol 3-kinase. Binds CD28 as one of the signals in T cells activation.
ROR-γt	Retinoic acid-related orphan receptor-γt. Transcription factor starting the expression of IL23R in T_H17 cells.
SHP1	SRC-homology-2-domain-containing protein tyrosine phosphatase 1. Plays a role in the suppression of APCs.

TCR	T cell receptor.
TGF-β	Transforming growth factor β. Suppressing cytokine secreted by Tregs.
T_H1	T helper 1 cell.
T_H2	T helper 2 cell.
T_H3	T helper 3 cell.
T_H17	T helper 17 cell.
TNF-α...	Tumor necrosis factor α.
T_R1	T regulatory 1 cell.
Tregs	Regulatory T cells.
VLA-4	Very late activation antigens 4. Adhesion molecule on T cells.
ZAP70	ζ-chain-associated protein kinase 70. Plays an essential role in the TCR. signalling and activation of T cells.

References

[1] O'Connor, A; Conforti, C; Boyd, E; Boer, RD; Vos, GD; Farrell, T; et al. Multiple sclerosis international federation. *About MS Introduction*. 2009.08.12. Available from http://www.msif.org/en/about_ms/index.html.

[2] Mills, KH. Regulatory t cells: friend or foe in immunity to infection? *Nature Reviews Immunology.,* 2004, 4(November), 841-855.

[3] Zozulya, AL; Wiendl, H. The role of regulatory T cells in multiple sclerosis. Nature clinical practice. *Neurology.,* 2008, 4(7), 384-398.

[4] Cava, AL. Natural Tregs and autoimmunity. *Frontiers in Bioscience.,* 2009, 14, 333-343.

[5] Murphy, K; Travers, P; Walport, M. *Janeway's Immunobiology*. 7th edition. New York: Garland Science, Taylor & Francis Group., 2008.

[6] Taylor, A; Verhagen, J; Blaser, K; Akdis, M; Akdis, CA. Mechanisms of immune suppression by interleukin-10 and transforming growth factor-beta: the role of T regulatory cells. *Immunology.,* 2006, 117, 433-442.

[7] Zhong, M; Weng, X; Liang, Z; Lu, S; Li, J; Chen, X; et al. Dimerization of soluble HLA-G by IgG-Fc fragment augments ILT2-mediated inhibition of T-cell alloresponse. *Transplantation.,* 2009, 87(1), 8-15.

[8] Huang, Y; Zozulya, AL; Weidenfeller, C; Schwab, N; Wiendl, H. T cell suppression by naturally occurring HLA-G-expressing regulatory CD4+ T cells is IL-10-dependent and reversible. *Journal of leukocyte biology.,* 2009, 86(August), 1-9.

[9] Bainbridge, DR; Ellis, SA; Sargent, IL. HLA-G suppresses proliferation of CD4+ T-lymphocytes. *Journal of Reproductive Immunology.,* 2000, 48, 17 - 26.

[10] Lila, N; Rouas-Freiss, N; Dausset, J; Carpentier, A; Carosella, ED. Soluble HLA-G protein secreted by allo-specific CD4+ T cells suppresses the allo-proliferative response: a CD4+ T cell regulatory mechanism. *Proc. Natl. Acad. Sci.,* U. S. A. 2001, 98(21), 12150-12155.

[11] Carosella, ED; HoWangYin, K; Favier, B; LeMaoult, J. HLA-G–dependent suppressor cells: diverse by nature; function; and significance. *Human Immunology.,* 2008, 69, 700-707.

[12] Feger, U; Tolosa, E; Huang, Y; Waschbisch, A; Biedermann, T; Melms, A; et al. HLA-G expression defines a novel regulatory T-cell subset present in human peripheral blood and sites of inflammation. *Blood*, 2007, 110(2), 568-577.
[13] Aractingi, S; Briand, N; Le Danff, C; Viguier, M; Bachelez, H; Michel, L; et al. HLA-G and NK Receptor are expressed in psoriatic skin. *American Journal of Pathology.*, 2001, 159(1), 71-77.
[14] Khosrotehrani, K; Le Danff, C; Reynaud-Mendel, B; Dubertret, L; Carosella, ED; Aractingi, S. HLA-G Expression in Atopic Dermatitis. *Journal of investigative dermatology.*, 2001, 117(3), 750-752.
[15] Wiendl, H; Feger, U; Mittelbronn, M; Jack, C; Schreiner, B; Stadelmann, C; et al. Expression of the immune-tolerogenic major histocompatibility molecule HLA-G in multiple sclerosis: implications for CNS immunity. *Brain*, 2005, 128, 2689-704.
[16] Wiendl, H; Behrens, L; Maier, S; Johnson, MA; Weiss, EH; Hohlfeld, R. Muscle fibers in inflammatory myopathies and cultured myoblasts express the nonclassical major histocompatibility antigen HLA-G. *Ann Neurol.*, 2000, 48, 679-684.
[17] Mitsdoerffer, M; Schreiner, B; Kieseier, BC; Neuhaus, O; Dichgans, J; Hartung, H; et al. Monocyte-derived HLA-G acts as a strong inhibitor of autologous CD4 T cell activation and is upregulated by interferon-beta in vitro and in vivo: rationale for the therapy of multiple sclerosis. *Journal of NeuroImmunology.*, 2005, 159, 155-164.
[18] Borges, L; Hsu, M; Fanger, N; Kubin, M; Cosman, D. A family of human lymphoid and myeloid lg-like receptors; some of which bind to MHC I Class I molecules. *Journal of Immunology.*, 1997, 159, 5192-5196.
[19] Colonna, M; Navarro, F; Bellón, T; Llano, M; García, P; Samaridis, J; et al. A common inhibitory receptor for major histocompatibility complex class I molecules on human lymphoid and myelomonocytic cells. *Journal of Experimental Medicine*, 1997, 186(11), 1809-1818.
[20] Cosman, D; Fanger, N; Borges, L. Human cytomegaiovirus; MHC class I and inhibitory signalling receptors: more questions than answers. *Immunological Reviews*., 1999, 168, 177-185.
[21] Wiendl, H. HLA-G in the nervous system. *Human Immunology.*, 2007, 68, 286-293.
[22] Bahri, R; Hirsch, F; Josse, A; Rouas-Freiss, N; Bidere, N; Vasquez, A; et al. Soluble HLA-G inhibits cell cycle progression in human alloreactive T lymphocytes. *Journal of Immunology.*, 2006, 176, 1331-1339.
[23] Sakaguchi, S. The origin of FOXP3-expressing CD4+ regulatory T cells: thymus or periphery. *Journal of Clinical Investigation*., 2003, 112(9), 1310-1312.
[24] Vignali, DA; Collison, LW; Workman, CJ. How regulatory T cells work. *Nature Reviews Immunology.*, 2008, 8, 523-532.
[25] Marie, JC; Letterio, JJ; Gavin, M; Rudensky, AY. TGF-beta1 maintains suppressor function and Foxp3 expression in CD4+CD25+ regulatory T cells. *Journal of Experimental Medicine*., 2005, 201(7), 1061-1067.
[26] Read, S; Greenwald, R; Izcue, A; Robinson, N; Mandelbrot, D; Francisco, L; et al. Blockade of CTLA-4 on CD4+CD25+ regulatory T cells abrogates their function In Vivo. *Journal of Immunology.*, 2006, 177, 4376-4383.
[27] Zheng, SG; Gray, JD; Ohtsuka, K; Yamagiwa, S; Horwitz, DA. Generation Ex Vivo of TGF-beta-producing regulatory T cells from CD4+CD25- precursors. *Journal of Immunology.*, 2002, 169, 4183-4189.

[28] Zheng, SG; Wang, JH; Gray, JD; Soucier, H; Horwitz, DA. Natural and induced CD4+CD25+ cells educate CD4+CD25- cells to develop suppressive activity: the role of IL-2; TGF-beta; and IL-10. *Journal of Immunology.,* 2004, 172, 5213-5221.

[29] Hori, S; Nomura, T; Sakaguchi, S. Control of regulatory T cell development by the transcription factor Foxp3. *Science.*, 2003, 299, 1057-1061.

[30] Fontenot, JD; Gavin, MA; Rudensky, AY. Foxp3 programs the development and function of CD4+CD25+ regulatory T cells. *Nature Immunology.,* 2003, 4(4), 330-336.

[31] Chen, W; Jin, W; Hardegen, N; Lei, K; Li, L; Marinos, N; et al. Conversion of peripheral CD4+CD25- naive T cells to CD4+CD25+ regulatory T cells by TGF-beta induction of transcription factor Foxp3. *Journal of Experimental Medicine.*, 2003, 198(12), 1875-1886.

[32] Bettelli, E; Carrier, Y; Gao, W; Korn, T; Strom, TB; Oukka, M; et al. Reciprocal developmental pathways for the generation of pathogenic effector TH17 and regulatory *T cells. Nature.*, 2006, 441, 235-238.

[33] Akdis, CA; Joss, A; Akdis, M; Faith, A; Blaser, K. A molecular basis for T cell suppression by IL-10, CD28- associated IL-10 receptor inhibits CD28 tyrosine phosphorylation and phosphatidylinositol 3-kinase binding. *The FASEB Journal.*, 2000, 14, 1666-1668.

[34] Collison, LW; Workman, CJ; Kuo, TT; Boyd, K; Wang, Y; Vignali, KM; et al. The inhibitory cytokine IL-35 contributes to regulatory T-cell function. *Nature*, 2007, 450, 566-569.

[35] Haskó, G; Linden, J; Cronstein, B; Pacher, P. Adenosine receptors: therapeutic aspects for inflammatory and immune diseases. *Nature reviews Drug discovery*, 2008, 7(September), 759-770.

[36] Cederbom, L; Hall, H; Ivars, F. CD4+CD25+ regulatory T cells down-regulate co-stimulatory molecules on antigen-presenting cells. Eur. *J. Immunol.*, 2000, 30, 1538-1543.

[37] Takahashi, T; Tagami, T; Yamazaki, S; Uede, T; Shimizu, J; Sakaguchi, N; et al. Immunologic self-tolerance maintained by CD25+CD4+ regulatory T cells constitutively expressing cytotoxic T lymphocyte-associated antigen 4. *Journal of Experimental Medicine*, 2000, 192(2), 303-309.

[38] Read, S; Malmström, V; Powrie, F. Cytotoxic T Lymphocyte–associated Antigen 4 plays an essential role in the function of CD25+CD4+ regulatory cells that control instestinal inflammation. *Journal of Experimental Medicine.*, 2000, 192(2), 295-302.

[39] Lee, K. Cortical areas differentially involved in multiplication and substraction: A functionel magnetic and correlation with a case of selective acalculia. *Annals of Neurology..,* 2000, 48, 657-684.

[40] Takahashi, T; Kuniyasu, Y; Toda, M; Sakaguchi, N; Itoh, M; Iwata, M; et al. Immunologic self-tolerance maintained by CD25+CD4+ naturally anergic and suppressive T cells: induction of autoimmune disease by breaking their anergic/suppressive state. *International Immunology.,* 1998, 10(12), 1969-1980.

[41] Thornton, AM; Shevach, EM. CD4+CD25+ immunoregulatory T cells suppress polyclonal T cell activation in vitro by inhibiting interleukin 2 production. *Journal of Experimental Medicine.*, 1998, 188(2), 287-296.

[42] Chen, W; Jin, W; Wahl, SM. Engagement of cytotoxic T lymphocyte-associated antigen 4 (CTLA-4) induces transforming growth factor beta (TGF-beta) production by murine CD4+ T cells. *Journal of Experimental Medicine*., 1998, 188(10), 1849-1857.

[43] Nakamura, K; Kitani, A; Fuss, I; Pedersen, A; Harada, N; Nawata, H; et al. TGF-beta1 plays an important role in the mechanism of CD4+CD25+ regulatory T cell activity in both humans and mice. *Journal of Immunology*., 2004, 172, 834-842.

[44] Nakamura, K; Kitani, A; Strober, W. Cell contact–dependent immunosuppression by CD4+CD25+ regulatory T cells is mediated by cell surface-bound transforming growth factor-beta. *Journal of Experimental Medicine*., 2001, 194(5), 629-644.

[45] Park, H; Paik, D; Jang, E; Hong, S; Youn, J. Acquisition of anergic and suppressive activities in transforming growth factor-beta-costimulated CD4+CD25- T cells. *International Immunology*., 2004, 16(8), 1203-1213.

[46] Sojka, DK; Hughson, A; Fowell, DJ. CTLA-4 is required by CD4+CD25+ Treg to control CD4+ T-cell lymphopenia-induced proliferation. *Eur. J. Immunol*., 2009, 39, 1544-1551.

[47] Itoh, M; Takahashi, T; Sakaguchi, N; Kuniyasu, Y; Shimizu, J; Otsuka, F; et al. Thymus and autoimmunity: production of CD25+CD4+ naturally anergic and suppressive T cells as a key function of the thymus in maintaining immunologic self-tolerance. *Journal of Immunology*, 1999, 162, 5317-5326.

[48] Wing, K; Onishi, Y; Prieto-Martin, P; Yamaguchi, T; Miyara, M; Fehervari, Z; et al. CTLA-4 control over Foxp3+ regulatory T cell function. Science. 2008, 322, 271-275.

[49] Schneider, H; Valk, E; da Rocha Dias, S; Wei, B; Rudd, CE. CTLA-4 up-regulation of lymphocyte function-associated antigen 1 adhesion and clustering as an alternate basis for coreceptor function. *Proc. Natl. Acad. Sci.,* U. S. A. 2005, 102(36), 12861-12866.

[50] McFarland, HF; Martin, R. Multiple sclerosis: a complicated picture of autoimmunity. *Nature Immunology*., 2007, 8(9), 913-919.

[51] Genain, CP; Cannella, B; Hauser, SL; Raine, CS. Identification of autoantibodies associated with myelin damage in multiple sclerosis. *Nature Medicine*., 1999, 5(2), 170-175.

[52] Lock, C; Hermans, G; Pedotti, R; Brendolan, A; Schadt, E; Garren, H; et al. Gene-microarray analysis of multiple sclerosis lesions yields new targets validated in autoimmune encephalomyelitis. *Nature Medicine*., 2002, 8(5), 500-508.

[53] Matusevicius, D; Kivisäkk, P; He, B; Kostulas, N; Özenci, V; Fredrikson, S; et al. Interleukin-17 mRNA expression in blood and CSF mononuclear cells is augmented in multiple sclerosis. *Multiple Sclerosis*., 1999, 5(2), 101-104.

[54] Wilson, NJ; Boniface, K; Chan, JR; Mckenzie, BS; Blumenschein, WM; Mattson, JD; et al. Development; cytokine profile and function of human interleukin 17–producing helper T cells. *Nature Immunology*., 2007, 8(9), 950-957.

[55] Acosta-Rodriguez, EV; Napolitani, G; Lanzavecchia, A; Sallusto, F. Interleukins 1beta and 6 but not transforming growth factor-beta are essential for the differentiation of interleukin 17–producing human T helper cells. *Nature Immunology*., 2007, 8(9), 942-949.

[56] Manel, N; Unutmaz, D; Littman, DR. The differentiation of human TH-17 cells requires transforming growth factor-beta and inducton of the nuclear receptor RORgamma-t. *Nature Immunology*., 2008, 9(6), 641-649.

[57] Volpe, E; Servant, N; Zollinger, R; Bogiatzi, SI; Hupe, P; Barillot, E; et al. A critical function for transforming growth factor-beta; interleukin 23 and pro-inflammatory cytokines in driving and modulating human TH-17 responses. *Nature Immunology.*, 2008, 9(6), 650-657.

[58] Lopez-Diego, RS; Weiner, HL. Novel therapeutic strategies for multiple sclerosis - a multifaceted adversary. *Nature reviews Drug discovery*., 2008, 7, 909-925.

[59] Awasthi, A; Kuchroo, VK. TH17 cells: from precursors to players in in ammation and infection. *International Immunology.*, 2009, 21(5), 489-498.

[60] Zhou, L; Lopes, JE; Chong, MM; Ivanov, II; Min, R; Victora, GD; et al. TGF-beta-induced Foxp3 inhibits TH17 cell differentiation by antagonizing RORgamma-t function. *Nature*., 2008, 453(May), 236-240.

[61] Bettelli, E; Oukka, M; Kuchroo, VK. TH-17 cells in the circle of immunity and autoimmunity. *Nature Immunology.*, 2007, 8(4), 345-350.

[62] Thakker, P; Leach, MW; Kuang, W; Benoit, SE; Leonard, JP; Marusic, S. IL-23 is critical in the induction but not in the effector phase of experimental autoimmune encephalomyelitis. *Journal of Immunology.*, 2007, 178, 2589-2598.

[63] Dong, C. TH17 cells in development: an updated view of their molecular identity and genetic programming. *Nature Reviews Immunology.*, 2008, 8(May), 337-348.

[64] Korn, T; Bettelli, E; Gao, W; Awasthi, A; Jäger, A; Storm, TB; et al. IL-21 initiates an alternative pathway to induce proinflammatory TH17 cells. *Nature*, 2007, 448(July), 484-488.

[65] Yang, XO; Panopoulos, AD; Nurieva, R; Chang, SH; Wang, D; Watowich, SS; et al. STAT3 regulates cytokine-mediated generation of inflammatory helper T cells. *Journal of Biological Chemistry*, 2007, 282(13), 9358 -9363.

[66] Martinez, GJ; Nurieva, RI; Yang, XO; Dong, C. Regulation and function of proinflammatory TH17 cells. *Annals of the New York Academy of Sciences*, 2008, 1143, 188-211.

[67] Langrish, CL; Chen, Y; Blumenschein, WM; Mattson, J; Basham, B; Sedgwick, JD; et al. IL-23 drives a pathogenic T cell population that induces autoimmune inflammation. *Journal of Experimental Medicine*, 2005, 201(2), 233-240.

[68] Murphy, CA; Langrish, CL; Chen, Y; Blumenschein, W; McClanahan, T; Kastelein, RA; et al. Divergent pro- and antiinflammatory roles for IL-23 and IL-12 in joint autoimmune inflammation. *Journal of Experimental Medicine*, 2003, 198(12), 1951-1957.

[69] Park, H; Li, Z; Yang, XO; Chang, SH; Nurieva, R; Wang, Y; et al. A distinct lineage of CD4 T cells regulates tissue inflammation by producing interleukin 17. *Nature Immunology.*, 2005, 6(11), 1133-1141.

[70] Mangan, PR; Harrington, LE; Quinn, DB; Helms, WS; Bullard, DC; Elson, CO; et al. Transforming growth factor-beta induces development of the TH17 lineage. *Nature*, 2006, 441(May), 231-234.

[71] Veldhoen, M; Hocking, RJ; Atkins, CJ; Locksley, RM; Stockinger, B. TGF-beta in the context of an inflammatory cytokine milieu supports de novo differentiation of IL-17-producing T cells. *Immunity*. 2006, 24(February), 179-189.

[72] Putheti, P; Pettersson, A; Soderstrom, M; Link, H; Huang, YM. Circulating CD4+CD25+ T regulatory cells are not altered in multiple sclerosis and unaffected by disease-modulating drugs. *Journal of clinical immunology.*, 2004, 24(2), 155-161.

[73] Feger, U; Luther, C; Poeschel, S; Melms, A; Tolosa, E; Wiendl, H. Increased frequency of CD4+CD25+ regulatory T cells in the cerebrospinal fluid but not in the blood of multiple sclerosis patients. *Clinical and Experimental Immunology.,* 2007, 147, 412-418.

[74] Huan, J; Culbertson, N; Spencer, L; Bartholomew, R; Burrows, GG; Chou, YK; et al. Decreased FOXP3 levels in multiple sclerosis patients. *Journal of Neuroscience Research.,* 2005, 81, 45-52.

[75] Venken, K; Hellings, N; Thewissen, M; Somers, V; Hensen, K; Rummens, J; et al. Compromised CD4+CD25high regulatory T-cell function in patients with relapsing-remitting multple sclerosis is correlated with a reduced frequency of FOXP3-positive cells and reduced FOXP3 expression at the single-cell level. *Immunology.,* 2007, 123, 79-89.

[76] Rønje, IH; Holm, NE; Bardenfleth, CL; Wæber, U; Klæbel, HH. Scleroseforeningen. *Fakta om sclerose. Behandling.* 2009.08.03 Available from: http://www. sclerose foreningen.dk/Fakta-om-Sclerose/Behandling.aspx.

[77] Linker, Ra; Kieseier, BC; Gold, R. Identification and development of new therapeutics for multiple sclerosis. *Trends in pharmacological sciences*, 2008, 29(11), 558-65.

[78] Weber, MS; Hohlfeld, R; Zamvil, SS. Mechanism of action of glatiramer acetate in treatment of multiple sclerosis. *Neurotherapeutics.,* 2007, 4(October), 647- 653.

[79] Johnson, WT; Nelson, JM; Lynch, CT; Mengel, RJ. MS national multiple sclerosis society. *About MS. Treatment;* Copaxone (Glatiramer Acetate). 2009.08.04 Available from: http://www.nationalmssociety.org/about-multiple-sclerosis/what-we-know-about-ms/treatments/medications/glatiramer-acetate/index.aspx.

[80] Rang, HP; Dale, MM; Ritter, JM; Flower, RJ. *Rang & Dale's Pharmacology.* 6th edition. Philadelphia: Churchill Livingstone Elsevier., 2007.

[81] *SPC tekst methotrexaat.*, 2009.08.04 Available from: http://db.cbg-meb.nl/IB-teksten/h100907.pdf, 2008.

[82] Sellebjerg, FT; Soerensen, PS. Behandling af multipel sklerose – nu og i fremtiden. *Ugeskrift for laeger.*, 2007, 169(12), 1121.

In: Regulatory T Cells
Editor: Ren S. Hayashi, pp. 79-98

ISBN: 978-1-61761-554-2

Chapter 4

Immunoregulatory Cells and Mechanisms of Immune Response

***Carol Aristimuño*[1], *Roseta Teijeiro*[1], *Clara de Andrés*[2] *and Silvia Sánchez-Ramón*[1*]**

[1]Department of Immunology Hospital General Universitario Gregorio Marañón, Madrid, Spain

[2]Neurology, Hospital General Universitario Gregorio Marañón, Madrid, Spain

Abstract

The immune system requires a homeostatic equilibrium between the mechanisms that assure self-tolerance, those that control the capacity to mount life-long immunity to pathogenic microbes or toxins, and those that attenuate effector mechanisms from inducing immune pathology. A growing body of evidence suggests that an imbalance in this homeostasis plays a major role in the pathophysiology of autoimmune disease.

A great diversity of regulatory cells types are being progressively identified, such as tolerogenic DCs, $CD4^+CD25^+$ T_{Reg}, regulatory natural killer (rNK) and $CD8^+CD28^-$ suppressor T-lymphocytes, among others. Among the different types of immunoregulatory cells, $CD4^+CD25^+$ (IL-2 receptor alpha chain) regulatory T $CD4^+$ cells (T_{Reg}) play a pivotal role in the tolerance to self-antigens and tissue grafts, and in the suppression of autoimmune reactions. T_{Reg} exert modulatory functions upon the activity of self-aggressive T cells not deleted in the thymus and are one of the pillars for maintaining the immune system in homeostatic balance. Indeed, these cells modulate the intensity and quality of immune reactions through attenuation of the cytolytic activities of reactive immune cells. The outcome of different contributions of all these complex mechanisms under variable inflammatory conditions, the diversity of immune

* Corresponding author: Department of Immunology, Hospital General Universitario Gregorio Marañón, Calle Doctor Esquerdo, 46, 28007 Madrid, Spain, Phone: +34 91 4265181, FAX: +34 91 5868018, E-mail: ssanchez.hgugm@salud.madrid.org

interactions between this constellation of cells conduct the evolution of the immune reaction. In this review, we summarize research characterizing the growing diversity of immunorregulatory cells subsets in the control of autoimmunity in rodents and humans. Here, we attempt to integrate the current experimental evidence on the major suppressive pathways of immunorregulation.

Keywords: Regulatory T cells; suppressive T cells; tolerogenic dendritic cells; IL-10; TGF-beta.

Immunological Identity, Tolerance, Autoimmunity and Autoimmune Disease

A germinal issue of Immunology is to understand how and why the immune response takes place, which remains still a matter of debate (Pradeu and Carosella, 2006). The central function of the immune system (IS) is to recognize and defend the self-integrity against external threats from pathogens and toxins. Thus, the generation of an efficient immune response against foreign antigens should be accompanied by a rigorous regulation of immune activation, in order to prevent collateral damage against self tissues, essential property of the IS that is defined as tolerance. A main characteristic of the adaptive immune response is the huge diversity of its specific repertoire of receptors to the antigen. However, the frontier that separate self and nonself is not always clear-cut, given the constantly changing nature of self and other particular circumstances. Among the later are included the food, non-pathogenic microbiota in gut or skin, air particles, etc; conditions such as pregnancy that requires tolerance versus an hemiallogeneic fetus, organ transplantation; or own antigens that have been modified by infections or tumors (Anaya et al., 2005). In this contextual framework, self could be defined as all those molecules of our organism that are generated physiologically throughout the growth and development, and those with which we interact every day.

Since Owen defined immune tolerance in 1945, many different immune cell subsets and different mechanisms of tolerance have been discovered, with distinct roles in diverse clinical conditions, such as gestation, autoimmune diseases, infections, organ transplantations, allergic diseases, neoplasias and paraneoplastic syndromes (Akbari et al., 2003; Aluvihare et al., 2004; Cobbold et al., 2003; Francois Bach, 2003; Qin, 2009). The mechanisms of "central tolerance" for T and B lymphocytes are generated very early in the development at the primary lymphoid organs, the thymus and the bone marrow, where most potentially threatening autoreactive lymphocytes with high avidity for the autoantigen are eliminated. Nevertheless, several clones of autoreactive T and B lymphocytes with low avidity for the autoantigen survive and are released to the bloodstream, whose activation will be regulated in the secondary lymphoid organs, in the bloodstream and in tissues by different mechanisms, globally referred as "peripheral tolerance". Among the later mechanisms are included the deletion (physical elimination of the immune cells during the process of maturation), the clonal anergy (clonal inactivation of mature lymphocytes by absence of costimulatory signals), the clonal ignorance (lymphocytes are silenced because the autoantigen is in small quantities or is sequestered in immunological privileged organs), and by the homeostatic control by several subsets of regulatory cells. In addition, many foreign antigens can behave as immunogenic or tolerogenic depending on their chemistry nature (protein, polysaccharide or lipid), dose and frequency, local environment and the nature of the antigen-presenting cell (APC) with which it enters in contact, as well as of the route of administration (Van Parijs and Abbas, 1998). The autotolerance is established and maintained through the inactivation of autoreactive $CD4^+$, $CD8^+$ and/or B effector cells by specific subsets of immune cells with "regulatory" abilities, namely tolerogenic dendritic cells (DCs), regulatory $CD4^+$ T-lymphocytes (T_{Reg}), CD8 suppressive cells, gamma-delta T-cells , natural killer (NK) cells, $CD3^+CD4^-CD8^-$ double negative cells.

According to this, we can distinguish a physiological autoimmunity, as in the case of the natural antibodies or the anti-idiotypic antibody network; local autoimmune regulation and repair process (Rolls et al., 2009) or the immune responses directed against infected or tumor-transformed own cells. Besides, there is a pathological autoimmunity leading, on one hand to autoimmune diseases, multifactorial clinical syndromes caused by the breakdown of immunological tolerance, in which excess of autoreactive activated T and/or B lymphocytes induce tissue damage; and on the other hand to paraneoplastic syndromes, in which the immune response is directed against antigens in self body tissues that are synthesized by distant tumor cells. In this review, we briefly delineate the main cell subsets interplaying in the immunoregulatory arm balance.

Immunogenic versus Tolerogenic Dendritic Cells

Recently, DCs have been described as the most powerful APCs and the only cells capable of inducing immunological primary responses on naïve T-lymphocytes, which can contribute to the peripheral tolerance on T lymphocytes by a bidirectional dialogue between both subsets (Adorini and Penna, 2009; Steinman et al., 2003; Wakkach et al., 2003). Therefore, the DCs play an essential dual role in opposite functions on the differentiation of counteracting T-lymphocytes, towards the induction and/or the regulation of the immune responses (Banchereau et al., 2003) (Figure 1). This functional plasticity turns them into the key cells of the immune response, which connects the innate and adaptive IS. Among other factors, the singularity of the DCs consists of their great capacity of recognition and capture of antigens when they are immature, their finely regulated migratory capability, their high expression of molecules of the major histocompatibility complex (MHC) and costimulatory molecules, and the secretion of polarizing cytokines once reached their functional maturity (Steinman et al., 1999). All these features make possible an interaction between the MHC on the DC surface and the clonotypic receptor of the T- lymphocyte (TCR) and then, in a suitable molecular environment, drives an efficient activation and polarization of naïve T-lymphocytes. In humans, two main subpopulations of DCs can be isolated in peripheral blood: the myeloid DCs (mDCs) and the plasmacytoid DCs (pCDs). Both populations lack specific markers of other lineages, show different phenotype, routes of development, tissue distribution, and function. The mDCs ($CD11c^+CD123^-$) are excellent APCs, which produce important quantities of IL-12 p70 and generate powerful Th1 (proinflammatory) and CTL (cytotoxic) responses. The pDCs ($CD11c^-CD123^+$(IL-3R)) are the principal producing cells of interferon of type I after infection of bacteria or/and viruses, induce Th2 responses, are tolerogenic in response to the CD40 or IL-3 molecules and after activation by virus or CD40 can generate Th1 responses (Fuchsberger et al., 2005; Zhang and Wang, 2005). During the activation, the $CD4^+$ T-lymphocytes can express the cytotoxic T-lymphocyte antigen-4 (CTLA-4) and programmed cell death-1 (PCD-1) molecules, which joins to the B7 molecule and programmed death ligand-1 (PD-L1) (B7-H1) on the DC, and generates signals that disable the production of IL-2 and progression of the cellular cycle of the T-lymphocyte (Fife and Bluestone, 2008). The CTLA-4 has demonstrated its ability to play a fundamental role in the induction of anergy. In fact, mutations in the genes that codify CTLA-4 or Fas (whose expression is essential for the induction of apoptosis), are associated with the occurrence of autoimmune diseases (Van Parijs and Abbas, 1998).

The ability to induce tolerance by DCs depends on their stage of functional maturation. Immature DCs induce tolerance due to presentation of the autoantigen without coestimulation, that is, in the absence of signal 2 at immune synapse (Figure 1). However, not only immature DC but also specialized subsets of tolerogenic DC can drive the differentiation of regulatory cells.

Tolerogenic DCs have a distinctive biological characteristic consisting of the production of the indolamine-2,3dioxigenase (IDO) enzyme, responsible for tryptofan degradation. This feature, in addition to inducing an expansion of the numbers of T_{Reg}, limits the need of tryptofan for pathogens, an essential aminoacid for their development. The expression of IDO has been detected in a constitutive way on the human regulatory pDCs $CD123^+$, and can be induced by classic stimuli of maturation of DCs, such as IFN-γ and LPS. In mice, T_{Reg} can induce the expression of functional activity of IDO on DCs and thus DC's immunosuppressive capacity (Fallarino et al., 2003; Munn et al., 2002).

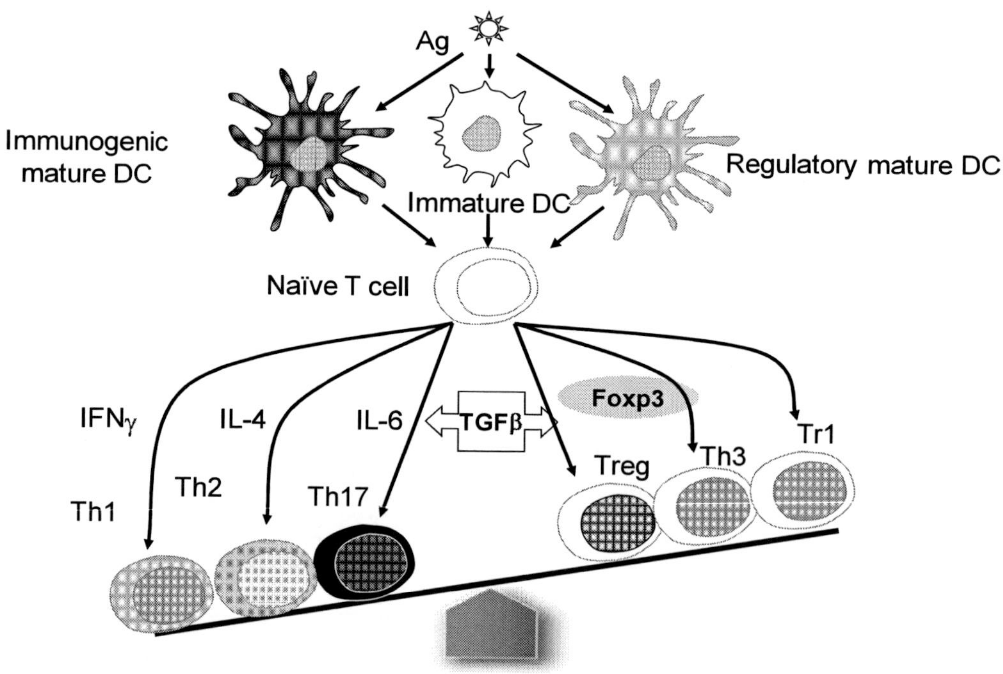

Figure 1. DC functions in the differentiation of T-lymphocytes, towards the induction and/or the regulation of the immune responses

Also, IL- 10 treatment can induce the differentiation of tolerogenic DCs (Tr-DCs), characterized by the expression of CD14, CDl Ic, CDl Ib, CD83, CD80, CD86, CD71 and

HLA-DR, but not CDIa. Tr-DC express immunoglobulin-like transcript (ILT-) 2, ILT-3, ILT-4, and the non classical MCH class I molecule HLA-G. Tr-DC secrete significantly higher levels of IL-10 compared to immature DC, whereas the amounts of IL- 12 are comparable to those produced by immature DC (Wu and Horuzsko, 2009). Autologous pDCs, in the presence of repetitive CpG's sequences (sequence 5′-purine-purine-CG-pirimidin-pirimidin-3′ without methylation) (Krieg et al., 1995), induce the transformation of naïve T-lymphocytes to cytokine-dependent T_{Reg}, which are anergic *in vitro* and express CTLA-4, CD25 and Foxp3 (Moseman et al., 2004).

Regulatory T Cells: Natural versus Induced

Regulatory T cells have emerged in the last years as key players in peripheral tolerance, in particular those with the $CD4^{+}CD25^{high}FoxP3^{+}$ phenotype, also known as T_{Reg}, which have shown their capability to control the immune response against autoantigens and alloantigens. Different studies support the concept that pregnancy is associated with an elevation in size of peripheral $CD4^{+}CD25^{+}$ T_{Reg} pool. In addition, T_{Reg} are recruited to the foetomaternal interface (Aluvihare et al., 2005), where they are probably playing a role in allowing fetal implantation and survival within the maternal uterus. Additionally, their contribution to the mechanisms mediating foetomaternal tolerance, in the particular context of an underlying autoimmune disease, may also play an important role in protection against autoimmune attacks (Sánchez-Ramón et al., 2005; Navarro et al., 2006; de Andrés et al., 2007). Such T_{Reg} expansion may contribute, in part, to the clinical amelioration of many autoimmune diseases during gestation.

The transcription nuclear factor Forkhead box p3 (FoxP3) is a negative regulator of T cell activation, and is involved in T_{Reg} lineage commitment (Fontenot et al., 2003; Hori et al., 2003; Sakaguchi et al., 2008). T_{Reg} are the best characterized and studied, comprising around 5% to 15% of $CD4^{+}$ T cells in peripheral blood (Gavin and Rudensky, 2003). A subset of $CD4^{+}$ T cells that exhibit the highest level of IL-2 receptor α-chain CD25, $CD4^{+}CD25^{+high}$, comprises the 1%–4% of $T_{Reg,}$ and have shown the highest suppressive activity *in vitro*, inhibiting completely the proliferation and secretion of cytokines by effector $CD4^{+}$ T cells (Baecher-Allan et al., 2001). The differentiation of most circulating T_{Reg} starts within the thymus, after the induction of FoxP3 expression with a subtype of TCRalfa-beta thymocytes, which shows high avidity for the autoantigen-MHC complexes (Apostolou et al., 2002; Sakaguchi et al., 2001). These cells are known as thymus-derived or natural T_{Reg} (nT_{Reg}), distinct from adaptive T_{Reg} that can be induced (iT_{Reg}), which include $FoxP3^{+}$ T_{Reg}; regulatory T helper type 3 (Th3 lymphocytes) and regulatory type 1 (Tr1); derived extra-thymically from circulating $CD4^{+}$ T cells, induced by specific antigens that can or can not express Foxp3 in particular conditions (Zheng et al., 2002).

Natural CD4+CD25+FoxP3 regulatory T cells (nT_{Reg})

The nT_{Reg} develop within a highly controlled thymic microenvironment (Josefowicz and Rudensky, 2009). A number of candidate markers for distinguishing nT_{Reg} cells have been reported. Some cell surface receptors characteristic of nT_{Reg} include the glucocorticoid-induced tumor necrosis factor receptor (GITR), CTLA-4 (CD152), galectin 1, CD38, CD62L,

OX-40L, CD103, TNFR2, and TGF-βR1, all expressed in classical activated T cells, with exception of the homing receptor CD103 (αE integrin). The neuropilin-1, receptor involved in the conduction of the axon, the angiogenesis and T cell activation; and the lymphocyte activation gen 3 (LAG-3), have been recently described as specific markers of nT_{Reg}. LAG-3 mRNA is selectively expressed in nT_{Reg}, but no LAG-3 protein is detected on the cell surface of *ex vivo* nT_{Reg}. Anti-LAG-3 antibodies could inhibit the suppression mediated by purified nT_{Reg} *in vitro* (Huang et al., 2004), which demonstrates that LAG-3 is involved in the regulatory function of nT_{Reg}. However, the lack of surface LAG-3 expression excludes the possibility of LAG-3 as a surface marker (Yi et al., 2006). The expression patterns of these molecules on mouse and human nT_{Reg} are summarized in Table 1.The discovery that T_{Reg} expressed the nuclear transcription factor FoxP3 suggested a T distinctive lineage for these cells. In humans, the disruption of FoxP3 gene function causes a syndrome characterized by immune disfunction, polyendocrinopathy, and X-linked enteropathy (IPEX syndrome). Recent studies have revealed that FoxP3, both in mice and humans, is a negative regulator of T lymphocyte activation, probably through the suppression of cytokines such as IL-2 (Hori and Sakaguchi, 2004). In human peripheral blood, the expression of FoxP3 is not specific of T_{Reg}, and can also be observed on DCs surface, in activated $CD4^{+}CD25^{-}$ T lymphocytes and even in $CD8^{+}$ T lymphocytes (Morgan et al., 2005).

Naturally occurring T_{Reg} are mitotically quiescent under basal conditions, displaying proliferative anergy. Innate T_{Reg} are a distinct thymic subset programmed for immunosuppressive tasks, which require antigenic stimulation to initiate proliferation *in vivo* (Kuniyasu et al., 2000; Gavin et al., 2002), but do not require engagement of major histocompatibility complex (MHC) antigens to trigger their suppressive activity (Bienvenu et al., 2005).

Thus, antigen-specific, TCR-mediated sensitization is required to induce proliferation, and to a lesser extent to activate the suppressive function of T_{Reg} *in vivo*. The nature of TCR engagement and the requirement for antigenic stimulation appear to be largely dependent on the environment under which T_{Reg} operate. On the one hand, activity of T_{Reg} *in vivo* at the site of inflammation (Chen et al., 2005; Green et al., 2002; Banz et al., 2002), and the *in vitro* demonstrated need to establish cell-to-cell contacts (Chen et al., 2005; Piccirillo et al., 2002) suggest an antigen specific mode of action of T_{Reg}. Indeed, antigen specific functions were implied by the dependence of T_{Reg} on TCR engagement to acquire full *in vivo* suppressive activity in transplant and autoimmune settings (Kingsley et al., 2002; Tang et al., 2004). Furthermore, T_{Reg} activation by APC elicits a reaction ten-fold more potent against the presented antigens than against third party antigens (Yamazaki et al., 2006). On the other hand, under various experimental conditions, antigen specificity is less accentuated, as professional T_{Reg} subsets do not always require further stimulation under selected conditions. Antigen specific activity of T_{Reg} is disputed in some transplant models and in immune colitis (Thornton et al., 2000). Likewise, polyclonal T_{Reg} effectively block autoimmunity, allograft rejection and graft versus host (GVH) reactions (Hoffmann et al., 2002). Efficient prevention of GVH by T_{Reg} stimulated against third party antigens (Steiner et al., 2006) and the lack of functional specificity in autoimmune disorders (Kuniyasu et al., 2000; Thornton et al., 2000) suggest that immune regulation can be achieved in an antigen-independent manner.

Table 1. Phenotypic characteristics of human and mouse nT_{Reg}

	Human	Mouse
Cell surface receptors	CD25high CD27 CD45ROhigh CD58 CD83 CD95high HLA-DRlow a4β1 a4β7 CCR4 CCR8 Granzyme A	CD11b/CD18high CD25intermediate~high CD30 CD45RBlow CD5high CD69 CD150low TNFRII OX40 4-1BB TRANCER CD103(aE) CCR2 CCR6 CCR7 CXCR3 LAG-3 Nrp1 PD-1 TLR4 TLR7 galectin-1 Ly6A/Ehigh TGFβR1 Granzyme B
Intracellular molecules		CD122high, CD132, CD28, CD38high, CD44high, CD62L, CD54, CD71, CD127, GITR, TLR5, TLR8, TGF-β CTLA4+, Foxp3+

Adapted from Huanfa Yi et al, 2006.

Taken together, these findings suggest that antigenic stimulation is required for initiation of T_{Reg} proliferation, whereas functional activation (suppressive function) does not necessarily require TCR engagement (Askenasy et al., 2008).

Mechanisms of action regulatory cells

Upon homing to the site of inflammation, T_{Reg} operate in concert with effector T-cells (T_{eff}) in modulation of the immune reaction. Three distinct mechanisms appear to mediate the suppressive effect of T_{Reg} on cytolytic cells, with variable contributions under different experimental conditions and inflammatory environments (Figure 2). All these mechanisms require close spatial proximity between the suppressor and suppressed cells, either in the target tissue or the regional lymph nodes (Askenasy, 2008). *The first mechanism* involves physical elimination of cytotoxic cells, by direct cell-to-cell contact at the site of inflammation. T_{Reg} directly kill the effector cells (Banz et al., 2002; Weber et al., 2006), with preferential use of perforin/granzyme by naturally occurring T_{Reg} (Grossman et al., 2004) and of Fas-ligand (FasL) by adaptive T_{Reg} (Weber et al., 2006; Gondek et al., 2005). This mechanism may be dominant in situations where the induction of tolerance depends on functional apoptotic pathways. In models of autoimmune diabetes, T_{Reg} use FasL to deplete the Th1 effector cells in lymphoid tissues, and FasL inhibition results in persistence of the diabetogenic cells (Weber et al., 2006). Fas-mediated apoptosis is critical both to the induction of tolerance via apoptosis of antigen-reactive lymphocytes, and also to the maintenance of tolerance through polarization of the immune response towards protective Th2 responses (Weber et al., 2006; Askenasy et al., 2005). However, it should be noted that suppressive functions of T_{Reg} do not always require physical elimination of T_{eff}, as demonstrated in other autoimmune models where T_{Reg} only repressed the cytolytic functions of T_{eff} (Thornton et al., 2000; Piccirillo et al., 2001). We have recently described how perforin-dependent suppression by T_{Reg} is enhanced by oestradiol (Valor et al, submitted manuscript).

The second mechanism involves inhibition of proliferation and/or cytokine production in pathogenic T cells (Thornton et al., 2000; Piccirillo et al., 2001). On one hand, T_{Reg} repress the cytolytic capacity of T_{eff} after antigen-specific sensitization by decreasing cytokine production, without affecting their proliferation (Wood et al., 2003). This regulatory mechanism may involve various cytokines and chemokine receptors, such as suppression of INF-γ-dependent CXCR3 expression in T_{eff}, reducing their ability to home to the pancreatic lymph nodes of diabetic mice (Sarween et al., 2004). The most important factor is inhibition of IL-2 production at the earliest stages of T_{eff} activation that results in anergy, before and without necessarily suppression of T_{eff} expansion (Kuniyasu, Y, 2000; Thornton, AM, 2000; Piccirillo, CA, 2001). On the other hand, the activity of T_{Reg} was associated with inhibition of T_{eff} proliferation. For instance, using T cells with a rearranged transgenic TCR, antigen-stimulated T_{Reg} blocked the migration, expansion and IL-2 production of T_{eff} (Duthoit et al., 2005). These data suggest that T_{Reg} can differentially suppress the cytolytic and proliferative activities of T_{eff} by inhibition of their cytokine production. Importantly, a direct inhibitory effect on T_{eff} can be induced prior to onset of cell proliferation. In addition to the contextual association of T_{Reg} with CD25 expression, IL-2 plays an important role in suppression of the immune reaction. Regulatory T cells attenuate IL-2 homeostasis in two ways: inhibition of IL-2 production and/or excessive IL-2 consumption. Both are significant mechanisms of T_{Reg}-mediated suppression of $CD4^+$ and $CD8^+$ T cells upon TCR stimulation (Kuniyasu et al., 2000; Thornton et al., 2000; Takahashi et al., 1998). Inhibition of IL-2 production in the earliest stages of cytotoxic T cell activation induces anergy, with (Duthoit et al., 2005; Inaba et al., 2006) and without (Wood et al., 2003) concomitant blockade of T_{eff} expansion *in vitro* and *in vivo*. The second pathway involves a paracrine depletion of IL-2 that impairs the cytolytic activity and expansion of reactive T_{eff} (Scheffold et al., 2005). Despite constitutive expression of CD25 by most $FoxP3^+$ T_{Reg}, naturally occurring T_{Reg} do not secrete IL-2 (Malek et al., 2004; Nelson, 2004; Fehervari et al., 2006), while adaptive T_{Reg} secrete this cytokine (Bluestone et al., 2003; Weber et al., 2006). Quantitatively, innate T_{Reg} depend primarily on IL-2 secreted by reactive T cells and to a lesser extent on IL-2 released by other cell types (de la Rosa et al., 2004). Thus, within the competition for IL-2, avid consumption of T_{Reg} (Scheffold et al., 2005) depletes the availability of this cytokine to the expanding clones of Teff. It should be noted that IL-2 induces expression of its receptor (including the α chain) as a positive feedback mechanism in activated T cells (Kim et al., 2001).

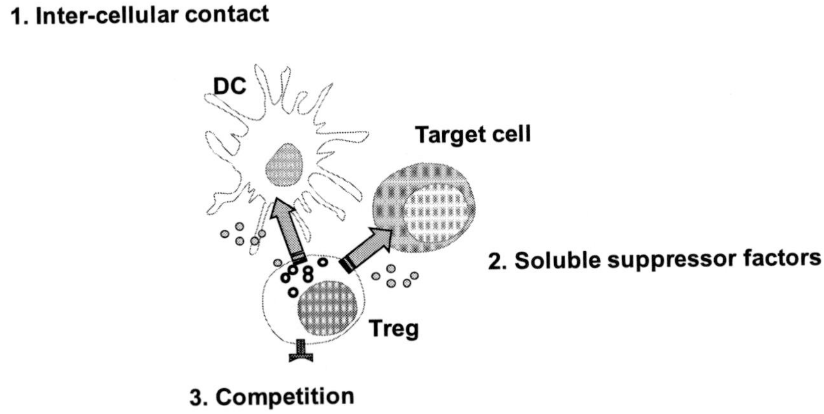

Figure 2. Mechanisms of action of regulatory T cells; Elimination of cytotoxic cells, by direct cell-to-cell contact (1). Inhibition of proliferation and/or cytokine production in target cells and direct secretion of cytokines (2). Competition for IL-2, avidly consumed by Treg cells (3) depleteting the availability of this cytokine to the expanding clones of effector cells

The third mechanism involves modulation of the cytokine environment at the site of inflammation through direct secretion of cytokines. $CD25^+CD4^+$ T_{Reg} produce immunoregulatory molecules such as TGF-β and IL-10, which directly affect the activity of cytotoxic T cells and APCs. TGF-β1 reduces cytokine secretion by activated $CD4^+$ T cells (Zheng et al., 2004), without limiting their capacity to expand and without inducing their apoptosis (Cottrez et al., 2001). TGF-β1 also induces IL-10 production in Th1 cells, which further inhibits cytokine production and directly attenuates T_{eff} function (Annacker et al., 2001). In a reciprocal interaction, IL-10 enhances the response of activated T cells to TGF-β1 through regulation of TGF receptor expression (Cottrez et al., 2001). Thus, combined effects the TGF-β and IL-10 inhibit the activity of T_{eff}, with minor effect on their expansion. In addition, the local cytokine composition indirectly affects the intensity of the immune reaction through modulation of the behavior of APC. IL-10 downregulates APC production of IL-12 (an inductive cytokine for Th1 differentiation) and thus, inhibits the differentiation and responses of Th1-type cells (Moore et al., 2001). In parallel, interaction of T cells with APC triggers IL-2 production in the latter, which acts to enhance reactive T cell proliferation. Therefore, IL-10 released by the T_{Reg} directly downmodulate the activity of APC (through IL-12), and indirectly dampens the intensity of entire immune reaction through inhibition of IL-2 production. The restrictive activities mediated by alterations in IL-2 concentrations in turn downregulate the activity of the T_{Reg}, which depend on this cytokine for sustained suppressive activity in the periphery (Malek et al., 2004; Nelson et al., 2004; Fehervari, et al., 2006).

Ito *et al* have recently reported the identification of two subsets of naturally occurring $FoxP3^+$ T_{Reg} according to their differential expression of the costimulatory receptor ICOS. Whereas the $ICOS^+FoxP3^+$ T_{Reg} used IL-10 to suppress DC function and TGF- β to suppress T cell function, the $ICOS^-FoxP3^+$ T_{Reg} used mainly TGF-β. The survival and proliferation of the two subsets of T_{Reg} were regulated by signaling through ICOS or CD28, respectively. These authors suggested that the molecules used by the $ICOS^+FoxP3^+$ and $ICOS^-FoxP3^+$ nT_{Reg} to mediate immunosuppression mirror those used by Tr1 cells and Th3 cells, respectively. Accordingly, selection of $FoxP3^+$ T_{Reg} in the thymus is probably coupled to their differentiation into the $ICOS^+FoxP3^+$ phenotype, which is imprinted with high IL-10-

producing capacity, and the ICOS⁻FoxP3⁺ subset is imprinted with high TGF-β-expression capacity (Ito et al., 2008).

Induced regulatory CD4⁺ T lymphocytes (iT$_{Reg}$)

Induced iT$_{Reg}$ are peripherally generated FoxP3 positive and negative cells that comprise at least two subsets of regulatory T cells, namely Th3 and Tr1. Unlike nT$_{Reg}$, iT$_{Reg}$ mediate suppression mainly by means of cytokines secretion (Berthelot and Maugars, 2004) (Table 2).

Table 2. Characterisitics of the various T CD4⁺ (Th3, Tr1 and T CD4⁺CD25⁺ʰⁱᵍʰ) subsets with suppressive effects and comparison with Th type 1 and Th2 cells subsets

	Th1	Th2	Th3	Tr1	CD25^{+high}
Origin	Lymph nodes	Lymph nodes	Periphery	Periphery	Thymus
IFN-γ secretion	++++	-	+/-	+	+/-
Il-4 secretion	-	++++	+/-	-	+/-
TGF-β secretion	+/-	+/-	++++	++	+/-
IL-10 secretion	-	++	+/-	++++	+/-
Growth stimuli	IL-2	IL-2/IL-4	IL-4/TGF-β	IL-10, IFN-α	IL-2+++
B-helper	IgG-2a	IgG1/IgE	IgA	-	-
Suppression upon	Th2	Th1	Th1 or 2	Th1	Th1 or 2
Proliferation	+++			+/- (IL-15)	IL-2
Affinity for self	+/-	+/-	+	+ ?	+++++

Adapted from Berthelot and Maugars, 2004.

Tr1: Interleukin-10 (IL-10)-secreting Tr1 CD4⁺ T cells

TrI cells are characterized by a unique cytokine secretion profile; upon TCR activation they secrete high levels of IL-10, significant amounts of IL-5, TGF-β and low levels of IFN-γ, and IL-2 but not IL-4. Tr1 cells suppress the immune response via cell–cell contact and/or via secretion of TGF-b and IL-10 (Roncarolo et al., 2001). Grossman *et al* showed that adaptive T$_{Reg}$ generated from human CD4⁺ T cells by colligation of the T cell coreceptor CD3 and the complement regulatory protein CD46 in the presence of IL-2 expressed predominantly granzyme B (Grossman et al., 2004). These adaptive T$_{Reg}$ cells demonstrated cytotoxicity against allogeneic tumor cell lines in a perforin-dependent, but MHC/TCR-independent, manner.

Tr1 cells have a limited growth potential (because of the high IL-10 levels), although they can divide in the presence of IL-15 (Roncarolo et al., 2001). They express CTLA-4, a feature required for TGF-β production by Tr1 cells. Tr1 cells do not seem more likely than other T cells to express CD25. Differentiation to Tr1 cells occurs after contact with APCs that produce large amounts of IL-10. This occurs mainly with relatively naïve DCs, when the antigen is present in low levels, and when contact between APCs and CD4⁺ T cells are repeated. An overabundance of Tr1 cells can favor the development of infections such as tuberculosis (Roncarolo et al., 2001) and Lyme disease. A number of infectious agents use this phenomenon to their advantage by producing molecules similar to IL-10; examples include the Epstein–Barr virus and the cytomegalovirus (Berthelot, 2004).

Th3 lymphocytes

This $CD4^+$ T cell subset secretes TGF-β (as well as IL-4 and IL-10), which stimulates plasma cells to secrete IgA and decreases the numbers of Th1 and Th2 cells. Th3-cell maturation requires the presence of TGF-β, IL-4, and IL-10, as well as inhibition of IL-12 and expression at the cell surface of CD86 and CTLA-4 (Weiner, 2001a; Weiner, 2001b). Th3 clones emerge mainly after penetration of a foreign antigen through the oral route, as the cytokine microenvironment in the bowel (high in TGF-β and containing DC subsets that induce suppressor responses (Weiner., 2001c)) is particularly prone to differentiation into Th3 cells. Rather than effector T cells, Th3 cells target the APCs that come into contact with effector T cells. Secretion of TGF-β by Th3 cells exerts a paracrine suppressor effect on APCs (Berthelot, 2004). The characteristics of the various $CD4^+$ T cell subsets with suppressor effects and comparisons with Th1 and Th2 cells are summarized in Table 2.

Regulatory/Suppressor $CD8^+$ T-Lymphocytes

The cells that were first suspected to have an immunosuppressive function were the $CD8^+$ T lymphocytes, as proposed by Gershon and Kondo (Gershon and Kondo, 1971; Filaci and Suciu-Foca, 2002). Antigen-specific $CD8^+$ T suppressor (Ts) lymphocytes have been isolated from mice and humans (Damle and Engleman, 1989; Sercarz and Krzych, 1991; Yuen et al., 1995) and evidence has been provided that Ts suppress autoimmune responses in mice with experimental autoimmune encephalomyelitis. Phenotypically, one subset of $CD8^+$ Ts cells are characterized by the lack of expression of the coestimulatory CD28 molecule (Damle and Engleman, 1989; Liu et al., 1998; Filaci et al., 2001). Filaci *et al*, found that $CD8^+$ Ts can be generated *in vitro* from $CD8^+CD28^-$ T lymphocytes, but not from $CD8^+CD28^+$ T cells (Filaci et al, 2004). Nonantigen-specific $CD8^+$ Ts inhibit T-cell proliferation of antigen-specific T lymphocytes. A key role in their generation is played by monocytes that secrete IL-10 after granulocyte macrophage-colony-stimulating factor (GM-CSF) stimulation. Cell-to-cell direct interaction between $CD8^+CD28^-$ T cells and monocytes does not play a role in the generation of $CD8^+$ Ts. In particular, two distinct functionally subpopulations of $CD8^+$ T suppressor lymphocytes ($CD8^+$ Ts) have been identified. The first characterized $CD8^+$ Ts subset induces an antigen-specific immune suppression through cell-to-cell contact with APCs after antigen presentation (Filaci et al., 2001, Ciubotariu et al., 1998 and Jiang et al., 1998). The second subset of $CD8^+$ Ts mediates a nonantigen specific suppression of T-cell proliferation via soluble factors such as IFNγ and IL-6 (Colovai et al., 2000; Liu et al., 1999). $CD8^+$ Ts suppress also the antigen-specific cytotoxic activity of CTL decreasing the expression of HLA class I molecules on target cells through IL-10 secretion. $CD8^+$ Ts have a $CD45RA^+$, $CD27^-$, $CCR7^-$, IL-10Rα$^+$ phenotype and a TCR Vβ chain repertoire overlapping that of autologous circulating $CD8^+$ T cells. This phenotype is typical of T lymphocytes previously expanded by antigen stimulation (Filaci et al, 2004). In parallel to these findings, our group has recently reported that $CD8^+CD25^+CD28^-$ T-cells are a relevant $CD8^+$ suppressor T-cells subset in multiple sclerosis (Aristimuño et al., 2008; Aristimuño et al., 2010).

Regulatory Natural Killer Cells

Natural killer (NK) cells represent a potent first line of defense in immunity. They preferentially attack and lyse cells that lack or express low levels of self-MHC class I proteins. However, accumulating data suggest that NK functional biology is more complex than previously thought. NK cells express many receptors, like killer cell Ig-like receptor (KIRs), both stimulatory and inhibitory. The net balance of activating and inhibitory signals resulting from interactions with target cells will determine the ultimate action to mediate killing or not. In addition to the receptors specific to self-MHC class I, various non-MHC stimulatory receptors on NK cells also participate in the process (Li et al., 2010). NK cell receptors responsible for detection of human leukocyte antigen (HLA) class I and class I-like proteins on potential target cells transmit inhibitory and activating signals that integrate to determine NK cell function. KIR 3binds to HLA class I molecules on the surface of the target cell, and it confers inhibitory signals to NK cells (Lanier, 1998). Upon its ligation by HLA class I molecules, KIR can deliver inhibitory signals via the immune-receptor tyrosine-based inhibitory motif. Therefore, NK cells can recognize the cells that do not express HLA class I molecules as cytotoxic target cells, and KIR plays a role in the cytotoxic target discrimination of NK cells (Colonna, 2005). Among inhibitory receptors, some are specific for different groups of MHC class I alleles, while others are still orphan receptors. On the contrary, various activating receptors are involved in the triggering of NK cell-mediated natural cytotoxicity: The expression of CD16 (FcγRIII), the low-affinity receptor for IgG and CD56, an isoform of neural cell adhesion molecule, and molecules that inform on the NK cell activation state, such as CD25, CD69, CD49d, activatory/inhibitory KIRs: CD94 (KLRD1), activatory KIR: CD161 (NKR-P1A), and inhibitory KIRs: CD158a (NKAT1, KIR2DL1) and CD158b (NKAT2, KIR2DL3) by IL-10-secreting NK cells are analyzed in different studies (Sambrook and Beck, 2007).

A growing work of evidence indicates that NK cells play positive or negative immune regulatory effects by secreting various cytokines or cell-to-cell contact and maintain immune homeostasis. The dysfunction or lack of NK cells are involved in the pathogenesis of many diseases (e.g., tumor, autoimmune diseases, infection, sterility, and recurrent spontaneous abortions), suggesting the existence of a “regulatory NK cell subset (NK-reg subset)”, which exerts important regulatory effects on innate and adaptive immunity (Zhang et al., 2006, Li et al., 2010).

Activated NK cells can be cytotoxic to autologous oligodendrocytes *in vitro* (Morse et al., 2001), and are detected in acute inflammatory CNS lesions (Matsumoto et al., 1998). However, depletion of NK cells results in increased severity of EAE, demonstrating their role as regulatory cells (Takahashi et al., 2001). These authors initially found that peripheral blood of multiple sclerosis patients in remission had increased proportions of NK cells that expressed high levels of CD95 (Fas), a receptor that can induce apoptosis (Takahashi et al., 2004). Thus NK cells subsets, analogous to CD8 T cells, have both cytotoxic and regulatory potential. It has been hypothesized that NK cells could fulfill a particularly important regulatory niche within the CNS, a compartment in which there is relatively low MHC expression. Recently, regulatory NK cells producing IL-10 and suppressing the proliferation of Th1 and Th2 lymphocytes have been described (Deniz et al., 2008).

Recent investigations have shown that not only T_{Reg}, but also regulatory NK cells, play very important roles in the maintenance of pregnancy (Saito et al., 2007; Higuma-Myojo et al., 2005). Veenstra van Nieuwenhowen *et al.* (Veenstra van Nieuwenhoven et al., 2002) showed an increase in the production of IL-10 by circulating NK cells during gestation, suggesting that these cells could play some important roles in the maintenance of pregnancy by regulation of maternal immune function.

Conclusions and Future Perspectives

At present, the potentiation of different immunoregulatory cells subsets are considered as a putative armamentarium for new therapeutical strategies in immune pathology, and even for individualized cell therapy. There still remain numerous open questions, in practical terms, by which this regulatory cell-based therapeutical strategy could be given, such as the most useful regulatory cell, the most reliable, effective and safe methods for their generation and expansion; the group of patients susceptible to this treatment and the best time-point of the disease to administer this kind of therapy. The next years promise hopeful and fascinating advances in their use.

Acknowledgments

This work was supported by grants from the Fondo de Investigation Sanitaria (FIS #PI040468), the Fundation Salud 2000, and the Fundation Alicia Koplowitz, Spain.

References

[1] Adorini, L. & Penna, G. (2009). Dendritic Cell Tolerogenicity in Immunomodulation by Vitamin D Receptor Agonists. *Hum Immunol.*

[2] Akbari, O., Stock, P., DeKruyff, R. H. & Umetsu, D. T. (2003). Role of regulatory T cells in allergy and asthma. *Curr Opin Immunol*, *15*, 627-633.

[3] Aluvihare, V. R., Kallikourdis, M. & Betz, A. G. (2004). Regulatory T cells mediate maternal tolerance to the fetus. *Nat Immunol*, *5*, 266-271.

[4] Aluvihare, V. R., Kallikourdis, M. & Betz, A. G. (2005). Tolerance, suppression and the fetal allograft. *J Mol Med*, *83*, 88-96.

[5] Anaya, J. M., Shoenfeld, Y., Correa, P. A., García Carrasco, M. & Cervera, R. (2005). *Autoinmunidad y Enfermedad Autoinmune.* Vol. *16,* Corporación para Investigaciones Biológicas, Barcelona.

[6] Annacker, O., Pimenta-Araujo, R., Burlen-Defranoux, O., Barbosa, T. C., Cumano, A. & Bandeira, A. (2001). CD25+ CD4+ T cells regulate the expansion of peripheral CD4 T cells through the production of IL-10. *J Immunol*, *166*, 3008-18.

[7] Apostolou, I., Sarukhan, A., Klein, L. & von Boehmer, H. (2002). Origin of regulatory T cells with known specificity for antigen. *Nat Immunol., 3,* 756-763.

[8] Aristimuno, C., Navarro, J., de Andres, C., Martinez-Gines, L., Gimenez-Roldan, S., Fernandez-Cruz, E. & Sanchez-Ramon, S. (2008). Expansion of regulatory CD8+ T-

lymphocytes and fall of activated CD8+ T-lymphocytes after i.v. methyl-prednisolone for multiple sclerosis relapse. *J Neuroimmunol., 204,* 131-135.

[9] Aristimuno, C., de Andres, C., Bartolome, M., de las Heras, V., Martinez-Gines, M. L., Arroyo, R., Fernandez-Cruz, E. & Sanchez-Ramon, S. (2010). IFNbeta-1a therapy for multiple sclerosis expands regulatory CD8+ T cells and decreases memory CD8+ subset: a longitudinal 1-year study. *Clin Immunol., 134,* 148-57.

[10] Askenasy, N. & Yolcu, E. S., et al. (2005). Induction of tolerance using Fas ligand: a double-edged immunomodulator. *Blood, 105,* 1396-404.

[11] Askenasy, N., Kaminitz, A. & Yarkoni, S. (2008). Mechanisms of T regulatory cell function. *Autoimmun Rev, 7,* 370-5.

[12] Baecher-Allan, C., Brown, J. A., Freeman, G. J. & Hafler, D. A. (2001). CD4+CD25high regulatory cells in human peripheral blood. *J Immunol, 167,* 1245-1253.

[13] Banchereau, J., Paczesny, S., Blanco, P., Bennett, L., Pascual, V., Fay, J. & Palucka, A. K. (2003). Dendritic cells: controllers of the immune system and a new promise for immunotherapy. *Ann N Y Acad Sci, 987,* 180-187.

[14] Banz, A., Pontoux, C. & Papiernik, M. (2002). Modulation of Fas-dependent apoptosis: a dynamic process controlling both the persistence and death of CD4 regulatory T cells and effector T cells. *J Immunol, 169,* 750-7.

[15] Berthelot, J. M. & Maugars, Y. (2004). Role for suppressor T cells in the pathogenesis of autoimmune diseases (including rheumatoid arthritis). Facts and hypotheses. *Joint Bone Spine, 71,* 374-380.

[16] Bienvenu, B., Martin, B., Auffray, C., Cordier, C., Becourt, C. & Lucas, B. (2005). Peripheral CD8+CD25+ T lymphocytes from MHC class II-deficient mice exhibit regulatory activity. *J Immunol, 175,* 246-53.

[17] Bluestone, J. A. & Abbas, A. K. (2003). Natural versus adaptive regulatory T cells. *Nat Rev Immunol, 3,* 253-7.

[18] Chen, Z., Herman, A. E., Matos, M., Mathis, D. & Benoist, C. (2005). Where CD4+CD25+ T reg cells impinge on autoimmune diabetes. *J Exp Med, 202,* 1387-97.

[19] Ciubotariu, R., Colovai, A. I., Pennesi, G., Liu, Z., Smith, D., Berlocco, P., Cortesini, R. & Suciu-Foca, N. (1998). Specific suppression of human CD4+ Th cell responses to pig MHC antigens by CD8+CD28- regulatory T cells. *J Immunol, 161,* 5193-202.

[20] Cobbold, S. P., Nolan, K. F., Graca, L., Castejon, R., Moine, A. L., Frewin, M., Humm, S., Adams, E., Thompson, S., Zelenika, D., Paterson, A., Yates, S., Fairchild, P. J. & Waldmann, H. (2003). Regulatory T cells and dendritic cells in transplantation tolerance: molecular markers and mechanisms. *Immunol Rev., 196,* 109-124.

[21] Colonna, M. (2005). Fine-tuning NK cell responses: it's a family affair. *Nat Immunol., 6,* 961-2.

[22] Colovai, A. I., Liu, Z., Ciubotariu, R., Lederman, S., Cortesini, R. & Suciu-Foca, N. (2000). Induction of xenoreactive CD4+ T-cell anergy by suppressor CD8+CD28- T cells. *Transplantation, 69,* 1304-10.

[23] Cottrez, F. & Groux., H. (2001). Regulation of TGF-beta response during T cell activation is modulated by IL-10. *J Immunol., 167,* 773-8.

[24] Damle, N. K. & Engleman, E. G. (1989). Antigen-specific suppressor T lymphocytes in man. *Clin Immunol Immunopathol., 53,* S17-24.

[25] de Andres, C., Aristimuno, C., de Las Heras, V., Martinez-Gines, M. L., Bartolome, M., Arroyo, R., Navarro, J., Gimenez-Roldan, S., Fernandez-Cruz, E. & Sanchez-Ramon, S. (2007). Interferon beta-1a therapy enhances CD4+ regulatory T-cell function: an ex vivo and in vitro longitudinal study in relapsing-remitting multiple sclerosis. *J Neuroimmunol., 182,* 204-11.

[26] de la Rosa, M., Rutz, S., Dorninger, H. & Scheffold, A. (2004). Interleukin-2 is essential for CD4+CD25+ regulatory T cell function. *Eur J Immunol., 34,* 2480-8.

[27] Deniz, G., Erten, G., Kucuksezer, U. C., Kocacik, D., Karagiannidis, C., Aktas, E., Akdis, C. A. & Akdis, M. (2008). Regulatory NK cells suppress antigen-specific T cell responses. *J Immunol, 180,* 850-7.

[28] Duthoit, C. T., Mekala, D. J., Alli, R. S. & Geiger, T. L. (2005). Uncoupling of IL-2 signaling from cell cycle progression in naive CD4+ T cells by regulatory CD4+CD25+ T lymphocytes. *J Immunol, 174,* 155-6.

[29] Fallarino, F., Grohmann, U., Hwang, K. W., Orabona, C., Vacca, C., Bianchi, R., Belladonna, M. L., Fioretti, M. C., Alegre, M. L. & Puccetti, P. (2003). Modulation of tryptophan catabolism by regulatory T cells. *Nat Immunol., 4,* 1206-1212.

[30] Fehervari, Z., Yamaguchi, T. & Sakaguchi, S. (2006). The dichotomous role of IL-2: tolerance versus immunity. *Trends Immunol, 27,* 109-11.

[31] Fife, B. T. & Bluestone, J. A. (2008). Control of peripheral T-cell tolerance and autoimmunity via the CTLA-4 and PD-1 pathways. *Immunol Rev., 224,* 166-182.

[32] Filaci, G., Bacilieri, S., Fravega, M., Monetti, M., Contini, P., Ghio, M., Setti, M., Puppo, F. & Indiveri, F. (2001). Impairment of CD8+ T suppressor cell function in patients with active systemic lupus erythematosus. *J Immunol, 166,* 6452-7.

[33] Filaci, G. & Suciu-Foca, N. (2002). CD8+ T suppressor cells are back to the game: are they players in autoimmunity? *Autoimmun Rev, 1,* 279-83.

[34] Filaci, G., Fravega, M., Negrini, S., Procopio, F., Fenoglio, D., Rizzi, M., Brenci, S., Contini, P., Olive, D., Ghio, M., Setti, M., Accolla, R. S., Puppo, F. & Indiveri, F. (2004). Nonantigen specific CD8+ T suppressor lymphocytes originate from CD8+CD28- T cells and inhibit both T-cell proliferation and CTL function. *Hum Immunol, 65,* 142-156.

[35] Fontenot, J. D., Gavin, M. A. & Rudensky, A. Y. (2003). Foxp3 programs the development and function of CD4+CD25+ regulatory T cells. *Nat Immunol., 4,* 330-336.

[36] Francois Bach, J. (2003). Regulatory T cells under scrutiny. Nat Rev Immunol *3,* 189-198.

[37] Fuchsberger, M., Hochrein, H. & O'Keeffe, M. (2005). Activation of plasmacytoid dendritic cells. *Immunol Cell Biol, 83,* 571-577.

[38] Gavin, M. A., Clarke, S. R., Negrou, E., Gallegos, A. & Rudensky, A. (2002). Homeostasis and anergy of CD4(+)CD25(+) suppressor T cells in vivo. *Nat Immunol., 3,* 33-41.

[39] Gavin, M. & Rudensky, A. (2003). Control of immune homeostasis by naturally arising regulatory CD4(+) T cells. *Curr Opin Immunol, 15,* 690-696.

[40] Gershon, R. K. & Kondo, K. (1971). Infectious immunological tolerance. *Immunology, 21,* 903-14.

[41] Gondek, D. C., Lu, L. F., Quezada, S. A., Sakaguchi, S. & Noelle, R. J. (2005). Cutting edge: contact-mediated suppression by CD4+CD25+ regulatory cells involves a granzyme B-dependent, perforin-independent mechanism. *J Immunol*, *174,* 1783-6.

[42] Green, E. A., Choi, Y. & Flavell, R. A. (2002). Pancreatic lymph node-derived CD4(+)CD25(+) Treg cells: highly potent regulators of diabetes that require TRANCE-RANK signals. *Immunity*, *16,* 183-91.

[43] Grossman, W. J., Verbsky, J. W., Barchet, W., Colonna, M., Atkinson, J. P. & Ley, T. J. (2004). Human T regulatory cells can use the perforin pathway to cause autologous target cell death. *Immunity*, *21,* 589-601.

[44] Higuma-Myojo, S., Sasaki, Y., Miyazaki, S., Sakai, M., Siozaki, A., Miwa, N. & Saito, S. (2005). Cytokine profile of natural killer cells in early human pregnancy. *Am J Reprod Immunol*, *54,* 21-9.

[45] Hoffmann, P., Ermann, J., et al., Edinger, M., Fathman, C. G. & Strober, S. (2002). Donor-type CD4(+)CD25(+) regulatory T cells suppress lethal acute graft-versus-host disease after allogeneic bone marrow transplantation. *J Exp Med, 196,* 389-99.

[46] Hori, S., Nomura, T. & Sakaguchi, S. (2003). Control of regulatory T cell development by the transcription factor Foxp3. *Science*, *299,* 1057-1061.

[47] Hori, S. & Sakaguchi, S. (2004). Foxp3: a critical regulator of the development and function of regulatory T cells. *Microbes Infect*, *6,* 745-751.

[48] Huang, C. T., Workman, C.J., Flies, D.M., Pan, X., Marson, A. L., Zhou, G., Hipkiss, E. L., Ravi, S., Kowalski, J., Levitsky, H. I., Powell, J. D., Pardoll, D. M., Drake, C. G. & Vignali, D. A. (2004). Role of LAG-3 in regulatory T cells. Immunity *21,* 503-13.

[49] Inaba, H. & Geiger, T. L. (2006). Defective cell cycle induction by IL-2 in naive T-cells antigen stimulated in the presence of refractory T-lymphocytes. *Int Immunol*, *18,* 1043-54.

[50] Ito, T., Hanabuchi, S., et al., Wang, Y. H., Park, W. R., Arima, K., Bover, L., Qin, F. X., Gilliet, M. & Liu, Y. J. (2008). Two functional subsets of FOXP3+ regulatory T cells in human thymus and periphery. *Immunity*, *28,* 870-80.

[51] Jiang, S., Tugulea, S., Pennesi, G., Liu, Z., Mulder, A., Lederman, S., Harris, P., Cortesini, R. & Suciu-Foca, N. (1998). Induction of MHC-class I restricted human suppressor T cells by peptide priming in vitro. *Hum Immunol*, *59,* 690-9.

[52] Josefowicz, S. Z. & Rudensky, A. (2009). Control of regulatory T cell lineage commitment and maintenance. *Immunity*, *30,* 616-625.

[53] Kim, H. P., Kelly, J. & Leonard, W. J. (2001). The basis for IL-2-induced IL-2 receptor alpha chain gene regulation: importance of two widely separated IL-2 response elements. *Immunity*, *15,* 159-72.

[54] Kingsley, C. I., Karim, M., Bushell, A. R. & Wood, K. J. (2002). CD25+CD4+ regulatory T cells prevent graft rejection: CTLA-4- and IL-10-dependent immunoregulation of alloresponses. *J Immunol*, *168,* 1080-6.

[55] Krieg, A. M., Yi, A. K., Matson, S., Waldschmidt, T. J., Bishop, G. A., Teasdale, R., Koretzky, G. A. & Klinman, D. M. (1995). CpG motifs in bacterial DNA trigger direct B-cell activation. *Nature*, *374,* 546-549.

[56] Kuniyasu, Y., Takahashi, T., Itoh, M., Shimizu, J., Toda, G. & Sakaguchi, S. (2000). Naturally anergic and suppressive CD25(+)CD4(+) T cells as a functionally and phenotypically distinct immunoregulatory T cell subpopulation. *Int Immunol*, *12,* 1145-55.

[57] Lanier, L. L. (1998). NK cell receptors. *Annu Rev Immunol, 16,* 359-93.

[58] Li, L., Liu, L. N., Feller, S., Allen, C., Shivakumar, R., Fratantoni, J., Wolfraim, L. A., Fujisaki, H., Campana, D., Chopas, N., Dzekunov, S. & Peshwa, M. (2010). Expression of chimeric antigen receptors in natural killer cells with a regulatory-compliant non-viral method. *Cancer Gene Ther, 17,* 147-54.

[59] Liu, Z., Tugulea, S., Cortesini, R. & Suciu-Foca, N. (1998). Specific suppression of T helper alloreactivity by allo-MHC class I-restricted CD8+CD28- T cells. *Int Immunol, 10,* 775-83.

[60] Liu, Z., Tugulea, S., Cortesini, R., Lederman, S. & Suciu-Foca, N. (1999). Inhibition of CD40 signaling pathway in antigen presenting cells by T suppressor cells. *Hum Immunol, 60,* 568-74.

[61] Malek, T. R. & Bayer, A. L. (2004). Tolerance, not immunity, crucially depends on IL-2. Nat *Rev Immunol, 4,* 665-74.

[62] Matsumoto, Y., Kohyama, K., Aikawa, Y., Shin, T., Kawazoe, Y., Suzuki, Y. & Tanuma, N. (1998). Role of natural killer cells and TCR gamma delta T cells in acute autoimmune encephalomyelitis. *Eur J Immunol., 28,* 1681-8.

[63] Moore, K. W., Waal Malefyt, R., Coffman, R. L. & O'Garra, A. (2001). Interleukin-10 and the interleukin-10 receptor. *Annu Rev Immunol, 19,* 683-765.

[64] Morgan, M. E., van Bilsen, J. H., Bakker, A. M., Heemskerk, B., Schilham, M. W., Hartgers, F. C., Elferink, B. G., van der Zanden, L., de Vries, R. R., Huizinga, T. W., Ottenhoff, T. H. & Toes, R. E. (2005). Expression of FOXP3 mRNA is not confined to CD4+CD25+ T regulatory cells in humans. *Hum Immunol, 66,* 13-20.

[65] Morse, R. H., Seguin, R., McCrea, E. L. & Antel, J. P. (2001). NK cell-mediated lysis of autologous human oligodendrocytes. *J Neuroimmunol., 116,* 107-15.

[66] Moseman, E. A., Liang, X., Dawson, A. J., Panoskaltsis-Mortari, A., Krieg, A. M., Liu, Y. J., Blazar, B. R. & Chen, W. (2004). Human plasmacytoid dendritic cells activated by CpG oligodeoxynucleotides induce the generation of CD4+CD25+ regulatory T cells. *J Immunol, 173,* 4433-4442.

[67] Munn, D. H., Sharma, M.D., Lee, J. R., Jhaver, K. G., Johnson, T. S., Keskin, D. B., Marshall, B., Chandler, P., Antonia, S. J., Burgess, R., Slingluff, C. L. & Jr., Mellor, A. L. (2002). Potential regulatory function of human dendritic cells expressing indoleamine *2,*3-dioxygenase. *Science, 297,* 1867-1870.

[68] Navarro, J., Aristimuno, C., Sanchez-Ramon, S., Vigil, D., Martinez-Gines, M. L., Fernandez-Cruz, E. & de Andres, C. (2006). Circulating dendritic cells subsets and regulatory T-cells at multiple sclerosis relapse: Differential short-term changes on corticosteroids therapy. *J Neuroimmunol., 176,* 153-61.

[69] Nelson, B. H. (2004). IL-*2,* regulatory T cells, and tolerance. *J Immunol, 172,* 3983-8.

[70] Piccirillo, C. A. & Shevach, E. M. (2001). Cutting edge: control of CD8+ T cell activation by CD4+CD25+ immunoregulatory cells. *J Immunol, 167,* 1137-40.

[71] Piccirillo, C. A., Letterio, J. J., Thornton, A. M., McHugh, R. S., Mamura, M., Mizuhara, H. & Shevach, E. M. (2002). CD4(+)CD25(+) regulatory T cells can mediate suppressor function in the absence of transforming growth factor beta1 production and responsiveness. *J Exp Med, 196,* 237-46.

[72] Pradeu, T. & Carosella, E. D. (2006). On the definition of a criterion of immunogenicity. *Proc Natl Acad Sci,* U S A *103,* 17858-17861.

[73] Qin, F. X. (2009). Dynamic behavior and function of Foxp3+ regulatory T cells in tumor bearing host. *Cell Mol Immunol, 6,* 3-13.
[74] Rolls, A., Shechter, R. & Schwartz, M. (2009). The bright side of the glial scar in CNS repair. *Nat Rev Neurosci, 10(3)*, 235-41.
[75] Roncarolo, M. G., Bacchetta, R., Bordignon, C., Narula, S. & Levings, M. K. (2001). Type 1 T regulatory cells. *Immunol Rev., 182,* 68-79.
[76] Saito, S., Shiozaki, A., Sasaki, Y., Nakashima, A., Shima, T. & Ito, M. (2007). Regulatory T cells and regulatory natural killer (NK) cells play important roles in feto-maternal tolerance. *Semin Immunopathol, 29,* 115-22.
[77] Sakaguchi, S., Sakaguchi, N., Shimizu, J., Yamazaki, S., Sakihama, T., Itoh, M., Kuniyasu, Y., Nomura, T., Toda, M. & Takahashi, T. (2001). Immunologic tolerance maintained by CD25+ CD4+ regulatory T cells: their common role in controlling autoimmunity, tumor immunity, and transplantation tolerance. *Immunol Rev., 182,* 18-32.
[78] Sakaguchi, S., Yamaguchi, T., Nomura, T. & Ono, M. (2008). Regulatory T cells and immune tolerance. *Cell, 133,* 775-787.
[79] Sambrook, J. G. & Beck, S. (2007). Evolutionary vignettes of natural killer cell receptors. *Curr Opin Immunol, 19,* 553-60.
[80] Sanchez-Ramon, S., Navarro, J., Aristimuno, C., Rodriguez-Mahou, M., Bellon, J. M., Fernandez-Cruz, E. & de Andres, C. (2005). Pregnancy-induced expansion of regulatory T-lymphocytes may mediate protection to multiple sclerosis activity. *Immunol Lett, 96,* 195-201.
[81] Sarween, N., Chodos, A., Raykundalia, C., Khan, M., Abbas, A. K. & Walker, L. S. (2004). CD4+CD25+ cells controlling a pathogenic CD4 response inhibit cytokine differentiation, CXCR-3 expression, and tissue invasion. *J Immunol, 173,* 2942-51
[82] Sercarz, E. & Krzych, U. (1991). The distinctive specificity of antigen-specific suppressor T cells. *Immunol Today, 12,* 111-8.
[83] Scheffold, A., Huhn, J. & Hofer, T. (2005). Regulation of CD4+CD25+ regulatory T cell activity: it takes (IL-) two to tango. *Eur J Immunol., 35,* 1336-41.
[84] Steiner, D., Brunicki, N., Blazar, B. R., Bachar-Lustig, E. & Reisner, Y. (2006). Tolerance induction by third-party "off-the-shelf" CD4+CD25+ Treg cells. *Exp Hematol, 34,* 66-71.
[85] Steinman, R. M., Hawiger, D. & Nussenzweig, M. C. (2003). Tolerogenic dendritic cells. *Annu Rev Immunol, 21,* 685-711.
[86] Takahashi, T., Kuniyasu, Y., Toda, M., Sakaguchi, N., Itoh, M., Iwata, M., Shimizu, J. & Sakaguchi, S. (1998). Immunologic self-tolerance maintained by CD25+CD4+ naturally anergic and suppressive T cells: induction of autoimmune disease by breaking their anergic/suppressive state. *Int Immunol, 10,* 1969-80.
[87] Takahashi, K., Miyake, S., Kondo, T., Terao, K., Hatakenaka, M., Hashimoto, S. & Yamamura, T. (2001). Natural killer type 2 bias in remission of multiple sclerosis. *J Clin Invest, 107,* R23-9.
[88] Takahashi, K., Aranami, T., Endoh, M., Miyake, S. & Yamamura, T. (2004). The regulatory role of natural killer cells in multiple sclerosis. *Brain, 127,* 1917-27.
[89] Tang, Q., Henriksen, K. J., Bi, M., Finger, E. B., Szot, G., Ye, J., Masteller, E. L., McDevitt, H., Bonyhadi, M. & Bluestone, J. A. (2004). In vitro-expanded antigen-specific regulatory T cells suppress autoimmune diabetes. *J Exp Med, 199,* 1455-65.

[90] Thornton, A. M. & Shevach, E. M. (2000). Suppressor effector function ofCD4+CD25+ immunoregulatory T cells is antigen nonspecific. *Journal of Immunology*, *164,* 183-90.

[91] Van Parijs, L. & Abbas, A. K. (1998). Homeostasis and self-tolerance in the immune system: turning lymphocytes off. *Science*, *280,* 243-248.

[92] Veenstra van Nieuwenhoven, A. L., Bouman, A., Moes, H., Heineman, M. J., de Leij, L. F., Santema, J. & Faas, M. M. (2002). Cytokine production in natural killer cells and lymphocytes in pregnant women compared with women in the follicular phase of the ovarian cycle. *Fertil Steril*, *77,* 1032-7.

[93] Wakkach, A., Fournier, N., Brun, V., Breittmayer, J. P., Cottrez, F. & Groux, H. (2003). Characterization of dendritic cells that induce tolerance and T regulatory 1 cell differentiation in vivo. *Immunity*, *18,* 605-617.

[94] Weber, S. E., Harbertson, J., Godebu, E., Mros, G. A, Padrick, R. C., Carson, B. D., Ziegler, S. F. & Bradley, L. M. (2006). Adaptive islet-specific regulatory CD4 T cells control autoimmune diabetes and mediate the disappearance of pathogenic Th1 cells in vivo. *J Immunol*, *176,* 4730-9.

[95] Weiner, H. L. (2001a). Induction and mechanism of action of transforming growth factor-beta-secreting Th3 regulatory cells. *Immunol Rev.*, *182,* 207-214.

[96] Weiner, H. L. (2001b). Oral tolerance: immune mechanisms and the generation of Th3-type TGF-beta-secreting regulatory cells. *Microbes Infect*, *3,* 947-954.

[97] Weiner, H. L. (2001c). The mucosal milieu creates tolerogenic dendritic cells and T(R)1 and T(H)3 regulatory cells. *Nat Immunol.*, *2,* 671-2.

[98] Wood, K. J. & Sakaguchi, S. (2003). Regulatory T cells in transplantation tolerance. *Nat Rev Immunol*, *3,* 199-210.

[99] Wu, J. & Horuzsko, A. (2009). Expression and function of immunoglobulin-like transcripts on tolerogenic dendritic cells. *Hum Immunol*, *70*, 353-356.

[100] Yamazaki, S., Inaba, K., Tarbell, K. V. & Steinman, R. M. (2006). Dendritic cells expand antigen-specific Foxp3+ CD25+ CD4+ regulatory T cells including suppressors of alloreactivity. *Immunol Rev.*, *212,* 314-29.

[101] Yi, H., Zhen, Y., Jiang, L., Zheng, J. & Zhao, Y. (2006). The phenotypic characterization of naturally occurring regulatory CD4+CD25+ T cells. *Cell Mol Immunol, 3,* 189-95.

[102] Yuen, M. H., Protti, M. P., Diethelm-Okita, B., Moiola, L., Howard, J. F. & Jr., Conti-Fine, B. M. (1995). Immunoregulatory CD8+ cells recognize antigen-activated CD4+ cells in myasthenia gravis patients and in healthy controls. *J Immunol*, *154,* 1508-20.

[103] Zhang, Z. & Wang, F. S. (2005). Plasmacytoid dendritic cells act as the most competent cell type in linking antiviral innate and adaptive immune responses. *Cell Mol Immunol*, *2,* 411-417.

[104] Zhang, C., Zhang, J. & Tian, Z. (2006). The regulatory effect of natural killer cells: do "NK-reg cells" exist? *Cell Mol Immunol*, *3*, 241-54.

[105] Zheng, S. G., Gray, J. D., Ohtsuka, K., Yamagiwa, S. & Horwitz, D. A. (2002). Generation ex vivo of TGF-beta-producing regulatory T cells from CD4+CD25- precursors. *J Immunol, 169,* 4183-4189.

[106] Zheng, S. G. & Wang, J. H. et al. (2004). Natural and induced CD4+CD25+ cells educate CD4+CD25- cells to develop suppressive activity: the role of IL-2, TGF-beta, and IL-10. *J Immunol*, *172,* 5213-21.

In: Regulatory T Cells
Editor: Ren S. Hayashi, pp. 99-116

ISBN: 978-1-61761-554-2

Chapter 5

Roles of Regulatory T Cells in Oral Tolerance and Food Allergy

Yusei Ohshima*
Department of Pediatrics, Faculty of Medical Sciences, University of Fukui

Abstract

Gastrointestinal mucosa is an important site for oral tolerance induction and allergic responses to exogenous innoxious antigens, such as food antigens. The failure or aberration of oral tolerance results in the development of allergic diseases, especially food allergy. Several factors, including mucosal barrier function, humoral factors, lymphocyte anergy, clonal deletion, and regulatory T cells, contribute to the development of oral tolerance. Accumulating clinical and experimental evidences have suggested that the decreased number and dysfunction of regulatory T cells are involved in the pathogenesis of food allergy and that the development of regulatory T cells is associated with outgrowing food allergy. Although the precise nature of regulatory T cells controlling oral tolerance to food Ags remains unknown, recent reports demonstrate that not only regulatory $CD4^+$ T cells but also subsets of $CD8^+$ T cells may function as Treg by inhibiting allergic responses. Increased understanding of the mechanisms of oral tolerance has made a paradigm shift of therapeutic strategies for food allergies from avoidance of foods to oral tolerance induction by augmentation of regulatory T cells. In this article, we review the roles of regulatory T cells in oral tolerance and the pathophysiology of food allergy, and discuss their therapeutic potential.

* Corresponding author: Department of Pediatrics, Faculty of Medical Sciences, University of Fukui, 23-3 Shimoaizuki, Matsuoka, Eiheiji-cho, Yoshida-gun Fukui, 910-1193, JAPAN, Tel +81-776-61-3111, FAX +81-776-61-8129, E-mail yohshima@u-fukui.ac.jp

Introduction

The mucosa of gastrointestinal tract is constantly exposed to food antigens (Ags) and commensal bacteria. The intestinal tract is also a major site of pathogen entry requiring an immune defense. Gut associated lymphoid tissue (GALT) is able to discriminate invading pathogens that require an immediate immune response from normal microbial flora or nutritive products. To prevent harmful bystander damage by inflammatory responses against the commensal bacteria and food Ags, GALT maintain a state of non-responsiveness. Despite the large extent of food antigenic exposure, only a small percentage of individuals have food allergies. This is due to development of oral tolerance. It is postulated that a breakdown in oral tolerance mechanisms or a failure of induction of oral tolerance results in food hypersensitivity, food allergy.

Oral tolerance refers to a state of active inhibition of immune responses to an Ag by means of prior exposure to the same Ag through the oral route [1]. Experimentally, oral tolerance can be induced by oral administration of either a single high dose of Ag or repeated low doses [2]. These low forms of tolerance, termed high-dose and low-dose oral tolerance, respectively, are mediated by two different mechanisms. A high dose of an oral Ag can induce clonal anergy or deletion, whereas repetitive feeding of low doses of Ags favors the generation of regulatory T cells [3-7]. Recent work by others and us suggests the two forms of tolerance may not be mutually exclusive and may occur simultaneously [8, 9]. Moreover, anergy and active suppression have been thought to not be distinct aspects of T-cell function [10]. Thus, regulatory T cells have attracted considerable attention as a key player in oral tolerance. In this article, we provide an overview of the roles of regulatory T cells in the mechanisms of oral tolerance and the pathophysiology of food allergy.

Processing and Sampling of Luminal Ags That Lead to Oral Tolerance

Ingested dietary Ags are subject to digestion by gastric juice and luminal digestive enzymes. The destruction of immunogenic epitopes of dietary Ags can lead to immunologic ignorance. Meanwhile, the dietary Ags that escape luminal digestion subsequently can be taken up by different cell types, such as microfold cells (M cells), dendritic cells (DCs) and epithelial cells, depending on their specific properties. The site of entry for Ags might also dictate the nature of the immune response [11].

M cells overlaying Peyer's patches (PPs) are inefficient at uptake of soluble protein Ags, but are specialized to take up particulate Ags [12]. The Ags taken up by M cells are in turn delivered into the subepithelial dome region of the PPs, an area rich in DCs, that in turn ingests the Ag and delivers it to the underlying B-cell follicles of the PPs [13].

DCs are present in different compartments of the alimentary tract, including intestinal LP (LP), isolated lymphoid follicles (ILFs), PPs, and mesenteric lymph nodes (MLNs) [14-16]. DCs can also intercalate between intestinal epithelial cells while preserving the integrity of the epithelial barrier through expression of tight junction proteins. They can send dendrites into the lumen and sample Ags directly from the lumen [17]. DCs in LP can acquire Ags, which the intestinal epithelial cells uptake and then transcytose [18]. The Ag-carrying DCs might then traffic through the lymphatics to MLNs [16, 19]. DCs have a pivotal role in directing the balance between tolerance and active immunity in the intestine. Generally, the development of pathogenic immunity or tolerance is largely determined by the DC phenotype presenting Ag to T cells in draining lymph nodes, including PPs and MLNs [20-23].

The intestinal epithelial cells, of which primary responsibility is nutrient absorption, play important roles in mucosal immunity by their barrier function and by transporting secretory IgA [24]. Moreover, the intestinal epithelial cells recognize and respond to bacterial and viral motifs by expressing nucleotide oligomerization domain (NOD) like receptors, Toll-like receptors and cytoplasmic nucleic acid sensors [24]. Upon detection of a pathogen, epithelial cells produce cytokines/chemokines that recruit DCs, monocytes, and neutrophils to the intestine, and thereby influence immune responses [23]. In the absence of infection, the intestinal epithelial cells normally produce large amounts of immunosuppressive cytokines, especially TGF–β or IL-10, and these cytokines probably also drive intestinal DCs to the tolerogenic phenotypes that preferentially induce regulatory T cells [25-27].

The intestinal epithelial cells take up dietary Ags that escape proteolysis in the intestinal lumen, through fluid-phase endocytosis, and might act as nonprofessional antigen-presenting cells. In contrast to professional APCs, the intestinal epithelial cells normally selectively activate $CD8^+$ T cells, playing a regulatory role in local suppression of immune responses [28, 29]. This process appears to be regulated by the non-classical class I molecule CD1d and the CD8 ligand gp180, an intestinal epithelial membrane glycoprotein [30, 31]. Moreover, intestinal epithelial cells were also shown to have the ability to assemble and release exosome-like structures that carry MHC class II with bound antigenic peptides sampled from the lumen. The exosome-like structures, "tolerosome" in the serum of animals shortly after feeding an antigen, were capable of inducing Ag-specific oral tolerance [32].

DCs Command the Induction of Oral Tolerance or Food Allergy

Intestinal DCs are not inherently tolerogenic. Under physiological conditions, they are "quiescent," capable of presenting Ags and inducing tolerance, but being sufficiently responsive to inflammatory stimuli to allow T-cell priming and protective immunity when necessary [33]. In steady state, LP-resident $CX3CR1^+$ DCs derived from circulating blood monocytes, specifically the $Ly6C^{hi}$ monocyte subset, can uniquely penetrate the intestinal epithelium to sense or sample gut luminal content are generated [34, 35]. In contrast, LP-resident $CD103^+$ $CD11b^+$ DCs and probably GALT resident $CD103^+$ $CD11b^+$ DCs, are derived from dedicated circulating DC precursors, i.e. pre-conventional DCs [36].

$CD103^+$ DCs comprise about 30%–50% of intestinal DCs and are rather homogenously distributed throughout the LP of colon and small intestine, and MLNs of healthy mice [37, 38]. Only $CD103^+$ $CD11b^+$ DCs, but not $CX3CR1^+$ DCs, migrate in a CCR7-dependent manner to the MLNs to induce differentiation of T cells into effector or regulatory cells [39]. LP-resident $CD103^+$ $CD11b^+$ DCs seem to be conditioned by the unique tissue environment of the LP and the overlying lumen content-exposed epithelium. Upon their arrival in the draining MLNs, retinoic acid, a metabolite of vitamin A, produced by $CD103^+$ $CD11b^+$ DCs can promote the differentiation of naïve T cells to Ag-specific regulatory forkhead box protein 3 $(Fox3)^+CD4^+$ T-cell in a TGF-β-dependent manner and thereby contributing to the 'default' peripheral immune tolerance process to food proteins and most other non-inflammatory environmental Ags [21, 25, 40].

$CD11c^+$ DCs in PPs from orally tolerant mice exhibit an immature phenotype and express higher levels of indoleamine 2, 3-dioxygnenase (IDO) compared to those from control mice [41]. Functionally, the IDO-expressing DCs inhibit Ag-specific T cell proliferation and pro-inflammatory Th1 cytokine production, and more importantly, induce the generation of Ag-specific $CD4^+CD25^+Foxp3^+$ Treg cells from highly purified $CD4^+CD25$- T cells in an IDO-dependent manner. IDO-expressing DCs can directly activate resting $CD4^+CD25^+FOXP3^+$ Treg cells for potent suppressor activity [42]. Both cytotoxic T-lymphocyte antigen-4 (CTLA-4) and glucocorticoid-inducible tumor necrosis factor receptor related protein (GITR) expressed on $CD4^+CD25^+$ Treg cells can induce IDO expression in DCs [43]. In this context, IDO-producing DCs and $CD4^+CD25^+$ Treg cells may form a positive feedback loop in the induction of oral tolerance.

Plasmacytoid DCs (pDCs) may also have an important role in sustaining tolerance to innoxious Ags at mucosal sites, as they can induce development of anergic T cells or T cells with regulatory functions *in vitro* [44]. Mice treated with Flit3 ligand, which expanded DCs, especially pDCs population *in vivo*, exhibited more efficient oral tolerance [45, 46]. On the other hand, systemic depletion of pDCs by antibodies prevented induction of oral tolerance [47]. The main mechanism by which pDCs contribute to oral tolerance seems to be by inducing anergy and/or deletion of Ag-specific T cells and is independent of their capacity to trigger $CD4^+$ Treg cell differentiation [47].

The role of DCs in food allergy has still not been fully investigated. Adoptive transfer of splenic and PP-derived DCs from the mice with IgE-dependent milk allergy can induce Ag-specific IgE responses in naive mice even without sensitizing the naive mice. When cultured in the presence of both Ag and Ag-specific T cells, splenic DCs and PP derived DCs from mice with milk allergy showed a reduced degree of apoptosis, compared with DCs from control animals without allergy [48].

$CD4^+$ Regulatory T Cells Act as a Key Player in the Development and Maintenance of Oral Tolerance

Regulatory T cells are thought to be crucial regulators of peripheral self-tolerance. They can suppress the effector functions of many cell types, including T cells, B cells, monocytes/macrophages and mast cells in a cell-contact dependent manner as well as via secretion of soluble mediators, such as IL-10 and TGF-β [49, 50]. Different types of regulatory T cells, e.g. $CD25^+CD4^+$ $Foxp3^+$ T cells, Th3 cells, and type 1 T regulatory (Tr1) cells are involved in oral tolerance.

$CD25^+CD4^+$ regulatory T (Treg) cells characteristically express Foxp3 [51]. Most $CD25^+CD4^+Foxp3^+$ Treg cells probably originate in the thymus, where their development requires signals that activate NF-κB via the PKCq-Carma-1-IKK2 axis and limit activation of the PI3-kinase-Akt-mTOR pathway [52]. After exiting the thymus, some part of the naturally occurring Treg (nTreg) cells migrate to the gastrointestinal tract where they can recognize intestinal Ags and prevent inappropriate immune responses. Since the Treg cells isolated from SPF-housed mice are better suppressors than those from germ-free mice, commensal bacterial flora affects the function and pool of the Treg cells in the peripheral lymphoid organs, allowing efficient regulation of the alimentary tract [53]. Naive $CD4^+$T cells can differentiate into "adaptive" or ''induced'' $Foxp3^+$ Treg (iTreg) cells when activated by transient TCR stimulation or TCR stimulation in the presence of TGF-β and IL-2 and the absence of inflammatory cytokines that promote effector T cell differentiation [54-56]. *In vivo*, transfusion of Ag under non-immunogenic conditions, homeostatic proliferation, or chronic inflammation can also generate iTreg cells [57]. Although nTreg and iTreg cells may express different gene transcripts and epigenetic markers, they are currently difficult to distinguish on a single-cell basis [58, 59]. The noninflammed gut seems to be a permissive site for the accumulation of iTreg cells, which together with nTreg cells collaborate to reinforce intestinal homeostasis.

Th3 cells were initially described as a population of regulatory T cells generated following the induction of oral tolerance to fed protein Ag, and produce TGF-β with various amounts of IL-4 and IL-10 [2, 60]. TGF-β not only serves as a class-switching factor of IgA, but also induces Foxp3 in naïve $CD4^+$ T cells [61]. $Foxp3^+$ T cells have been shown to be induced following oral exposure to protein Ag via a TGF-β-dependent mechanism [62]. Thus, in the intestine, it seems likely that the $Foxp3^+$ regulatory T cells contain a substantial portion of iTreg cells, corresponding Th3 cells.

TGF-β is important for the maintenance and effector function of both $Foxp3^+$ Treg cells and Tr1-like cells in the intestine [63-67]. Latency-associated peptide (LAP) is the amino-terminal domain of the TGF-β precursor peptide. The LAP remains non-covalently associated with the TGF-β peptide after cleavage, and forms the latent complex [68]. The presence of membrane bound TGF-β or LAP on the surface of regulatory T cells has linked TGF-β with the suppressive function of regulatory T cells [68]. Although the intestine harbors a high concentration of TGF-β, the majority is thought to exist in an inactive form. αvβ6 and αvβ8 integrins expressed by intestinal epithelial cells and DCs, respectively, may play a key role in the activation of TGF-β [69-71]. Additionally, thrombospondin and the proprotein convertase Furin expressed by T cells might also activate TGF-β [72-74]. $CD4^+$ T cells, DCs, and intestinal epithelial cells all collaborate to activate TGF-β and maintain T cell tolerance in the intestine. To support the notion that TGF-β is a key molecule for oral tolerance, oral administration of Ags together with exogenous recombinant TGF-β enhances oral tolerance by augmenting reduction of Ag-specific IgE and IgG1 antibodies, and T-cell reactivity to the Ag [75].

Tr1 cells preferentially secrete IL-10, which also drives generation of these cells [76, 77]. IL-10 was shown to suppress Ag-specific immune responses and prevent colitis in a mouse model [76]. IL-10 can be produced by $CD4^+$ T cells in the dome of PPs and the villi of tolerant T cell receptor transgenic mice [78].The presence of IL-10 in PP cells from tolerant mice after β-lactoglobulin feeding but not in sensitized mice with β-lactoglobulin–induced anaphylaxis [79].

$CD4^+$ T cell populations with regulatory functions in the intestine include $Foxp3^+$ Treg cells and Foxp3- Tr1-like cells [80, 81]. In the colon, most of the $Foxp3^+$ Treg cells in the LP produce IL-10 and Tr1-like cells are less frequent. In the small intestine, Tr1-like cells are frequently found in the intraepithelial layer and both Tr1-like cells and $Foxp3^+$ Treg cells in the LP produce IL-10. In the draining MLN, fewer IL-10-producing cells are found than in the intestinal tissue and the major regulatory population is the $Foxp3^+$ Treg cell. Thus, the composition of $Foxp3^+IL\text{-}10^-$, $Foxp3^+IL\text{-}10^+$, and Tr1-like cells varies greatly between intestinal compartments.

Oral administration of Ag can expand Ag-specific regulatory T cell populations in PPs, MLNs, and peripheral lymphoid tissues [63, 82] [83]. Although oral tolerance might also involve the expansion of reactive nTreg cells, this phenomenon involves the induction of Foxp3$^+$ iTreg cells from the conventional T cell pool and predominantly occurs in the GALT [38, 40, 82]. Orally induced Treg cells can produce either or both of TGF-β and IL-10 and may confer suppressor activity ('infectious tolerance') on various CD4$^+$ T cells by inducing the expression of the Foxp3 gene, which appears to be critical for the generation and suppressive function of many but not all regulatory T cells (e.g. IL-10-producing Tr1 cells appear to maintain a Foxp3- status) [81]. Orally induced regulatory T cells act both locally in the gut and systemically [84]. Indeed, orally induced ovalbumin-specific Foxp3$^+$ Treg cells can suppress lung inflammation in a model of asthma driven by Th2 effector cells [85].

CD8$^+$ T Cells-Mediated Suppression: Regulatory CD8$^+$ T Cells

In the oral tolerance induced by fed Ag, CTL, helper T cell and antibody responses were suppressed. The suppression could be adoptively transferred with total splenocytes but not with spleen suspensions depleted of CD8$^+$ cells [86]. Although CD8$^+$ DCs rather than regulatory CD8$^+$ T cells might be involved in the phenomenon, splenocytes depleted of cells expressing CD75s, a marker expressed by several T-cell lines and hybridomas, did not transfer tolerance to adoptive hosts neither, supporting the conclusion that CD8$^+$ Treg were involved in the oral tolerance [87, 88].

In contrast to thymus-derived CD8$^+$ cells, which express the CD8αβ hererodimer, the majority of CD8+ cells in the intestinal epithelium express a CD8αα homodimer. Most TCRαβ CD8αα cells develop in the intestine [89]. TCRαβCD8αα intraepithelial lymphocytes (IELs) may have regulatory capacity *in vivo*. Indeed, SCID mice injected with CD4$^+$CD45RB^{high} T cells develop colitis, which could be prevented by injection of TCRαβCD8αα IELs [90, 91]. IL-10 production by these Treg appeared to be involved in protection from colitis. However the physiologic role of TCRαβCD8αα IELs is still unknown.

Intestinal epithelial cells can induce the proliferation of CD8$^+$CD28- peripheral T cell subset through gp180 *in vitro* [26]. The CD8$^+$ T cells activated by intestinal epithelial cells express CD101 and CD103, and display a regulatory function. CD8$^+$ T cells with regulatory activity are found in the LP of healthy individuals, but not in patients with inflammatory bowel disease, indicating that these cells are involved in mucosal tolerance [92]. CD8$^+$ T cells may function as a regulatory T cells by secreting TGF-β [93]. Alternatively, IL-4 or IL-10 produced by antigen-primed CD8$^+$ T-cell may be associated with tolerance induction [94, 95]

Although CD4$^+$ T-cell populations are thought to be involved in the pathophysiology of food allergy and oral tolerance, the role of CD8$^+$ T cells remains uncertain. We have recently demonstrated that adoptive transfer of splenic CD8$^+$ T cells from Ag-primed mice, but not from non-primed mice, suppressed the development of allergic diarrhea in the Ag-sensitized mice that had a significantly reduced number of conventional TCRαβ$^+$CD8$^+$ T cells [96]. The CD8$^+$ T cells exert the regulatory functions through both IL-10 dependent and independent mechanisms. These data suggest that systemic immunization with Ag simultaneously induce regulatory $CD8^+$ T cells, as well as oral Ag administration and that subsets of $CD8^+$ might play an important role in the regulation of oral tolerance and food allergy.

The Roles of NKT Cells in Oral Tolerance Induction

NKT cells are a unique lineage of T cells expressing invariant Vα14-Jα18 TCRα-chain that share properties with both NK cells and memory T cells [97, 98]. This subset of lymphocytes may be either $CD4^+$ or double negative and is CD1d reactive. Although NKT cells can generate both Th1 and Th2 responses, it has been shown that NKT cells are essential for Ag-specific suppression of the delayed-type hypersensitivity response mediated by both oral and portal venous tolerance [99]. NKT cells can produce cytokines immediately after exposure to activating signals and can determine the differentiation of Treg cells [100, 101]. DCs may play an important role in the activation of Treg cells by stimulation of NKT cells [102-104]. In a model of experimental colitis, induction of oral tolerance is associated with an increased number and altered function of NKT cells [105-107]. However, the roles of NKT cells in the pathophysiology of food allergy are unknown.

Operations Centers for the Induction of Oral Tolerance

Orally administered Ag passes from the intestinal lumen via either the villus epithelial cells or M cells to the organized lymphoid tissues of the PPs and MLNs. It can also be taken up via active sampling of DCs [108]. Oral tolerance cannot be induced in CCR7-deficient mice that display impaired migration of DCs from the intestine to the MLNs, suggesting that the relevant antigen is transported in a cell-bound fashion and that Ag transport via afferent lymphatics into the draining MLN is obligatory for oral tolerance [109, 110]. Moreover, oral tolerance is absent in lymphotoxin–α knockout mice, which lack MLNs [111] [112]. Reduced $CD3^+$ expression and defects in TCR signaling mediate the anergic state of MLN $CD4^+$ T cells, and thereby play a role in the maintenance of oral tolerance [113]. In this context, MLNs seem to be a key site for the induction of tolerance to dietary Ags.

Actively tolerant mice, which were fed Ag, were found to have a higher number of Ag–specific IgA-secreting cells in PPs, as well higher fecal Ag-specific IgA titers, than anaphylactic mice that were actively sensitized to the same antigen [114]. An increase of Ag-induced IL-10 and TGF-β production by T cells from PPs was observed in tolerant mice. Compared with the mice lacking PPs, the PP of intact mice contained $CD4^+CD25^+$ $Foxp3^+$ Treg cells that produce high levels of IL-10, and inhibited the development of allergic diarrhea [115]. How essential PPs are in oral tolerance induction is still debatable, since tolerance to soluble antigen could be induced in mice lacking PPs [111].

It was estimated that approximately 2% of intact proteins of ingested dietary Ags reach the intestinal lymph and portal circulation under physiologic conditions [116]. Discriminating immune responses between transported nutrients and translocated harmful microbes into the portal circulation may occur in the liver. Thus, the liver is a candidate site for the induction of oral tolerance to dietary Ags. Direct evidence for a role of the liver in tolerance is provided by the observation that portocaval anastomosis, which prevents drainage into the liver of Ags from the gut, abrogates oral tolerance to proteins or haptens [117]. There are several explanations for the tolerogenic environment in the liver. First, the liver contains abundant NK cells and NKT cells, which are relevant for intrahepatic T-cell trapping and killing, resulting in clonal deletion [98, 106, 107, 118]. Liver-derived DCs produce IL-10, and express low levels of costimulatory molecules, and thereby are inherently tolerogenic when compared with skin DCs. Local secretion of IL-10 and TGF-β by Kupffer cells and hepatocytes can skew DC function toward the induction of regulatory T cells [119]. Liver resident pDCs appear to induce anergy or deletion of Ag-specific T cells relatively rapidly after oral Ag administration [47]. Liver sinusoidal endothelial cells can cross-present Ags to $CD8^+$ T cells, resulting in apoptosis of $CD8^+$ T cells and induce IL-4 and IL-10 production by $CD4^+$ T cells [120]. Finally, hepatocytes themselves can function as Ag presenting cells to activate Treg cells and naive T cells, leading to Ag-specific tolerance [121] [122].

Regulatory t Cells and Patients with Food Allergy

Children with mutations in Foxp3 develop the immunedysregulation, polyendocrinopathy, enteropathy, X-linked (IPEX) syndrome [123]. Patients with this disease have very high Th2 responses, high IgE levels and sever allergic diarrhea. Infants who later developed egg allergy by 12 months of age had lower Treg function than infants who were tolerant to foods, suggesting that a defect in Treg cells may predispose to food allergy [124]. Children with food allergy were found to have fewer TGF-β1–expressing duodenal lymphocytes in both epithelial and LP compartments compared to control [125]. Milk-specific lymphocyte clones established from the duodenal mucosa of children with milk-induced allergic gastroenteropathies failed to release TGF-β and IL-10, when cultured *in vitro* in the presence of milk [126]. These circumstantial evidences suggest that the impairment of regulatory T cell induction and function cause the development of food allergy.

It is generally thought that the majority of children with egg or cow's milk allergy develop tolerance by late childhood [127]. The mechanisms involved in outgrowing food allergy have been explained by the development of digestive function and physical barrier function of the alimentary tracts, and by the maturation of mucosal immunity, such as secretary IgA production. The acquisition of tolerance could be attributed to a decrease in the uptake of intact epitopes derived from ingested dietary Ags. Recently, however, accumulating evidence demonstrates that regulatory T cells also play important role in outgrowing food allergy. Cow's milk-specific T cell clones of children with persistent cow's milk allergy were Th2 skewed, whereas those of cow's milk-tolerant control subjects were characterized by the production of IL-10 [128]. Children who tolerate extensively heated milk products outgrow their milk allergy earlier than those who react to heated milk products. Heated milk-tolerant children have higher numbers of allergen-specific $Foxp3^+CD25^{high}CD27^+CTLA4^+$ Treg cells than milk allergy children [129]. Children who outgrew their cow milk allergy had higher frequencies of circulating $CD4^+CD25^+$ Treg cells and decreased *in vitro* proliferative responses to bovine β-lactoglobulin in peripheral blood mononuclear cells (PBMCs) compared with children who maintained clinically active allergy [130]. These data collectively suggest that mucosal induction of tolerance against dietary Ags is associated with the appearance of circulating iTreg cells and/or Tr1 cells that are capable of suppressing the effector T-cells generated.

Regulatory t Cells as a Therapeutic Target for Food Allergy

A number of experimental approaches for the prevention and treatment of food allergy have been reported. Early studies in murine models aimed at promoting Th1 skewing, but recent ones seem to focus on induction or modulation of regulatory T cells. Pretreatment with a transfected IL-10-producing *Lactococcus lactis* before sensitization can induce IL-10 secretion by PP cells and mitigated food-induced anaphylaxis [131]. Mucosal administration of conjugate of cholera toxin subunit B (CTB) to Ag or gene fusion proteins have been found to induce tolerance with superior efficiency compared with administration of Ag alone, suppressing allergies in animal models [132]. The mucosal administration of Ag-CTB conjugate results in an increase in Ag-specific regulatory $CD4^+$ T cells, via two separate pathways: one using B cells as APCs and leading to a strong expansion of $Foxp3^+$ Treg cells and one being B cell-independent and associated with development of Foxp3- regulatory T cells that express membrane latency-associated peptide and TGF-β and/or IL-10.

Classical subcutaneous Ag-specific immunotherapy exhibited severe adverse reactions in food allergy patients, so that more recently, oral immunotherapy, and oral desensitization has been introduced in the therapy for cow's milk, egg, or peanut allergy [133]. Throughout oral immunotherapy for peanut allergy, peanut-specific IgE decreased, whereas IgG4 and secretion IL-10, IL-5, IFN-γ and TNFα from PBMCs increased [134]. A transient increase of peanut-specific $Foxp3^+$ T cells was observed. Since IL-10 is a class switch factor of IgG4, IL-10 could support an initial increase in Treg cells and Ag-specific IgG4 production.

Concluding Remarks

Animal models have demonstrated that regulatory T cells play important roles in the pathophysiology of food allergy as well as the induction and maintenance of oral tolerance. However, why a certain individual develops a food allergy against specific foods remains an enigma. Since the majority of patients with food allergy are children, it is difficult to gather specimens from patients' gastrointestinal tracts. Whether the results obtained from PBMC could reflect the immunological status of GALT is unknown. To solve the enigma, many further studies based on patients' data are needed. It will be important to elucidate the underlying immunological mechanisms of oral immunotherapy and the roles of regulatory T cells in outgrowing food allergy.

References

[1] Holmgren, J; Czerkinsky, C. Mucosal immunity and vaccines. *Nature medicine.*, 2005 Apr; 11(4 Suppl), S45-53.

[2] Friedman, A; Weiner, HL. Induction of anergy or active suppression following oral tolerance is determined by antigen dosage. Proceedings of the National *Academy of Sciences of the United States of America.*, 1994, Jul 5, 91(14), 6688-92.

[3] Whitacre, CC; Gienapp, IE; Orosz, CG; Bitar, DM. Oral tolerance in experimental autoimmune encephalomyelitis. III. Evidence for clonal anergy. *J Immunol.*, 1991, Oct 1, 147(7), 2155-63.

[4] Barone, KS; Jain, SL; Michael, JG. Effect of in vivo depletion of CD4+ and CD8+ cells on the induction and maintenance of oral tolerance. *Cellular Immunology.*, 1995, Jun, 163(1), 19-29.

[5] Chen, Y; Inobe, J; Marks, R; Gonnella, P; Kuchroo, VK; Weiner, HL. Peripheral deletion of antigen-reactive T cells in oral tolerance. *Nature*, 1995, Jul 13, 376(6536), 177-80.

[6] Chen, Y; Inobe, J; Weiner, HL. Induction of oral tolerance to myelin basic protein in CD8-depleted mice: both CD4+ and CD8+ cells mediate active suppression. *J Immunol.*, 1995, Jul 15, 155(2), 910-6.

[7] Garside, P; Steel M; Liew FY; Mowat AM. CD4+ but not CD8+ T cells are required for the induction of oral tolerance. *International immunology.* 1995 Mar; 7(3), 501-4.

[8] Omata N; Ohshima Y; Yasutomi M; Yamada A; Karasuyama H; Mayumi M. Ovalbumin-specific IgE modulates ovalbumin-specific T-cell response after repetitive oral antigen administration. *The Journal of allergy and clinical immunology.* 2005 Apr; 115(4), 822-7.

[9] Sun JB; Czerkinsky C; Holmgren J. Sublingual 'oral tolerance' induction with antigen conjugated to cholera toxin B subunit generates regulatory T cells that induce apoptosis and depletion of effector T cells. *Scandinavian journal of immunology.* 2007 Aug-Sep; 66(2-3), 278-86.

[10] von Boehmer, H. Mechanisms of suppression by suppressor T cells. *Nature immunology*, 2005, Apr; 6(4), 338-44.

[11] Chehade, M; Mayer, L. Oral tolerance and its relation to food hypersensitivities. The *Journal of allergy and clinical Immunology*, 2005, Jan 115(1), 3-12, quiz 3.

[12] Sicinski, P; Rowinski, J; Warchol, JB; Jarzabek, Z; Gut, W; Szczygiel, B; et al. Poliovirus type 1 enters the human host through intestinal M cells. *Gastroenterology*, 1990, Jan, 98(1), 56-8.

[13] Shreedhar, VK; Kelsall, BL; Neutra, MR. Cholera toxin induces migration of dendritic cells from the subepithelial dome region to T- and B-cell areas of Peyer's patches. *Infection and immunity*, 2003, Jan, 71(1), 504-9.

[14] Pavli, P; Woodhams, CE; Doe, WF; Hume, DA. Isolation and characterization of antigen-presenting dendritic cells from the mouse intestinal lamina propria. *Immunology*, 1990 May; 70(1), 40-7.

[15] Ruedl, C; Rieser, C; Bock, G; Wick, G; Wolf, H. Phenotypic and functional characterization of CD11c+ dendritic cell population in mouse Peyer's patches. *European journal of immunology*, 1996, Aug; 26(8), 1801-6.

[16] Scheinecker, C; McHugh, R; Shevach, EM; Germain, RN. Constitutive presentation of a natural tissue autoantigen exclusively by dendritic cells in the draining lymph node. *The Journal of experimental medicine*., 2002, Oct 21, 196(8), 1079-90.

[17] Rescigno, M; Urbano, M; Valzasina, B; Francolini, M; Rotta, G; Bonasio, R; et al. Dendritic cells express tight junction proteins and penetrate gut epithelial monolayers to sample bacteria. *Nature immunology*, 2001, Apr; 2(4), 361-7.

[18] Rimoldi, M; Chieppa, M; Salucci, V; Avogadri, F; Sonzogni, A; Sampietro, GM; et al. Intestinal immune homeostasis is regulated by the crosstalk between epithelial cells and dendritic cells. *Nature immunology*, 2005, May, 6(5), 507-14.

[19] Liu, LM; MacPherson, GG. Lymph-borne (veiled) dendritic cells can acquire and present intestinally administered antigens. *Immunology*, 1991 Jul, 73(3), 281-6.

[20] Ohshima, Y; Omata, N; Motoko, Y; Mayumi, M. The role of dendritic cells in Th1/Th2 balance: A novel therapeutic target of allergic diseases. *Allergol Int*, 2004, 53, 219-26.

[21] Kelsall, BL; Leon, F. Involvement of intestinal dendritic cells in oral tolerance; immunity to pathogens; and inflammatory bowel disease. *Immunological reviews*, 2005, Aug, 206, 132-48.

[22] Niess, JH; Reinecker, HC. Dendritic cells: the commanders-in-chief of mucosal immune defenses. *Current opinion in gastroenterology*, 2006, Jul; 22(4), 354-60.

[23] Coombes, JL; Powrie, F. Dendritic cells in intestinal immune regulation. *Nature reviews*, 2008, Jun; 8(6), 435-46.

[24] Dahan, S; Roth-Walter, F; Arnaboldi, P; Agarwal, S; Mayer, L. Epithelia: lymphocyte interactions in the gut. *Immunological reviews*., 2007, Feb, 215, 243-53.

[25] von Boehmer, H. Oral tolerance: is it all retinoic acid? *The Journal of experimental medicine*, 2007, Aug 6, 204(8), 1737-9.

[26] Allez, M; Brimnes, J; Dotan, I; Mayer, L. Expansion of CD8+ T cells with regulatory function after interaction with intestinal epithelial cells. *Gastroenterology*, 2002, Nov, 123(5), 1516-26.

[27] Dotan, I; Allez, M; Nakazawa, A; Brimnes, J; Schulder-Katz, M; Mayer, L. Intestinal epithelial cells from inflammatory bowel disease patients preferentially stimulate CD4+ T cells to proliferate and secrete interferon-gamma. *American journal of physiology*, 2007, Jun; 292(6), G1630-40.

[28] Mayer, L; Shlien, R. Evidence for function of Ia molecules on gut epithelial cells in man. *The Journal of experimental medicine*, 1987, Nov 1, 166(5), 1471-83.

[29] Bland, PW; Warren, LG. Antigen presentation by epithelial cells of the rat small intestine. II. Selective induction of suppressor T cells. *Immunology*, 1986, May, 58(1), 9-14.

[30] Panja, A; Blumberg, RS; Balk, SP; Mayer, L. CD1d is involved in T cell-intestinal epithelial cell interactions. *The Journal of experimental medicine*, 1993, Sep 1, 178(3), 1115-9.

[31] Yio, XY; Mayer, L. Characterization of a 180-kDa intestinal epithelial cell membrane glycoprotein; gp180. A candidate molecule mediating t cell-epithelial cell interactions. *The Journal of biological chemistry*, 1997, May 9, 272(19), 12786-92.

[32] Karlsson, M; Lundin, S; Dahlgren, U; Kahu, H; Pettersson, I; Telemo, E. "Tolerosomes" are produced by intestinal epithelial cells. *European journal of immunology*, 2001, Oct; 31(10), 2892-900.

[33] Collison, LW; Workman, CJ; Kuo, TT; Boyd, K; Wang, Y; Vignali, KM; et al. The inhibitory cytokine IL-35 contributes to regulatory T-cell function. *Nature*, 2007, Nov 22, 450(7169), 566-9.

[34] Bogunovic, M; Ginhoux, F; Helft, J; Shang, L; Hashimoto, D; Greter, M; et al. Origin of the lamina propria dendritic cell network. *Immunity*, 2009, Sep 18, 31(3), 513-25.

[35] Varol, C; Vallon-Eberhard, A; Elinav, E; Aychek, T; Shapira, Y; Luche, H; et al. Intestinal lamina propria dendritic cell subsets have different origin and functions. *Immunity*, 2009, Sep 18, 31(3), 502-12.

[36] Liu, K; Victora, GD; Schwickert, TA; Guermonprez, P; Meredith, MM; Yao, K; et al. In vivo analysis of dendritic cell development and homeostasis. *Science* (New York; NY., 2009, Apr 17, 324(5925), 392-7.

[37] Annacker, O; Burlen-Defranoux, O; Pimenta-Araujo, R; Cumano, A; Bandeira, A. Regulatory CD4 T cells control the size of the peripheral activated/memory CD4 T cell compartment. *J Immunol*, 2000, Apr 1, 164(7), 3573-80.

[38] Sun, CM; Hall, JA; Blank, RB; Bouladoux, N; Oukka, M; Mora, JR; et al. Small intestine lamina propria dendritic cells promote de novo generation of Foxp3 T reg cells via retinoic acid. *The Journal of experimental medicine*, 2007, Aug 6, 204(8), 1775-85.

[39] Schulz, O; Jaensson, E; Persson, EK; Liu, X; Worbs, T; Agace, WW; et al. Intestinal CD103+; but not CX3CR1+; antigen sampling cells migrate in lymph and serve classical dendritic cell functions. *The Journal of experimental medicine*, 2009, Dec 21, 206(13), 3101-14.

[40] Coombes, JL; Siddiqui, KR; Arancibia-Carcamo, CV; Hall, J; Sun, CM; Belkaid, Y; et al. A functionally specialized population of mucosal CD103+ DCs induces Foxp3+ regulatory T cells via a TGF-beta and retinoic acid-dependent mechanism. *The Journal of experimental medicine*, 2007, Aug 6, 204(8), 1757-64.

[41] Park, MJ; Min, SY; Park, KS; Cho, YG; Cho, ML; Jung, YO; et al. Indoleamine 2, 3-dioxygenase-expressing dendritic cells are involved in the generation of CD4+CD25+ regulatory T cells in Peyer's patches in an orally tolerized; collagen-induced arthritis mouse model. *Arthritis research & therapy*, 2008, 10(1), R11.

[42] Sharma, MD; Baban, B; Chandler, P; Hou, DY; Singh, N; Yagita, H; et al. Plasmacytoid dendritic cells from mouse tumor-draining lymph nodes directly activate mature Tregs via indoleamine 2, 3-dioxygenase. *The Journal of clinical investigation*, 2007, Sep, 117(9), 2570-82.

[43] Puccetti, P; Grohmann, U. IDO and regulatory T cells: a role for reverse signaling and non-canonical NF-kappaB activation. *Nature reviews*, 2007, Oct, 7(10), 817-23.

[44] Contractor, N; Louten, J; Kim, L; Biron, CA; Kelsall, BL. Cutting edge: Peyer's patch plasmacytoid dendritic cells (pDCs) produce low levels of type I interferons: possible role for IL-10, TGFbeta; and prostaglandin E2 in conditioning a unique mucosal pDC phenotype. *J Immunol*, 2007, Sep 1, 179(5), 2690-4.

[45] Gilliet, M; Boonstra, A; Paturel, C; Antonenko, S; Xu, XL; Trinchieri, G; et al. The development of murine plasmacytoid dendritic cell precursors is differentially regulated by FLT3-ligand and granulocyte/macrophage colony-stimulating factor. *The Journal of experimental medicine*, 2002 Apr 1, 195(7), 953-8.

[46] Viney, JL; Mowat, AM; O'Malley, JM; Williamson, E; Fanger, NA. Expanding dendritic cells in vivo enhances the induction of oral tolerance. *J Immunol*, 1998, Jun 15, 160(12), 5815-25.

[47] Goubier, A; Dubois, B; Gheit, H; Joubert, G; Villard-Truc, F; Asselin-Paturel, C; et al. Plasmacytoid dendritic cells mediate oral tolerance. *Immunity*, 2008, Sep 19, 29(3), 464-75.

[48] Man, AL; Bertelli, E; Regoli, M; Chambers, SJ; Nicoletti, C. Antigen-specific T cell-mediated apoptosis of dendritic cells is impaired in a mouse model of food allergy. The *Journal of allergy and clinical immunology*, 2004, May, 113(5), 965-72.

[49] Roncarolo, MG; Battaglia, M. Regulatory T-cell immunotherapy for tolerance to self antigens and alloantigens in humans. *Nature reviews*., 2007, Aug, 7(8), 585-98.

[50] Gri, G; Piconese, S; Frossi, B; Manfroi, V; Merluzzi, S; Tripodo, C; et al. CD4+CD25+ regulatory T cells suppress mast cell degranulation and allergic responses through OX40-OX40L interaction. *Immunity*, 2008, Nov 14, 29(5), 771-81.

[51] Sakaguchi, S. Naturally arising CD4+ regulatory t cells for immunologic self-tolerance and negative control of immune responses. *Annual review of immunology*, 2004, 22, 531-62.

[52] Feuerer, M; Hill, JA; Mathis, D; Benoist, C. Foxp3+ regulatory T cells: differentiation; specification; subphenotypes. *Nature immunology*, 2009, Jul; 10(7), 689-95.

[53] Singh, B; Read, S; Asseman, C; Malmstrom, V; Mottet, C; Stephens, LA; et al. Control of intestinal inflammation by regulatory T cells. *Immunological reviews*, 2001, Aug, 182, 190-200.

[54] Sauer, S; Bruno, L; Hertweck, A; Finlay, D; Leleu, M; Spivakov, M; et al. T cell receptor signaling controls Foxp3 expression via PI3K; Akt; and mTOR. *Proceedings of the National Academy of Sciences of the United States of America*., 2008, Jun 3, 105(22), 7797-802.

[55] Chen, W; Jin, W; Hardegen, N; Lei, KJ; Li, L; Marinos, N; et al. Conversion of peripheral CD4+CD25- naive T cells to CD4+CD25+ regulatory T cells by TGF-beta induction of transcription factor Foxp3. *The Journal of experimental medicine*., 2003, Dec 15, 198(12), 1875-86.

[56] Maynard, CL; Weaver, CT. Intestinal effector T cells in health and disease. *Immunity*, 2009, Sep, 18, 31(3), 389-400.

[57] Curotto de Lafaille, MA; Lafaille, JJ. Natural and adaptive foxp3+ regulatory T cells: more of the same or a division of labor? *Immunity*, 2009, May, 30(5), 626-35.

[58] Hill, JA; Feuerer, M; Tash, K; Haxhinasto, S; Perez, J; Melamed, R; et al. Foxp3 transcription-factor-dependent and -independent regulation of the regulatory T cell transcriptional signature. *Immunity*, 2007, Nov, 27(5), 786-800.

[59] Wei, G; Wei, L; Zhu, J; Zang, C; Hu-Li, J; Yao, Z; et al. Global mapping of H3K4me3 and H3K27me3 reveals specificity and plasticity in lineage fate determination of differentiating CD4+ T cells. *Immunity*, 2009, Jan 16, 30(1), 155-67.

[60] Strobel, S; Mowat, AM. Oral tolerance and allergic responses to food proteins. *Current opinion in allergy and clinical immunology*, 2006, Jun, 6(3), 207-13.

[61] Chen, W; Konkel, JE. TGF-beta and 'adaptive' Foxp3(+) regulatory T cells. *Journal of molecular cell biology*., Feb; 2(1), 30-6.

[62] Mucida, D; Park, Y; Kim, G; Turovskaya, O; Scott, I; Kronenberg, M; et al. Reciprocal TH17 and regulatory T cell differentiation mediated by retinoic acid. *Science* (New York; NY. 2007, Jul 13, 317(5835), 256-60.

[63] Chen, Y; Kuchroo, VK; Inobe, J; Hafler, DA; Weiner, HL. Regulatory, T. cell clones induced by oral tolerance: suppression of autoimmune encephalomyelitis. *Science* (New York; NY. 1994, Aug 26, 265(5176), 1237-40.

[64] Li, MO; Sanjabi, S; Flavell, RA. Transforming growth factor-beta controls development; homeostasis; and tolerance of T cells by regulatory T cell-dependent and -independent mechanisms. *Immunity*, 2006, Sep, 25(3), 455-71.

[65] Marie, JC; Liggitt, D; Rudensky, AY. Cellular mechanisms of fatal early-onset autoimmunity in mice with the T cell-specific targeting of transforming growth factor-beta receptor. *Immunity*, 2006, Sep, 25(3), 441-54.

[66] Marie, JC; Letterio, JJ; Gavin, M; Rudensky, AY. TGF-beta1 maintains suppressor function and Foxp3 expression in CD4+CD25+ regulatory T cells. *The Journal of experimental medicine*, 2005, Apr 4, 201(7), 1061-7.

[67] Powrie, F; Carlino, J; Leach, MW; Mauze, S; Coffman, RL. A critical role for transforming growth factor-beta but not interleukin 4 in the suppression of T helper type 1-mediated colitis by CD45RB(low) CD4+ T cells. *The Journal of experimental medicine*, 1996, Jun 1, 183(6), 2669-74.

[68] Gandhi, R; Anderson, DE; Weiner, HL. Cutting Edge: Immature human dendritic cells express latency-associated peptide and inhibit T cell activation in a TGF-beta-dependent manner. *J Immunol*, 2007, Apr 1, 178(7), 4017-21.

[69] Lacy-Hulbert, A; Smith, AM; Tissire, H; Barry, M; Crowley, D; Bronson, RT; et al. Ulcerative colitis and autoimmunity induced by loss of myeloid alphav integrins. Proceedings of the *National Academy of Sciences of the United States of America*, 2007, Oct 2, 104(40), 15823-8.

[70] Munger, JS; Huang, X; Kawakatsu, H; Griffiths, MJ; Dalton, SL; Wu, J; et al. The integrin alpha v beta 6 binds and activates latent TGF beta 1: a mechanism for regulating pulmonary inflammation and fibrosis. *Cell*, 1999, Feb 5, 96(3), 319-28.

[71] Travis, MA; Reizis, B; Melton, AC; Masteller, E; Tang, Q; Proctor, JM; et al. Loss of integrin alpha(v)beta8 on dendritic cells causes autoimmunity and colitis in mice. *Nature*, 2007, Sep 20, 449(7160), 361-5.

[72] Crawford, SE; Stellmach, V; Murphy-Ullrich, JE; Ribeiro, SM; Lawler, J; Hynes, RO; et al. Thrombospondin-1 is a major activator of TGF-beta1 in vivo. *Cell*, 1998 Jun, 26, 93(7), 1159-70.

[73] Oida, T; Zhang, X; Goto, M; Hachimura, S; Totsuka, M; Kaminogawa, S; et al. CD4+CD25- T cells that express latency-associated peptide on the surface suppress CD4+CD45RBhigh-induced colitis by a TGF-beta-dependent mechanism. *J Immunol*, 2003, Mar 1, 170(5), 2516-22.

[74] Pesu, M; Watford, WT; Wei, L; Xu, L; Fuss, I; Strober, W; et al. T-cell-expressed proprotein convertase furin is essential for maintenance of peripheral immune tolerance. *Nature*, 2008, Sep 11, 455(7210), 246-50.

[75] Ando, T; Hatsushika, K; Wako, M; Ohba, T; Koyama, K; Ohnuma, Y; et al. Orally administered TGF-beta is biologically active in the intestinal mucosa and enhances oral tolerance. *The Journal of allergy and clinical immunology*, 2007, Oct; 120(4), 916-23.

[76] Groux, H; O'Garra, A; Bigler, M; Rouleau, M; Antonenko, S; de Vries, JE; et al. A CD4+ T-cell subset inhibits antigen-specific T-cell responses and prevents colitis. *Nature*, 1997, Oct 16, 389(6652), 737-42.

[77] Faria, AM; Weiner, HL. Oral tolerance: therapeutic implications for autoimmune diseases. *Clinical & developmental immunology*, 2006, Jun-Dec, 13(2-4), 143-57.

[78] Gonnella, PA; Chen, Y; Inobe, J; Komagata, Y; Quartulli, M; Weiner, HL. In situ immune response in gut-associated lymphoid tissue (GALT) following oral antigen in TCR-transgenic mice. *J Immunol*, 1998, May 15, 160(10), 4708-18.

[79] Frossard, CP; Tropia, L; Hauser, C; Eigenmann, PA. Lymphocytes in Peyer patches regulate clinical tolerance in a murine model of food allergy. *The Journal of allergy and clinical immunology*, 2004, May; 113(5), 958-64.

[80] Maynard, CL; Harrington, LE; Janowski, KM; Oliver, JR; Zindl, CL; Rudensky, AY; et al. Regulatory T cells expressing interleukin 10 develop from Foxp3+ and Foxp3- precursor cells in the absence of interleukin 10. *Nature immunology*, 2007, Sep; 8(9), 931-41.

[81] Vieira, PL; Christensen, JR; Minaee, S; O'Neill, EJ; Barrat, FJ; Boonstra, A; et al. IL-10-secreting regulatory T cells do not express Foxp3 but have comparable regulatory function to naturally occurring CD4+CD25+ regulatory T cells. *J Immunol*, 2004, May 15, 172(10), 5986-93.

[82] Zhang, X; Izikson, L; Liu, L; Weiner, HL. Activation of CD25(+)CD4(+) regulatory T cells by oral antigen administration. *J Immunol*, 2001, Oct 15, 167(8), 4245-53.

[83] Hauet-Broere, F; Unger, WW; Garssen, J; Hoijer, MA; Kraal, G; Samsom, JN. Functional CD25- and CD25+ mucosal regulatory T cells are induced in gut-draining lymphoid tissue within 48 h after oral antigen application. *European journal of immunology*, 2003, Oct, 33(10), 2801-10.

[84] Zheng, SG; Wang, JH; Gray, JD; Soucier, H; Horwitz, DA. Natural and induced CD4+CD25+ cells educate CD4+CD25- cells to develop suppressive activity: the role of IL-2, TGF-beta; and IL-10. *J Immunol*, 2004, May 1, 172(9), 5213-21.

[85] Curotto de Lafaille, MA; Kutchukhidze, N; Shen, S; Ding, Y; Yee, H; Lafaille, JJ. Adaptive Foxp3+ regulatory T cell-dependent and -independent control of allergic inflammation. *Immunity*, 2008, Jul 18, 29(1), 114-26.

[86] Ke, Y; Kapp, JA. Oral antigen inhibits priming of CD8+ CTL; CD4+ T cells; and antibody responses while activating CD8+ suppressor T cells. *J Immunol*, 1996, Feb 1, 156(3), 916-21.

[87] Kapp, JA; Pierce, CW; Webb, DR; Devens, B; Godfrey, W; Fukuse, S; et al. Characterization of the epitope recognized by a mAb that reacts differentially with murine suppressor T cells. *International immunology*, 1995, Aug, 7(8), 1319-30.

[88] Zimring, JC; Levery, SB; Kniep, B; Kapp, LM; Fuller, M; Kapp, JA. CD75s is a marker of murine CD8(+) suppressor T cells. *International immunology*, 2003, Nov; 15(11), 1389-99.

[89] Rocha, B. The extrathymic T-cell differentiation in the murine gut. *Immunological reviews*, 2007, Feb, 215, 166-77.

[90] Izcue, A; Coombes, JL; Powrie, F. Regulatory T cells suppress systemic and mucosal immune activation to control intestinal inflammation. *Immunological reviews*, 2006, Aug, 212, 256-71.

[91] Poussier, P; Ning, T; Banerjee, D; Julius, M. A unique subset of self-specific intraintestinal T cells maintains gut integrity. *The Journal of experimental medicine*, 2002, Jun 3, 195(11), 1491-7.

[92] Brimnes, J; Allez, M; Dotan, I; Shao, L; Nakazawa, A; Mayer, L. Defects in CD8+ regulatory T cells in the lamina propria of patients with inflammatory bowel disease. *J Immunol.,* 2005, May 1, 174(9), 5814-22.

[93] Miller, JF; Kurts, C; Allison, J; Kosaka, H; Carbone, F; Heath, WR. Induction of peripheral CD8+ T-cell tolerance by cross-presentation of self antigens. *Immunological reviews.*, 1998, Oct; 165:267-77.

[94] Lee, MS; Sawyer, S; Arnush, M; Krahl, T; von Herrath, M; Oldstone, MA; et al. Transforming growth factor-beta fails to inhibit allograft rejection or virus-induced autoimmune diabetes in transgenic mice. *Transplantation*, 1996, Apr 15, 61(7), 1112-5.

[95] Von Herrath, MG; Yokoyama, M; Dockter, J; Oldstone, MB; Whitton, JL. CD4-deficient mice have reduced levels of memory cytotoxic T lymphocytes after immunization and show diminished resistance to subsequent virus challenge. *Journal of virology*, 1996, Feb, 70(2), 1072-9.

[96] Yamada, A; Ohshima, Y; Yasutomi, M; Ogura, K; Tokuriki, S; Naiki, H; et al. Antigen-primed splenic CD8+ T cells impede the development of oral antigen-induced allergic diarrhea. *The Journal of allergy and clinical immunology*, 2009, Apr; 123(4), 889-94.

[97] Bendelac, A; Rivera, MN; Park, SH; Roark, JH. Mouse CD1-specific NK1 T cells: development; specificity; and function. *Annual review of immunology*, 1997, 15, 535-62.

[98] Godfrey, DI; MacDonald, HR; Kronenberg, M; Smyth, MJ; Van Kaer, L. NKT cells: what's in a name? *Nature reviews*. 2004, Mar, 4(3), 231-7.

[99] McFadden, JP; White, JM; Basketter, DA; Kimber, I. Does hapten exposure predispose to atopic disease? The hapten-atopy hypothesis. *Trends in immunology*, 2009, Feb; 30(2), 67-74.

[100] Nowak, M; Stein-Streilein, J. Invariant NKT cells and tolerance. *International reviews of immunology*, 2007, Jan-Apr, 26(1-2), 95-119.

[101] La Cava, A; Van Kaer, L; Fu Dong, S. CD4+CD25+ Tregs and NKT cells: regulators regulating regulators. *Trends in immunology.,* 2006, Jul; 27(7), 322-7.

[102] Fujii, S; Shimizu, K; Hemmi, H; Steinman, RM. Innate Valpha14(+) natural killer T cells mature dendritic cells; leading to strong adaptive immunity. *Immunological reviews.,* 2007, Dec; 220:183-98.

[103] Dougan, SK; Kaser, A; Blumberg, RS. CD1 expression on antigen-presenting cells. *Current topics in microbiology and immunology*, 2007, 314:113-41.

[104] Tupin, E; Kinjo, Y; Kronenberg, M. The unique role of natural killer T cells in the response to microorganisms. *Nat Rev Microbiol.*, 2007, Jun, 5(6), 405-17.

[105] Trop, S; Samsonov, D; Gotsman, I; Alper, R; Diment, J; Ilan, Y. Liver-associated lymphocytes expressing NK1.1 are essential for oral immune tolerance induction in a murine model. Hepatology (*Baltimore; Md*, 1999 Mar; 29(3), 746-55.

[106] Ilan, Y. Immune downregulation leads to upregulation of an antiviral response: a lesson from the hepatitis B virus. *Microbes and infection / Institut Pasteur*. 2002, Nov; 4(13), 1317-26.

[107] Shibolet, O; Alper, R; Avraham, Y; Berry, EM; Ilan, Y. Immunomodulation of experimental colitis via caloric restriction: role of Nk1.1+ T cells. *Clinical immunology* (Orlando; Fla. 2002 Oct; 105(1), 48-56.

[108] Garside, P; Millington, O; Smith, KM. The anatomy of mucosal immune responses. *Annals of the New York Academy of Sciences*, 2004, Dec; 1029, 9-15.

[109] Worbs, T; Bode, U; Yan, S; Hoffmann, MW; Hintzen, G; Bernhardt, G; et al. Oral tolerance originates in the intestinal immune system and relies on antigen carriage by dendritic cells. *The Journal of experimental medicine*, 2006, Mar 20, 203(3), 519-27.

[110] Macpherson, AJ; Smith, K. Mesenteric lymph nodes at the center of immune anatomy. *The Journal of experimental medicine*, 2006, Mar 20, 203(3), 497-500.

[111] Kunkel, D; Kirchhoff, D; Nishikawa, S; Radbruch, A; Scheffold, A. Visualization of peptide presentation following oral application of antigen in normal and Peyer's patches-deficient mice. *European journal of immunology*, 2003, May; 33(5), 1292-301.

[112] Kraus, TA; Brimnes, J; Muong, C; Liu, JH; Moran, TM; Tappenden, KA; et al. Induction of mucosal tolerance in Peyer's patch-deficient; ligated small bowel loops. *The Journal of clinical investigation*, 2005, Aug; 115(8), 2234-43.

[113] Yi, HJ; Lee, CG; Kwon, HK; So, JS; Sahoo, A; Hwang, JS; et al. Defect in TCR-CD3zeta signaling mediates T cell hypo-responsiveness in mesenteric lymph node. *Molecular immunology*., 2008, Aug; 45(14), 3748-55.

[114] Frossard, CP; Hauser, C; Eigenmann, PA. Antigen-specific secretory IgA antibodies in the gut are decreased in a mouse model of food allergy. *The Journal of allergy and clinical immunology*, 2004, Aug; 114(2), 377-82.

[115] Takayama, N; Igarashi, O; Kweon, MN; Kiyono, H. Regulatory role of Peyer's patches for the inhibition of OVA-induced allergic diarrhea. *Clinical immunology* (Orlando; Fla. 2007, May; 123(2), 199-208.

[116] Warshaw, AL; Walker, WA; Isselbacher, KJ. Protein uptake by the intestine: evidence for absorption of intact macromolecules. *Gastroenterology*, 1974, May; 66(5), 987-92.

[117] Yang, R; Liu, Q; Grosfeld, JL; Pescovitz, MD. Intestinal venous drainage through the liver is a prerequisite for oral tolerance induction. *Journal of pediatric surgery*, 1994, Aug; 29(8), 1145-8.

[118] Stanford, M; Whittall, T; Bergmeier, LA; Lindblad, M; Lundin, S; Shinnick, T; et al. Oral tolerization with peptide 336-351 linked to cholera toxin B subunit in preventing relapses of uveitis in Behcet's disease. *Clinical and experimental immunology*, 2004, Jul; 137(1), 201-8.

[119] Sugiura, K; Kato, K; Hashimoto, F; Jin, T; Amoh, Y; Yamamoto, Y; et al. Induction of donor-specific T cell anergy by portal venous injection of allogeneic cells. *Immunobiology*, 1997, Nov; 197(5), 460-77.

[120] Gotsman, I; Beinart, R; Alper, R; Rabbani, E; Engelhardt, D; Ilan, Y. Induction of oral tolerance towards hepatitis B envelope antigens in a murine model. *Antiviral research*, 2000, Oct; 48(1), 17-26.

[121] Limmer, A; Ohl, J; Wingender, G; Berg, M; Jungerkes, F; Schumak, B; et al. Cross-presentation of oral antigens by liver sinusoidal endothelial cells leads to CD8 T cell tolerance. *European journal of immunology*, 2005, Oct; 35(10), 2970-81.

[122] Mehal, WZ; Juedes, AE; Crispe, IN. Selective retention of activated CD8+ T cells by the normal liver. *J Immunol*, 1999, Sep 15, 163(6), 3202-10.

[123] Wildin, RS; Ramsdell, F; Peake, J; Faravelli, F; Casanova, JL; Buist, N; et al. X-linked neonatal diabetes mellitus; enteropathy and endocrinopathy syndrome is the human equivalent of mouse scurfy. *Nature genetics*, 2001, Jan; 27(1), 18-20.

[124] Smith, M; Tourigny, MR; Noakes, P; Thornton, CA; Tulic, MK; Prescott, SL. Children with egg allergy have evidence of reduced neonatal CD4(+)CD25(+)CD127(lo/-)

regulatory T cell function. *The Journal of allergy and clinical immunology*, 2008, Jun; 121(6), 1460-6, 6 e1-7.

[125] Perez-Machado, MA; Ashwood, P; Thomson, MA; Latcham, F; Sim, R; Walker-Smith, JA; et al. Reduced transforming growth factor-beta1-producing T cells in the duodenal mucosa of children with food allergy. *European journal of immunology*, 2003, Aug; 33(8), 2307-15.

[126] Beyer, K; Castro, R; Birnbaum, A; Benkov, K; Pittman, N; Sampson, HA. Human milk-specific mucosal lymphocytes of the gastrointestinal tract display a TH2 cytokine profile. *The Journal of allergy and clinical immunology*, 2002, Apr; 109(4), 707-13.

[127] Savage, JH; Matsui, EC; Skripak, JM; Wood, RA. The natural history of egg allergy. *The Journal of allergy and clinical immunology*, 2007, Dec; 120(6), 1413-7.

[128] Tiemessen, MM; Van Ieperen-Van Dijk, AG; Bruijnzeel-Koomen, CA; Garssen, J; Knol, EF; Van Hoffen, E. Cow's milk-specific T-cell reactivity of children with and without persistent cow's milk allergy: key role for IL-10. *The Journal of allergy and clinical immunology*, 2004, May; 113(5), 932-9.

[129] Shreffler, WG; Wanich, N; Moloney, M; Nowak-Wegrzyn, A; Sampson, HA. Association of allergen-specific regulatory T cells with the onset of clinical tolerance to milk protein. *The Journal of allergy and clinical immunology*, 2009, Jan; 123(1), 43-52 e7.

[130] Karlsson, MR; Rugtveit, J; Brandtzaeg, P. Allergen-responsive CD4+CD25+ regulatory T cells in children who have outgrown cow's milk allergy. *The Journal of experimental medicine*, 2004, Jun 21, 199(12), 1679-88.

[131] Frossard, CP; Steidler, L; Eigenmann, PA. Oral administration of an IL-10-secreting Lactococcus lactis strain prevents food-induced IgE sensitization. *The Journal of allergy and clinical immunology*, 2007, Apr; 119(4), 952-9.

[132] Sun, JB; Czerkinsky, C; Holmgren, J. Mucosally induced immunological tolerance; regulatory T cells and the adjuvant effect by cholera toxin B subunit. *Scandinavian journal of immunology*, Jan; 71(1), 1-11.

[133] Tang, ML. Oral immunotherapy for food allergy. *Current allergy and asthma reports*, 2009, Jan; 9(1), 43-9.

[134] Jones, SM; Pons, L; Roberts, JL; Scurlock, AM; Perry, TT; Kulis, M; et al. Clinical efficacy and immune regulation with peanut oral immunotherapy. *The Journal of allergy and clinical immunology*, 2009, Aug; 124(2), 292-300, e1-97.

In: Regulatory T Cells
Editor: Ren S. Hayashi, pp. 117-135

ISBN: 978-1-61761-554-2

Chapter 6

The Role of Regulatory T Cells in GVL and GVHD

Abdullah Alsuliman, Paula Garland and Katayoun Rezvani*
Imperial College, Hammersmith Campus, London, United Kingdom

Abstract

The recognition and elimination of malignant cells by donor derived immune effector cells, the so-called graft-versus-leukemia effect (GVL), is key to the success of Allogeneic Hematopoietic Stem Cell Transplantation (AHSCT) in hematological malignancies. The ongoing challenge limiting the success of AHSCT is the maximization of this powerful GVL effect, while minimizing attack of the donor immune cells against the host normal tissue, the so-called graft-versus-host disease (GVHD). The role of CD4+CD25+FOXP3+ regulatory T cells (Treg) in the promotion of peripheral tolerance to self and thus the aversion of autoimmunity is now widely accepted. Furthermore, increasing evidence support their role in modulating alloresponses in the stem cell transplant setting, demonstrating that Tregs are critical players both in the engraftment of donor stem cells and the abrogation of allo-reactive responses that can result in GVHD. Importantly evidence from both murine models and human data suggests that Tregs dampen the alloresponse to host, thereby ameliorating the severity of GVHD without an apparent loss of useful GVL effect. This observation has sparked much research interest into both the underlying mechanisms by which Tregs mediate this effect and ways in which their function can be harnessed and translated into therapies to improve AHSCT outcomes. In this review we seek to summarize the existing data on the role of regulatory T cells in the stem cell transplant setting and examine how their manipulation might be incorporated as a useful technique to reduce GVHD following allogeneic stem cell transplantation.

* Corresponding author: Katayoun Rezvani MD, PhD, MRCP, FRCPath, Clinical senior lecturer and consultant haematologist, Rm 4N7a, 4th floor Commonwealth Bldg, Imperial College, Hammersmith Campus, Du Cane Road, London W12 0NN. Tel: +44 (0) 208 383 2175. K.rezvani@imperial.ac.uk

Allogeneic Stem Cell Transplantation

Allogeneic Hematopoietic Stem Cell Transplantation (AHSCT) remains the only curative option for numerous malignant hematological conditions. The development of reduced intensity conditioning regimens and improved supportive care have broadened the indications for the procedure and in Europe alone, over 24,000 stem cell transplants are currently performed each year [1]. In addition to the treatment of hematological malignancies, AHSCT is increasingly employed in a number of non-malignant conditions including inherited hemaglobinopathies, metabolic disorders and congenital immunodeficiency syndromes. However, the most common indication continues to be the treatment of malignant disease where, in contrast to the non-malignant situation, harnessing the powerful anti-tumor capacity of the donor immune system is the cornerstone of long term disease eradication and prevention of relapse.

The success of AHSCT is influenced by a number of variables including the underlying condition, disease status, patient / donor age and importantly the degree of HLA matching between donor and recipient [1;2]. In addition, a number of complications hamper its beneficial effect and increase procedure related morbidity and mortality including infection, conditioning regimen toxicity and graft versus host disease (GVHD) [3]. The control of these and other procedure related complications coupled with the ability to maximize immune mediated, graft derived anti-tumor effect are key to improving transplant outcomes.

Graft versus Leukemia and Graft versus Host Disease

The beneficial graft versus tumor / leukemia effect (GVL) was observed in early murine experiments in the 1950s where it was first demonstrated that the eradication of leukemia was possible [4]. However this success was accompanied by a fatal, wasting diarrheal syndrome, that would now be recognized as graft versus host disease (GVHD). In contrast to the setting of solid organ transplantation, where the major concern is rejection of the donor organ by the recipient immune system, AHSCT reconstitutes an active, donor derived immune system which, on the one hand, is able to recognize the recipient's residual leukemia cells as foreign and eradicate them, but on the other, has the capacity to recognize the host tissues as foreign, resulting in recipient directed immune attack –GVHD. Some of the most compelling evidence for the power of anti-leukemia alloresponses comes from the efficacy of donor lymphocyte infusion (DLI), introduced into clinical practice in the early 1990s, in the enhancement of anti-tumor immune responses and treatment of post-transplant disease relapse [5;6]. In addition, the ability of reduced intensity conditioned (non-myeloablative) transplants to eradicate hematological malignancy relies largely on immunologic GVL reactions.

Both GVL and GVHD involve the recognition of self-antigens presented on the recipient's antigen presenting cells (APC), which accounts for the overlap in their occurrence and the frequent enhancement in the effect of one at the expense of adverse outcomes from the other. This situation and the overlapping biology of the immunological mechanisms at play has been clearly demonstrated from the observation of post transplant clinical outcomes: hematological disease relapse was found to be more common in those without any evidence of GVHD and conversely allograft T cell depletion resulted in a better GVHD profile but worse relapse rates [7]. Nevertheless there is evidence that these two effects may be teased out from one another and it is the achievement of this effect that is the current pursuit of much stem cell transplant research.

The clinical syndrome of GVHD manifests as tissue injury mediated both by inflammatory cytokines and cytotoxic effecter T cells. The syndrome has been divided into two phases: the acute phase, occurring earlier than 100 days post-allograft and the chronic phase, occurring after 100 days. Acute GVHD primarily affects the skin, liver and gastrointestinal tract, whereas chronic GVHD typically has a multisystem, classically autoimmune-syndrome pattern and may have a wider distribution of affected tissues (Table 1). Temporal division of the two manifestations is however problematic. Acute GVHD is not necessarily followed by the chronic form and conversely chronic GVHD can arise de novo without prior acute GVHD. In addition late-onset GVHD with the clinical pattern of acute GVHD is recognized [8]. These problems led to a working group consensus statement in 2005 from the National Institutes of Health that proposed new criteria for diagnosis and classification of chronic GVHD for clinical trials. According to the consensus criteria, clinical manifestations, rather than time after transplantation, should be used in clinical trials to distinguish chronic GVHD from late acute GVHD, which includes persistent, recurrent, or late-onset acute GVHD [9].

The most important antigens in the pathogenesis of GVHD are the human leukocyte antigens (HLAs) [10]. These proteins are highly polymorphic and expressed on almost all nucleated cells in the body. The degree of mismatch between donor and recipient HLA proteins is directly related to the frequency of GVHD [1;10]. However, GVHD still occurs in fully HLA matched transplants indicating the importance of other proteins, termed minor histocompatibility antigens (mHA), in the generation of the disease. In contrast to HLA molecules, mHAs manifest a variety of tissue distributions. Some such as HY and HA-3 are expressed ubiquitously whereas others such as HA-1 and HA-2 are restricted to hematopoietic tissues [11]. The potential exploitation of this tissue separation to extract useful GVL effect from unwanted GVHD is one avenue of active current research.

Table 1. Clinical Patterns of Acute and Chronic GVHD

Organ involved	Acute GVHD	Chronic GVHD
Skin	Maculopapular rash, ulceration	Abnormal pigmentation, alopecia, sclerosis
Upper gastrointestinal tract	Nausea and vomiting	Xerostomia, ulceration, esophageal strictures
Lower gastrointestinal tract	Watery diarrhea (>500ml), abdominal pain, ileus	Weight loss
Liver	Cholestatic hyperbilirubinemia	Jaundice, transaminitis
Eyes		Sicca syndrome, conjunctivitis
Lungs	-	Obliterative bronchiolitis, restrictive defects, pleural effusions
Bone marrow	-	cytopenias
Musculoskeletal	-	Myositis, fasciitis, contractures

The initiation of GVHD is characterized by massive cytokine storm, induced by the conditioning regimen itself. Damaged host tissue produce proinflammatory cytokines and chemokines, resulting in APC activation and increased cross-presentation of polypeptide antigens to donor immune cells. This may be further enhanced by the presence of infection, the underlying hematological malignancy itself and the systemic translocation of additional inflammatory stimuli across damaged mucosa [12;13]. These activated APCs then stimulate donor T cells to proliferate and produce large amounts of cytokines, particularly interferon-gamma (IFN-γ), interleukin-2 (IL-2) and tumor necrosis factor-alpha (TNF-α), which in turn promote the stimulation of cellular mediators including cytotoxic T lymphocytes (CTL) and natural killer (NK) cells. These mediators destroy target tissues via cell lysis pathways such as Fas-Fas Ligand, perforin and granzyme, resulting in clinical GVHD [8;14;15].

The mainstay of prophylaxis and treatment of GVHD consists of pharmacological agents to interrupt the cascade of inflammation. The calcineurin inhibitors (e.g. cyclosporine and tacrolimus) and methotrexate constitute the usual standard of care for prophylaxis, whilst high dose corticosteroids remain front-line in the management of active GVHD. However failure of prophylaxis, steroid refractory GVHD and complications associated with immunosuppressive regimens continue to thwart the success of AHSCT [15]. Severe GVHD has a poor prognosis, with <25% 5 year survival for disease grade III or above despite maximum intervention [1;15]. Harnessing the power of regulatory T cells for the improved control of GVHD and consequent reduction in transplant related morbidity and mortality is an enticing strategy, although to be a successful approach it will be essential to ensure that donor engraftment, anti-tumor allo-response and protection against infection are not adversely affected.

Regulatory T Cells

Maintaining immune tolerance to self antigens is a crucial role of the immune system in preventing autoimmune reactions. A variety of cells are involved in this process including naturally occurring regulatory T cells (Tregs) [16-19], IFN-γ/IL-10-secreting T regulatory type 1 (Tr1) [20;21], transforming growth factor-β-secreting Th3 regulatory cells [22], natural killer T cells (NKT) [23], and double negative cells ($CD4^{-}CD8^{-}$) [24], of which naturally occurring regulatory T cells have been the principal focus of transplantation tolerance studies.

Although studies have suggested that naturally occurring regulatory T cells are generated as a distinct population in the thymus, $CD4^{+}CD25^{hi}Foxp3^{+}$ Tregs, that are identical to the thymus-derived population, can also be generated after antigen-induced proliferation of CD4+ T cells in peripheral tissues in both mice and humans [25-27]. $CD4^{+}CD25^{hi}Foxp3^{+}$ Tregs constitute 5-10% of total CD4+ T cells and play a pivotal role in immune tolerance by inhibiting auto-reactive T cell populations [28;29]. As such they are critical in the maintenance of normal immune homeostasis [30]. In addition to their ability to suppress autoimmune reactivity they can suppress reactions directed against non-self antigens [16]. Naturally, Tregs are anergic when stimulated *in vitro*. However, co-stimulation with anti-CD28 and high concentrations of IL-2 has been shown to break this hypo-responsiveness. By contrast, *in vivo* studies have shown the ability of these cells to accumulate and proliferate at the site of inflammation upon second antigen recall [25].

Regulatory T cells can be identified phenotypically by the expression of CD4 and the IL-2 receptor α chain CD25 [31]. In addition they are characterized by negative or low expression of CD127, the IL-7 receptor α chain [32]. More recently, the forkhead/winged helix transcription factor protein (FoxP3), which is constitutively expressed in Tregs and involved in their development and suppressive functions, has been employed as an intracellular marker for Tregs [33;34]. Mutation or disruption in the FoxP3 gene leads to a fatal multi-organ autoimmune disease in both human and mice, namely immunodysregulation polyendocrinopathy enteropathy X-linked syndrome (IPEX) and the scurfy mouse phenotype, respectively [35;36], demonstrating the critical role of this transcription factor in Treg functionality. More detailed characterization of Treg cell subpopulations based on CD45RA and FoxP3 expression has also been reported [37].

Other markers that are commonly used for Treg identification, include cytotoxic T lymphocytes antigen-4(CTLA-4), glucocorticoid-induced tumor necrosis factor receptor (GITR) and more recently tumor necrosis factor receptor 2 (TNFR2) [38-40]. Although FoxP3 and CTLA-4 are among the most reliable markers for the identification of regulatory T cells, they cannot be employed for the selection of functional Tregs due to their intracellular expression. To overcome this problem, several groups have focused on finding a reliable surface marker. Recently, CD39, the ectoenzyme degrading ATP to AMP, and CD49d, the α-chain of the integrin VLA-4 (α4β1), have been used along with $CD4^+$ $CD25^{hi}$ and $CD127^{low}$ as markers to facilitate the isolation of functionally active $Foxp3^+$ Tregs [41-43]. Other markers include the tumor-necrosis factor receptor family member, CD27 [44] and HLA-DR [45], although both are limited in their selectivity.

The exact mechanisms by which CD4+CD25+Foxp3+ regulatory cells exert their anti-inflammatory and immunoregulatory functions are not fully understood. To date efforts to identify these mechanisms have cast some light on the topic but progress has been hampered by conflicting data. The majority of *in vitro* studies have shown the requirement for cell contact for Tregs to exert their suppressive effect [46]. However, a key role of immunoregulatory cytokines, such as IL-10, TGF-β and IL-4 in T cell mediated suppressive activities has also been demonstrated in both *in vitro* and *in-vivo* neutralization studies [47-49]. In addition to these two main mechanisms, Edinger et al. have shown perforin dependency of these cells in the control of allo-reactions in GVHD [50]. In a recent study by Cai et al. granzyme-B was reported to be a key player in the suppressive action of Tregs against anti-tumor T effector T-cells (mediating GVL), but not important in suppressing allo-reactive effector T-cells in GVHD, suggesting that the manipulation of granzyme-B could allow separation of the two phenomena [51;52].

A variety of effecter cells, most notably those of the adaptive immune system such as $CD4^+$ and $CD8^+$ T-cells have been shown to be the main targets for Treg suppressive activity [53-58]. Interestingly, it has been shown that once activated, T regulatory suppressor functions are nonspecific [59], therefore it may also be possible for these cells to control innate immunity such as NK function. Conversely, recent work has shown that there may in fact be counter-regulation of regulatory T cells by innate immune mechanisms [60].

The manipulation of regulatory T cells in the allogeneic setting has the potential to achieve two opposing, although not necessarily mutually exclusive effects; their depletion to enhance antitumor immunity [61-63] or harnessing their regulatory capacity to achieve protection from autoimmunity, graft rejection and critically, GVHD. The segregation and optimization of these two effects is at the forefront of current allogeneic stem cell transplant research.

Regulatory T Cells and GVHD/GVL: Lessons from Animal Models

Several studies using animals models have demonstrated the prominent role of regulatory T cells in the prevention of pathological GVHD and more interestingly, in the separation of GVHD from favorable GVL. Depletion of $CD4^{+}CD25^{+}$ cells has been shown to accelerate and exacerbate GVHD whereas the adoptive transfer of these cells can rescue mice from such harmful effects [64]. Furthermore, $CD4^{+}CD25^{+}$ cells can prevent GVHD related damage of the thymus and peripheral lymphoid tissues, which in turn facilitates the development and expansion of an enhanced T-cell repertoire, resulting in improved reconstitution of functional immune responses following AHSCT [65]. Regulatory T cells can also prevent the production of pro-inflammatory cytokines, such as IFN-γ, IL-17, IL-1β and G-CSF that are mediators of chronic GVHD [66]. Different subpopulations of Tregs have been shown to have different regulatory potency. $CD4^{+}CD25^{+}CD62L^{+}$were shown to be more effective at preventing GVHD, whereas their counterpart CD62L- fraction did not demonstrate the same anti-GVHD capacity. This differential effect may be related to the ability of CD62L+ Treg subpopulation to home to secondary lymphoid organs more efficiently and therefore more effectively exert their protective function against aGVHD [67;68].

It is possible to break the anergic status of Tregs *in vitro* using TCR ligation, co-stimulation and cytokines. *Ex-vivo* expanded Tregs are even more suppressive than their fresh counterparts [64;68]. Recently, Beyersdorf et al showed that Tregs isolated from animals that were pre-treated with superagonistic anti-CD28 displayed enhanced suppressive activity and following adoptive transfer protected recipients from acute GVHD without abrogating the beneficial GVL effect of CD28 pre-activated effector T cells [69]. The suppressive ability of regulatory T cells has been shown by some groups to be antigen-specific, as third party recipients are not protected from GVHD [70;71]. Conversely, others have shown that once activated, regulatory T cells act in a non-specific manner [59]. CD4+CD25+ cells can also suppress CD8-mediated GVHD across minor histocompatibility barriers [68] and have even been shown to be suppressive in xenogenic antigen-mediated GVHD [72].

The paucity of regulatory T-cells in the peripheral blood and their unknown specificity have been a challenge to employing these cells in transplantation tolerance. However, several groups have successfully expanded polyclonal, as well as antigen-specific regulatory T cells using several different expansion strategies. Taylor and colleagues have evaluated different methods for polyclonal expansion, including immobilized anti-CD3 with low dose IL-2 in the presence of TGF-β (yielding the highest frequency and most suppressive population), and soluble anti-CD3 with APC or high concentration IL-2 [64]. Stimulating Tregs with anti-CD3/CD28 microbeads with IL-2 [73;74] or with IL-15 [75] has also been shown to result in successful expansion of polyclonal regulatory T cells. An alternative approach is the use of lentivirus to ectopically express Foxp3 in $CD4^{+}CD25^{-}$ T cells. These engineered regulatory T cells were shown to be phenotypically and functionally similar to natural T regulatory cells and able to suppress GVHD while preserving GVL in a murine EL4/DsRed leukemia model [76].

Although polyclonal regulatory T cells have been shown to possess sustainable phenotypic and functional characteristics of regulatory T cells [77] and have the ability to delay or protect from GVHD, their effect on beneficial immune responses is not fully understood. Therefore, the idea of expanding regulatory T cells with a particular specificity, so-called recipient allo-antigen specific, has emerged in transplantation models. The advantage of using antigen-specific over polyclonal regulatory T cells has been proven in different settings such as diabetes [78] and solid organ transplantation. Trenado A et al have shown that recipient-specific regulatory T cells, generated by expanding donor regulatory T cells with recipient's APCs, were able to prevent GVHD while preserving GVL [71]. The specificity of this protection was further assessed using a graft rejection model whereby mice were transplanted with syngeneic bone marrow (BM) and two different allogeneic BMs along with syngeneic splenocytes (stimulator), in the presence of allo-activated CD4+CD25+ Tregs. Only BM from which APCs were used to activate the Tregs was protected from rejection while the third party BM was not [70]. Furthermore, expanded antigen-specific regulatory T cells were even more suppressive than polyclonal ones. As is the case with adoptive immunity, specificity is preferable. Many protocols have been and continue to be developed, aiming to generate antigen-specific regulatory T cells. However, further research is needed to achieve the required specificity to enable the use of these powerful cells to be translated to human applications.

Translation of Treg Use into the Human Clinical Setting

The translation from murine Treg models into the human clinical setting remains in its infancy. Early work in the field yielded conflicting data, partly because of difficulties in the use of markers to accurately identify true regulatory T-cells and discriminate them from activated T-cell effectors. Indeed some early studies reported a correlation between GVHD occurrence and *high* Treg frequency, defined as CD4+CD25+ T-cells, both in the graft [79] and in the recipient [80]. Others failed to show a correlation between $CD4^{+}CD25^{+}$ content and outcome in the setting of donor lymphocyte infusion (DLI); although the addition of a further discriminatory marker defining Treg as $CD25^{hi}$ $CD45RB^{low}$, did show a relationship between higher DLI $CD4^{+}CD25^{hi}CD45RB^{low}$ cells and lower severity of GVHD [81]. However, these early works used CD25+ as the only marker for identifying Treg which is also expressed on activated T-cells, rendering these data difficult to interpret due to the probable overlap between regulatory and activated T cell populations in these studies.

The use of Foxp3 as a marker for the identification of regulatory T cells served to aid progress in the field and human observational studies demonstrated that decreased Foxp3 expression, measured by flow cytometry and PCR techniques, correlates with a heightened incidence of both acute and chronic GVHD post allogeneic stem cell transplant [82;83]. The number of transfused Treg within the graft has been shown to also affect overall survival post-transplant [84]. Our group and others have shown an association between the Treg content of the graft and GVHD [85-86]. Furthermore, we showed that high frequencies and absolute numbers of $Foxp3^{+}CD4^{+}$ T-cells in the donor graft and the recipient blood during immune reconstitution early after AHSCT, correlate with a lower incidence of GVHD [85]. However, when we analyzed CD25+ T-cell content, which had previously been employed as a marker for regulatory T cells independently, there was a direct correlation between

$CD4^{+}CD25^{+}$ and *increased* incidence of GVHD, further confirming the inadequacy of CD25+ as a stand-alone marker for Tregs. Even following the selective depletion of CD25+ T-cells in the graft using anti-CD25 immunotoxin, we observed that the CD4+Foxp3+ regulatory T-cell compartment reconstitutes promptly and undergoes further in-vivo expansion which may then confer additional protection against GVHD [86]. These data were further supported by a recent study reporting that the balance between Tregs and T-effector cells is important in determining outcome from GVHD [87]. In contrast, others have failed to show a correlation between Foxp3 expressing Tregs and the incidence of GVHD [88]; the discrepancy in these results is probably related to the heterogeneity of the patient groups in the different studies.

Recent published data of the comparative examination of pathological specimen have also shown an association between the Treg frequencies in the affected organ and severity of GVHD. Rieger et al. demonstrated a post-transplant shift in the T-cell repertoire during GVHD with reduced numbers of peripheral blood $CD4+CD25^{hi}$ regulatory T-cells, accompanied by a reduction in gastrointestinal (GI) mucosal Foxp3 expression [89]. Others have shown that the number of Tregs is higher in the skin of patients with aGVHD that are responding to therapy compared to those with resistant aGVHD, illustrating the importance of their role, even in established disease [90]. These data have led to a proposal to consider Treg frequencies as diagnostic and prognostic biomarkers for aGVHD [91].

The potential suppressive effect of Treg on useful anti-tumor responses in humans remains largely unknown. Whereas in some murine studies, the beneficial protection from aGVHD has been reported without the loss of anti-tumor immune responses, clinical data reported in the literature are conflicting. Nadal et al. examined a cohort of patients transplanted for chronic myeloid leukemia (CML) and found that, although an increment in Tregs may have an advantageous effect on graft survival in the early post-transplant phase, this was associated with a loss of GVL effect and a consequent increased relapse rate. Furthermore, their data failed to support a role for Tregs against a GVHD [92]. The weight of published data, however, favors regulatory T cells as potentially useful therapy in GVHD.

Regulatory T-Cells as an Immunomodulatory Tool

Various strategies may be employed to enhance the Treg population, with the aim of exploiting their powerful immunomodulatory effect to dampen pathogenic alloresponses.

Cytokine Therapy

Interleukin-2 (IL-2) is an essential cytokine in the maintenance of the pool of Treg cells *in vivo* and is also critical to their function. Furthermore, in the lymphopenic setting such as might be encountered immediately following allogeneic stem cell transplantation, IL-2 may preferentially promote the expansion of the Treg compartment [93]. IL-2 directly induces the up-regulation of Foxp3 gene expression through the phosphorylation of STAT proteins that subsequently translocate to the nucleus and bind to regulatory motifs in the Foxp3 gene [94]. Exogenous use of this cytokine has therefore been proposed as a potential immunomodulatory therapy to boost the number of Tregs post AHSCT. However, for preferential *in vivo* Treg expansion, the presence of adequate numbers of Treg cells as a starting pool is necessary.

Recently Zorn et al. reported the feasibility of a clinical study of low-dose IL-2 combined with infusions of donor CD4+ lymphocytes to expand Tregs *in vivo* [95-96]. Concerns remain that this approach may diminish the GVL action of effector T cells and carefully monitored, prospective clinical trials are required to explore the efficacy of this strategy further.

Selective Immunosuppression

Anti-thymocyte globulin has been used for the prophylaxis and treatment of GVHD as well as wider clinical applications that harness its potent immunosuppressive action e.g. in the management of aplastic anaemia. More recently it has been suggested that ATG may preferentially deplete conventional T cells, while sparing their regulatory T-cell counterparts. However, work seeking to evaluate the notion that rabbit anti-thymocyte globulin (rATG) may provide protection from GVHD through the enhancement of regulatory T cells, has shown that although Foxp3 expression increases in human T cells exposed to rATG, the induction is transient rather than stable. Furthermore, these induced Foxp3 positive cells do not appear to have immunosuppressive capacity even at the time of maximal Foxp3 expression [97].

Rapamycin, an mTOR inhibitor, has been shown to effectively induce Tregs both *in vitro* and *in vivo* [95]. This agent has been used successfully in the prophylaxis and treatment of GVHD post AHSCT [98]. However, in contrast to other classical prophylactic agents such as cyclosporine and tacrolimus, it avoids inhibition of TCR mediated calcineurin activity, thereby producing a differential effect on regulatory and conventional effector T-cell populations and leading to an increase in the Treg : Teffector ratio [99]. Recent data from the murine setting have shown that post allograft, rapamycin results in an increase in the frequencies of Tregs in GVHD target organs, such as the skin, although circulating numbers of Tregs are unaffected [100]. Such pathological data in the human allograft setting is as yet unavailable, principally due to the difficulties in obtaining such samples from patients.

Extracorporeal Photopheresis (ECP)

ECP is used for the treatment of chronic graft versus host disease, although the exact mechanism for the efficacy of this therapy remains unclear. Leukapheresed peripheral blood mononuclear cells are treated with UVA radiation in the presence of a DNA alkylating agent that induces apoptosis. It is proposed that upon phagocytosis by antigen presenting cells, tolerogenic signals are induced resulting in reduced graft versus host disease. The precise nature of these tolerogenic signals and the cell phenotypes involved have been an area of active investigation. A number of groups have shown that induction or differentiation of regulatory T-cells is an important component of this tolerogenic response [101;102]. In addition to an increase in Treg frequencies, each successive cycle of treatment with ECP leads to an increase in the suppressive capacity of these cells [103;104]. Di Biaso et al. recently demonstrated that only patients responding to ECP have a rise in their fraction of circulating Tregs whereas patients who fail to respond to the therapy have no such rise in Treg numbers [105]. In addition, they identified a concurrent decrease in the CD4+ T cells producing interleukin-17 in patients responding to ECP, further supporting the importance of the recently described Th17 axis in the pathology of GVHD and its frequent reciprocal relationship with the regulatory T-cell axis [106;107].

Regulatory T-Cells as Cellular Therapy

The most established model of cellular therapy is in the setting of donor lymphocyte infusion (DLI). DLI is employed to augment the graft versus leukemia response post AHSCT, either to treat relapsed disease or pre-emptively to prevent relapse, particularly in the context of reduced intensity conditioning transplant regimens. Increasingly other cell-based therapies are being used in the clinic – mesenchymal stem cells have been employed as successful GVHD therapy [108] and illustrate the feasibility of expanding the necessary Good Manufacturing Practices (GMP) standards to a range of cellular products including regulatory T cells for therapeutic adoptive transfer.

Tackling the source of regulatory T-cells, given their low frequencies in peripheral blood, has been a major block in their development as a viable cellular therapy. Several methodologies have been under development to overcome this obstacle, either through polyclonal or antigen-specific expansion of these cells. The *in vitro* conversion of conventional T cells (CD4+CD25-) into polyconal cells with a Treg phenotype has been achieved using specific cytokines, such as retinoic acid [109] and/or TGF-β [110;111] in addition to IL-2, which plays a critical role in the expansion of Tregs. The use of agonistic antibodies against stimulatory and co-stimulatory molecules, mainly CD3 and CD28, has also been shown to be effective in expanding these cells [64;77;112;113]. Anti-CD28 superagonist was shown to preferentially expand the Treg compartment [114], however translation of this approach in a recent clinical trial failed to yield significant preferential Treg generation [115]. Several studies have also focused on the expansion of antigen-specific regulatory T cells, employing cellular mediators to promote Treg expansion. Yamazaki and colleagues have directly expanded Tregs using dendritic cell stimulation [116]. Allogeneic B cells [117] or CD40-activated B cells [118] have also been suggested as stimulators to generate and expand alloantigen-specific Tregs. CD40-activated B cells were shown to be more efficient at generating Tregs with more suppressive activity, compared to immature dendritic cells [119]. An alternative methodology has been the *ex-vivo* selection of CD4+CD25+ natural regulatory T-cells and their *in vitro* expansion. Regulatory T cells are selected according to CD4+CD25+ phenotype for either direct adoptive transfer or *in vitro* expansion; this raises the issue of activated conventional T cell contamination of the collected product, which might in turn skew the benefits expected. It is therefore important to further characterize these cells phenotypically to achieve accurate selection and expansion of a pure regulatory T cell product.

The technology is now ripe for transfer into the human clinical setting, in carefully conducted and closely monitored trials – only then will it become possible to glean answers to questions on optimal dose, timing, compatibility with other immunosuppressive drugs and critically safety, in terms of relapse of underlying disease and susceptibility to infection.

Adoptive Treg Transfer – Clinical Trials

The first human study of adoptive transfer of *ex vivo* expanded CD4+CD25+CD127- T regulatory cells in the treatment of GVHD was reported in October 2009 in two patients, using cells expanded from buffy coats taken from family members [120]. The first patient, who was being treated for chronic GVHD, achieved significant alleviation of symptoms and reduction of pharmacologic immunosuppression. However the second patient, who was being treated for grade IV acute GvHD, experienced only transient improvement. A number of phase I clinical studies, currently ongoing, have employed both freshly isolated donor Tregs administered alongside DLI, and Tregs expanded from umbilical cord blood administered in the double cord blood transplant setting [121;122]. Recently, Di Ianni and colleagues conducted a phase I/II clinical trial involving twenty-two patients with hematological malignancies, receiving a haploidentical AHSCT [123]. Patients received freshly isolated CD4+CD25hi regulatory T cells after transplant conditioning therapy followed by infusion of mobilized CD34+ stem cells and conventional T cells from their haploidentical donor. The infused Tregs :Teffector ratio was established at 1:1.5 and no post-transplant prophylaxis against GVHD was used. Strikingly, only 2 of 26 evaluable patients developed GVHD. The authors concluded that in the setting of haploidentical transplantation, the success of which has been thwarted by infection-related deaths due to aggressive immunosupression needed to prevent GVHD, infusion of freshly purified Tregs may allow safe administration of high doses of T cells, providing protection from GVHD and facilitating robust immune reconstitution.

Conclusion

Although there is a significant amount of conflict within the published data in the human setting, including lack of specific markers for identifying Tregs, differences in sampling techniques/protocols and heterogenous conditioning and transplant settings, the weight of published human data support the powerful property of regulatory T cells in the control of allo-responses post transplant [85;86;124;125].

The translation of regulatory T cell therapy from bench to bedside is now within reach, with a number of monitored clinical trials ongoing. It is hoped that these trials will be able to discern whether Tregs can actually deliver the beneficial outcomes they seem to promise, i.e. protection from GVHD without compromise in terms of disease relapse or infection rates.

References

[1] ESH-EBMT. Haematopoietic Stem Cell Transplantation; *The EBMT Handbook*; 5th Edition., 2008.

[2] Ringden, O; Barrett, AJ; Zhang, MJ et al. Decreased treatment failure in recipients of HLA-identical bone marrow or peripheral blood stem cell transplants with high CD34 cell doses. *Br.J.Haematol.*, 2003, 121, 874-885.

[3] Gratwohl, A; Brand, R; Frassoni, F; et al. Cause of death after allogeneic haematopoietic stem cell transplantation (HSCT) in early leukaemias: an EBMT analysis of lethal infectious complications and changes over calendar time. *Bone Marrow Transplant.,* 2005, 36, 757-769.

[4] Barnes, DW; Loutit, JF. Treatment of murine leukaemia with x-rays and homologous bone marrow. II. *Br.J.Haematol.,* 1957, 3, 241-252.

[5] Helg, C; Starobinski, M; Jeannet, M; Chapuis, B. Donor lymphocyte infusion for the treatment of relapse after allogeneic hematopoietic stem cell transplantation. *Leuk.Lymphoma.,* 1998, 29, 301-313.

[6] Deol, A; Lum, LG. Role of donor lymphocyte infusions in relapsed hematological malignancies after stem cell transplantation revisited. *Cancer Treat.Rev.,* 2010

[7] Alyea, EP. Modulating graft-versus-host disease to enhance the graft-versus-leukemia effect. *Best.Pract.Res.Clin.Haematol.,* 2008, 21, 239-250.

[8] Mielcarek, M; Martin, PJ; Leisenring, W; et al. Graft-versus-host disease after nonmyeloablative versus conventional hematopoietic stem cell transplantation. *Blood.,* 2003, 102, 756-762.

[9] Filipovich, AH; Weisdorf, D; Pavletic, S; et al. National Institutes of Health consensus development project on criteria for clinical trials in chronic graft-versus-host disease: I. Diagnosis and staging working group report. *Biol.Blood Marrow Transplant.,* 2005, 11, 945-956.

[10] Ferrara, JL; Levine, JE; Reddy, P; Holler, E. Graft-versus-host disease. *Lancet.,* 2009, 373, 1550-1561.

[11] Loiseau, P; Busson, M; Balere, ML; et al. HLA Association with hematopoietic stem cell transplantation outcome: the number of mismatches at HLA-A; -B; -C; -DRB1, or - DQB1 is strongly associated with overall survival. *Biol.Blood Marrow Transplant.,* 2007, 13, 965-974.

[12] Pidala, J; Anasetti, C. Can antigen-specific regulatory T cells protect against graft vs. host disease and spare anti-malignancy alloresponse? *Haematologica.,* 2009.

[13] Xun, CQ; Thompson, JS; Jennings, CD; Brown, SA; Widmer, MB. Effect of total body irradiation; busulfan-cyclophosphamide; or cyclophosphamide conditioning on inflammatory cytokine release and development of acute and chronic graft-versus-host disease in H-2-incompatible transplanted SCID mice. *Blood.,* 1994, 83, 2360-2367.

[14] Hill, GR; Crawford, JM; Cooke, KR; et al. Total body irradiation and acute graft-versus-host disease: the role of gastrointestinal damage and inflammatory cytokines. *Blood.,* 1997, 90, 3204-3213.

[15] Jacobsohn, DA; Hallick, J; Anders, V; et al. Infliximab for steroid-refractory acute GVHD: a case series. *Am.J.Hematol.,* 2003, 74, 119-124.

[16] Sakaguchi, S. Naturally arising Foxp3-expressing CD25+CD4+ regulatory T cells in immunological tolerance to self and non-self. *Nat.Immunol.,* 2005, 6, 345-352.

[17] Hoffmann, P; Ermann, J; Edinger, M; Fathman, CG; Strober, S. Donor-type CD4(+)CD25(+) regulatory T cells suppress lethal acute graft-versus-host disease after allogeneic bone marrow transplantation. *J.Exp.Med.,* 2002, 196, 389-399.

[18] Cohen, JL; Trenado, A; Vasey, D; Klatzmann, D; Salomon, BL. CD4(+)CD25(+) immunoregulatory T Cells: new therapeutics for graft-versus-host disease. *J.Exp.Med.,* 2002, 196, 401-406.

[19] Albert, MH; Liu, Y; Anasetti, C; Yu, XZ. Antigen-dependent suppression of alloresponses by Foxp3-induced regulatory T cells in transplantation. *Eur.J.Immunol.*, 2005, 35, 2598-2607.

[20] Groux, H; O'Garra, A; Bigler, M; et al. A CD4+ T-cell subset inhibits antigen-specific T-cell responses and prevents colitis. *Nature.*, 1997, 389, 737-742.

[21] Levings, MK; Sangregorio, R; Galbiati, F; et al. IFN-alpha and IL-10 induce the differentiation of human type 1 T regulatory cells. *J.Immunol.*, 2001, 166, 5530-5539.

[22] Weiner, HL. Induction and mechanism of action of transforming growth factor-beta-secreting Th3 regulatory cells. *Immunol.Rev.*, 2001, 182, 207-214.

[23] Seino, KI; Fukao, K; Muramoto, K; et al. Requirement for natural killer T (NKT) cells in the induction of allograft tolerance. *Proc.Natl.Acad.Sci.*, U.S.A 2001, 98, 2577-2581.

[24] Zhang, J; Gao, JX; Salojin, K; et al. Regulation of fas ligand expression during activation-induced cell death in T cells by p38 mitogen-activated protein kinase and c-Jun NH2-terminal kinase. *J.Exp.Med.*, 2000, 191, 1017-1030.

[25] Vukmanovic-Stejic, M; Agius, E; Booth, N; et al. The kinetics of CD4+Foxp3+ T cell accumulation during a human cutaneous antigen-specific memory response in vivo. *J.Clin.Invest*, 2008, 118, 3639-3650.

[26] Cobbold, SP; Castejon, R; Adams, E; et al. Induction of foxP3+ regulatory T cells in the periphery of T cell receptor transgenic mice tolerized to transplants. *J.Immunol.*, 2004, 172, 6003-6010.

[27] Apostolou, I; Von, BH. In vivo instruction of suppressor commitment in naive T cells. *J.Exp.Med.*, 2004, 199, 1401-1408.

[28] Sakaguchi, S; Sakaguchi, N; Shimizu, J; et al. Immunologic tolerance maintained by CD25+ CD4+ regulatory T cells: their common role in controlling autoimmunity; tumor immunity; and transplantation tolerance. *Immunol.Rev.*, 2001, 182, 18-32.

[29] Takahashi, T; Kuniyasu, Y; Toda, M; et al. Immunologic self-tolerance maintained by CD25+CD4+ naturally anergic and suppressive T cells: induction of autoimmune disease by breaking their anergic/suppressive state. *Int.Immunol.*, 1998, 10, 1969-1980.

[30] Dudda, JC; Perdue, N; Bachtanian, E; Campbell, DJ. Foxp3+ regulatory T cells maintain immune homeostasis in the skin. *J.Exp.Med.*, 2008, 205, 1559-1565.

[31] Sakaguchi, S; Sakaguchi, N; Asano, M; Itoh, M; Toda, M. Immunologic self-tolerance maintained by activated T cells expressing IL-2 receptor alpha-chains (CD25). Breakdown of a single mechanism of self-tolerance causes various autoimmune diseases. *J.Immunol.*, 1995, 155, 1151-1164.

[32] Liu, W; Putnam, AL; Xu-Yu, Z; et al. CD127 expression inversely correlates with FoxP3 and suppressive function of human CD4+ T reg cells. *J.Exp.Med.*, 2006, 203, 1701-1711.

[33] Fontenot, JD; Gavin, MA; Rudensky, AY. Foxp3 programs the development and function of CD4+CD25+ regulatory T cells. *Nat.Immunol.*, 2003, 4, 330-336.

[34] Hori, S; Nomura, T; Sakaguchi, S. Control of regulatory T cell development by the transcription factor Foxp3. *Science*, 2003, 299, 1057-1061.

[35] Bennett, CL; Christie, J; Ramsdell, F; et al. The immune dysregulation; polyendocrinopathy; enteropathy; X-linked syndrome (IPEX) is caused by mutations of FOXP3. *Nat.Genet.*, 2001, 27, 20-21.

[36] Brunkow, ME; Jeffery, EW; Hjerrild, KA; et al. Disruption of a new forkhead/winged-helix protein; scurfin; results in the fatal lymphoproliferative disorder of the scurfy mouse. *Nat.Genet.*, 2001, 27, 68-73.

[37] Miyara, M; Yoshioka, Y; Kitoh, A; et al. Functional delineation and differentiation dynamics of human CD4+ T cells expressing the FoxP3 transcription factor. *Immunity.*, 2009, 30, 899-911.

[38] Takahashi, T; Tagami, T; Yamazaki, S; et al. Immunologic self-tolerance maintained by CD25(+)CD4(+) regulatory T cells constitutively expressing cytotoxic T lymphocyte-associated antigen 4. *J.Exp.Med.,* 2000, 192, 303-310.

[39] Shimizu, J; Yamazaki, S; Takahashi, T; Ishida, Y; Sakaguchi, S. Stimulation of CD25(+)CD4(+) regulatory T cells through GITR breaks immunological self-tolerance. *Nat.Immunol.,* 2002, 3, 135-142.

[40] Chen, X; Subleski, JJ; Hamano, R; et al. Co-expression of TNFR2 and CD25 identifies more of the functional CD4+FOXP3+ regulatory T cells in human peripheral blood. *Eur.J.Immunol.,* 2010, 40, 1099-1106.

[41] Borsellino, G; Kleinewietfeld, M; Di, MD; et al. Expression of ectonucleotidase CD39 by Foxp3+ Treg cells: hydrolysis of extracellular ATP and immune suppression. *Blood.,* 2007, 110, 1225-1232.

[42] Kleinewietfeld, M; Starke, M; Di, MD; et al. CD49d provides access to "untouched" human Foxp3+ Treg free of contaminating effector cells. *Blood.,* 2009, 113, 827-836.

[43] Mandapathil, M; Lang, S; Gorelik, E; Whiteside, TL. Isolation of functional human regulatory T cells (Treg) from the peripheral blood based on the CD39 expression. *J.Immunol Methods*, 2009, 346, 55-63.

[44] Ruprecht, CR; Gattorno, M; Ferlito, F; et al. Coexpression of CD25 and CD27 identifies FoxP3+ regulatory T cells in inflamed synovia. *J.Exp.Med.,* 2005, 201, 1793-1803.

[45] Baecher-Allan, C; Wolf, E; Hafler, DA. MHC class II expression identifies functionally distinct human regulatory T cells. *J.Immunol.,* 2006, 176, 4622-4631.

[46] Nakamura, K; Kitani, A; Strober, W. Cell contact-dependent immunosuppression by CD4(+)CD25(+) regulatory T cells is mediated by cell surface-bound transforming growth factor beta. *J.Exp.Med.,* 2001, 194, 629-644.

[47] Seddon, B; Mason, D. Regulatory T cells in the control of autoimmunity: the essential role of transforming growth factor beta and interleukin 4 in the prevention of autoimmune thyroiditis in rats by peripheral CD4(+)CD. *J.Exp.Med.,* 1999, 189, 279-288.

[48] Powrie, F; Carlino, J; Leach, MW; Mauze, S; Coffman, RL. A critical role for transforming growth factor-beta but not interleukin 4 in the suppression of T helper type 1-mediated colitis by CD45RB(low) CD4+ T cells. *J.Exp.Med.,* 1996, 183, 2669-2674.

[49] Asseman, C; Mauze, S; Leach, MW; Coffman, RL; Powrie, F. An essential role for interleukin 10 in the function of regulatory T cells that inhibit intestinal inflammation. *J.Exp.Med.,* 1999, 190, 995-1004.

[50] Edinger, M; Hoffmann, P; Ermann, J; et al. CD4+CD25+ regulatory T cells preserve graft-versus-tumor activity while inhibiting graft-versus-host disease after bone marrow transplantation. *Nat.Med.*, 2003, 9, 1144-1150.

[51] Cai, SF; Cao, X; Hassan, A; Fehniger, TA; Ley, TJ. Granzyme B is not required for regulatory T cell-mediated suppression of graft-versus-host disease. *Blood.,* 2010, 115, 1669-1677.

[52] Cao, X; Cai, SF; Fehniger, TA; et al. Granzyme B and perforin are important for regulatory T cell-mediated suppression of tumor clearance. *Immunity.*, 2007, 27, 635-646.

[53] Pandiyan, P; Zheng, L; Ishihara, S; Reed, J; Lenardo, MJ. CD4+CD25+Foxp3+ regulatory T cells induce cytokine deprivation-mediated apoptosis of effector CD4+ T cells. *Nat.Immunol.,* 2007, 8, 1353-1362.

[54] Levings, MK; Sangregorio, R; Roncarolo, MG. Human cd25(+)cd4(+) t regulatory cells suppress naive and memory T cell proliferation and can be expanded in vitro without loss of function. *J.Exp.Med.,* 2001, 193, 1295-1302.

[55] Kim, J; Kim, HJ; Choi, WS; et al. Maintenance of CD8+ T-cell anergy by CD4+CD25+ regulatory T cells in chronic graft-versus-host disease. *Exp.Mol.Med.*, 2006, 38, 494-501.

[56] Simon, AK; Jones, E; Richards, H; et al. Regulatory T cells inhibit Fas ligand-induced innate and adaptive tumour immunity. *Eur.J.Immunol.,* 2007, 37, 758-767.

[57] Piccirillo, CA; Shevach, EM. Naturally-occurring CD4+CD25+ immunoregulatory T cells: central players in the arena of peripheral tolerance. *Semin.Immunol.*, 2004, 16, 81-88.

[58] Iikuni, N; Lourenco, EV; Hahn, BH; La, CA. Cutting edge: Regulatory T cells directly suppress B cells in systemic lupus erythematosus. *J.Immunol.,* 2009, 183, 1518-1522.

[59] Thornton, AM; Shevach, EM. Suppressor effector function of CD4+CD25+ immunoregulatory T cells is antigen nonspecific. *J.Immunol.,* 2000, 164, 183-190.

[60] Kim, JI; Lee, MK; Moore, DJ; et al. Regulatory T-cell counter-regulation by innate immunity is a barrier to transplantation tolerance. *Am.J.Transplant.*, 2009, 9, 2736-2744.

[61] Heier, I; Hofgaard, PO; Brandtzaeg, P; Jahnsen, FL; Karlsson, M. Depletion of CD4+ CD25+ regulatory T cells inhibits local tumour growth in a mouse model of B cell lymphoma. *Clin.Exp.Immunol.*, 2008, 152, 381-387.

[62] Imai, H; Saio, M; Nonaka, K; et al. Depletion of CD4+CD25+ regulatory T cells enhances interleukin-2-induced antitumor immunity in a mouse model of colon adenocarcinoma. *Cancer Sci.*, 2007, 98, 416-423.

[63] Kline, J; Brown, IE; Zha, YY; et al. Homeostatic proliferation plus regulatory T-cell depletion promotes potent rejection of B16 melanoma. *Clin.Cancer Res.,* 2008, 14, 3156-3167.

[64] Taylor, PA; Lees, CJ; Blazar, BR. The infusion of ex vivo activated and expanded CD4(+)CD25(+) immune regulatory cells inhibits graft-versus-host disease lethality. *Blood.,* 2002, 99, 3493-3499.

[65] Nguyen, VH; Shashidhar, S; Chang, DS; et al. The impact of regulatory T cells on T-cell immunity following hematopoietic cell transplantation. *Blood.,* 2008, 111, 945-953.

[66] Chen, X; Vodanovic-Jankovic, S; Johnson, B; et al. Absence of regulatory T-cell control of TH1 and TH17 cells is responsible for the autoimmune-mediated pathology in chronic graft-versus-host disease. *Blood.,* 2007, 110, 3804-3813.

[67] Ermann, J; Hoffmann, P; Edinger, M; et al. Only the CD62L+ subpopulation of CD4+CD25+ regulatory T cells protects from lethal acute GVHD. *Blood.,* 2005, 105, 2220-2226.

[68] Jones, SC; Murphy, GF; Korngold, R. Post-hematopoietic cell transplantation control of graft-versus-host disease by donor CD425 T cells to allow an effective graft-versus-leukemia response. *Biol.Blood Marrow Transplant.,* 2003, 9, 243-256.

[69] Beyersdorf, N; Ding, X; Hunig, T; Kerkau, T. Superagonistic CD28 stimulation of allogeneic T cells protects from acute graft-versus-host disease. *Blood.,* 2009, 114, 4575-4582.

[70] Joffre, O; Gorsse, N; Romagnoli ,P; Hudrisier, D; van Meerwijk, JP. Induction of antigen-specific tolerance to bone marrow allografts with CD4+CD25+ T lymphocytes. *Blood.,* 2004, 103, 4216-4221.

[71] Trenado, A; Charlotte, F; Fisson, S; et al. Recipient-type specific CD4+CD25+ regulatory T cells favor immune reconstitution and control graft-versus-host disease while maintaining graft-versus-leukemia. *J.Clin.Invest.,* 2003, 112, 1688-1696.

[72] Cao, T; Soto, A; Zhou, W; et al. Ex vivo expanded human CD4+CD25+Foxp3+ regulatory T cells prevent lethal xenogenic graft versus host disease (GVHD). Cell *Immunol.* 2009, 258, 65-71.

[73] Trenado, A; Sudres, M; Tang, Q; et al. Ex vivo-expanded CD4+CD25+ immunoregulatory T cells prevent graft-versus-host-disease by inhibiting activation/differentiation of pathogenic T cells. *J.Immunol.,* 2006, 176, 1266-1273.

[74] Taylor, PA; Panoskaltsis-Mortari, A; Swedin, JM; et al. L-Selectin(hi) but not the L-selectin(lo) CD4+25+ T-regulatory cells are potent inhibitors of GVHD and BM graft rejection. *Blood.,* 2004, 104, 3804-3812.

[75] Karakhanova, S; Munder, M; Schneider, M; et al. Highly efficient expansion of human CD4+CD25+ regulatory T cells for cellular immunotherapy in patients with graft-versus-host disease. *J.Immunother.*, 2006, 29, 336-349.

[76] Cao, J; Chen, C; Zeng, L; et al. Engineered regulatory T cells prevent graft-versus-host disease while sparing the graft-versus-leukemia effect after bone marrow transplantation. *Leuk.Res.*, 2009.

[77] Godfrey, WR; Ge, YG; Spoden, DJ; et al. In vitro-expanded human CD4(+)CD25(+) T-regulatory cells can markedly inhibit allogeneic dendritic cell-stimulated MLR cultures. *Blood.,* 2004, 104, 453-461.

[78] Tang, Q; Henriksen, KJ; Bi, M; et al. In vitro-expanded antigen-specific regulatory T cells suppress autoimmune diabetes. *J.Exp.Med.,* 2004, 199, 1455-1465.

[79] Stanzani, M; Martins, SL; Saliba, RM; et al. CD25 expression on donor CD4+ or CD8+ T cells is associated with an increased risk for graft-versus-host disease after HLA-identical stem cell transplantation in humans. *Blood.,* 2004, 103, 1140-1146.

[80] Clark, FJ; Gregg, R; Piper, K; et al. Chronic graft-versus-host disease is associated with increased numbers of peripheral blood CD4+CD25high regulatory T cells. *Blood.,* 2004, 103, 2410-2416.

[81] Mutis, T; Aarts-Riemens, T; Verdonck, LF. The association of CD25 expression on donor CD8+ and CD4+ T cells with graft-versus-host disease after donor lymphocyte infusions. *Haematologica.,* 2005, 90, 1389-1395.

[82] Miura, Y; Thoburn, CJ; Bright, EC; et al. Association of Foxp3 regulatory gene expression with graft-versus-host disease. *Blood.,* 2004, 104, 2187-2193.

[83] Zorn, E; Kim, HT; Lee, SJ; et al. Reduced frequency of FOXP3+ CD4+CD25+ regulatory T cells in patients with chronic graft-versus-host disease. *Blood.,* 2005, 106, 2903-2911.

[84] Vela-Ojeda, J; Montiel-Cervantes, L; Granados-Lara, P; et al. Role of CD4+CD25+(high)Foxp3+CD62L+ Regulatory T Cells and Invariant NKT Cells in Human Allogeneic Hematopoietic Stem Cell Transplantation. *Stem Cells Dev.*, 2010.

[85] Rezvani, K; Mielke, S; Ahmadzadeh, M; et al. High donor FOXP3-positive regulatory T-cell (Treg) content is associated with a low risk of GVHD following HLA-matched allogeneic SCT. *Blood.,* 2006, 108, 1291-1297.

[86] Mielke, S; Rezvani, K; Savani, BN; et al. Reconstitution of FOXP3+ regulatory T cells (Tregs) after CD25-depleted allotransplantation in elderly patients and association with acute graft-versus-host disease. *Blood.,* 2007, 110, 1689-1697.

[87] Matthews, K; Lim, Z; Afzali, B; et al. Imbalance of effector and regulatory CD4 T cells is associated with graft-versus-host disease after hematopoietic stem cell transplantation using a reduced intensity conditioning regimen and alemtuzumab. *Haematologica.,* 2009, 94, 956-966.

[88] Meignin, V; Peffault de, LR; Zuber, J; et al. Numbers of Foxp3-expressing CD4+CD25high T cells do not correlate with the establishment of long-term tolerance after allogeneic stem cell transplantation. *Exp.Hematol.*, 2005, 33, 894-900.

[89] Rieger, K; Loddenkemper, C; Maul, J; et al. Mucosal FOXP3+ regulatory T cells are numerically deficient in acute and chronic GvHD. *Blood.,* 2006, 107, 1717-1723.

[90] Fondi, C; Nozzoli, C; Benemei, S; et al. Increase in FOXP3+ regulatory T cells in GVHD skin biopsies is associated with lower disease severity and treatment response. *Biol.Blood Marrow Transplant.,* 2009, 15, 938-947.

[91] Magenau, JM; Qin, X; Tawara, I; et al. Frequency of CD4(+)CD25(hi)FOXP3(+) Regulatory T Cells has Diagnostic and Prognostic Value as a Biomarker for Acute Graft-Versus-Host-Disease. *Biol.Blood Marrow Transplant.,* 2010

[92] Nadal, E; Garin, M; Kaeda, J; et al. Increased frequencies of CD4(+)CD25(high) T(regs) correlate with disease relapse after allogeneic stem cell transplantation for chronic myeloid leukemia. *Leukemia*, 2007, 21, 472-479.

[93] Zhang, H; Chua, KS; Guimond, M; et al. Lymphopenia and interleukin-2 therapy alter homeostasis of CD4+CD25+ regulatory *T cells. Nat.Med.*, 2005, 11, 1238-1243.

[94] Wuest, TY; Willette-Brown, J; Durum, SK; Hurwitz, AA. The influence of IL-2 family cytokines on activation and function of naturally occurring regulatory T cells. *J.Leukoc.Biol.*, 2008, 84, 973-980.

[95] Thornton, AM; Donovan, EE; Piccirillo, CA; Shevach, EM. Cutting edge: IL-2 is critically required for the in vitro activation of CD4+CD25+ T cell suppressor function. *J.Immunol.,* 2004, 172, 6519-6523.

[96] Zorn, E; Mohseni, M; Kim, H; et al. Combined CD4+ donor lymphocyte infusion and low-dose recombinant IL-2 expand FOXP3+ regulatory T cells following allogeneic hematopoietic stem cell transplantation. *Biol.Blood Marrow Transplant.,* 2009, 15, 382-388.

[97] Broady, R; Yu, J; Levings, MK. ATG-induced expression of FOXP3 in human CD4(+) T cells in vitro is associated with T-cell activation and not the induction of FOXP3(+) T regulatory cells. *Blood.,* 2009, 114, 5003-5006.

[98] Ghez, D; Rubio, MT; Maillard, N; et al. Rapamycin for refractory acute graft-versus-host disease. *Transplantation*, 2009, 88, 1081-1087.

[99] Coenen, JJ; Koenen, HJ; van, RE; et al. Rapamycin; not cyclosporine; permits thymic generation and peripheral preservation of CD4+ CD25+ FoxP3+ T cells. *Bone Marrow Transplant.,* 2007, 39, 537-545.

[100] Palmer, JM; Chen, BJ; Deoliveira, D; Le, ND; Chao, NJ. Novel mechanism of rapamycin in GVHD: increase in interstitial regulatory T cells. *Bone Marrow Transplant.,* 2010, 45, 379-384.

[101] Lamioni, A; Parisi, F; Isacchi, G; et al. The immunological effects of extracorporeal photopheresis unraveled: induction of tolerogenic dendritic cells in vitro and regulatory T cells in vivo. *Transplantation*, 2005, 79, 846-850.

[102] Maeda, A; Schwarz, A; Kernebeck, K; et al. Intravenous infusion of syngeneic apoptotic cells by photopheresis induces antigen-specific regulatory T cells. *J.Immunol.,* 2005, 174, 5968-5976.

[103] Rao, V; Saunes, M; Jorstad, S; Moen, T. Cutaneous T cell lymphoma and graft-versus-host disease: a comparison of in vivo effects of extracorporeal photochemotherapy on Foxp3+ regulatory T cells. *Clin.Immunol.*, 2009, 133, 303-313.

[104] Schmitt, S; Johnson, TS; Karakhanova, S; et al. Extracorporeal photophoresis augments function of CD4+CD25+FoxP3+ regulatory T cells by triggering adenosine production. *Transplantation*, 2009, 88, 411-416.

[105] Di, BI; Di, ML; Bugarin, C; et al. Regulatory T cells and extracorporeal photochemotherapy: correlation with clinical response and decreased frequency of proinflammatory T cells. *Transplantation*, 2009, 87, 1422-1425.

[106] Carvalho, A; Cunha, C; Di, IM; et al. Prognostic significance of genetic variants in the IL-23/Th17 pathway for the outcome of T cell-depleted allogeneic stem cell transplantation. *Bone Marrow Transplant.,* 2010

[107] Dander, E; Balduzzi, A; Zappa, G; et al. Interleukin-17-producing T-helper cells as new potential player mediating graft-versus-host disease in patients undergoing allogeneic stem-cell transplantation. *Transplantation.,* 2009, 88, 1261-1272.

[108] Dazzi, F; Marelli-Berg, FM. Mesenchymal stem cells for graft-versus-host disease: close encounters with T cells. *Eur.J.Immunol.,* 2008, 38, 1479-1482.

[109] Wang, J; Huizinga, TW; Toes, RE. De novo generation and enhanced suppression of human CD4+CD25+ regulatory T cells by retinoic acid. *J.Immunol.,* 2009, 183, 4119-4126.

[110] Chen, W; Jin, W; Hardegen, N; et al. Conversion of peripheral CD4+CD25- naive T cells to CD4+CD25+ regulatory T cells by TGF-beta induction of transcription factor Foxp3. *J.Exp.Med.,* 2003, 198, 1875-1886.

[111] Park, HB; Paik, DJ; Jang, E; Hong, S; Youn, J. Acquisition of anergic and suppressive activities in transforming growth factor-beta-costimulated CD4+. *Int.Immunol.*, 2004, 16, 1203-1213.

[112] Hoffmann, P; Eder, R; Kunz-Schughart, LA; Andreesen, R; Edinger, M. Large-scale in vitro expansion of polyclonal human CD4(+)CD25high regulatory T cells. *Blood.,* 2004, 104, 895-903.

[113] Taylor, PA; Friedman, TM; Korngold, R; Noelle, RJ; Blazar, BR. Tolerance induction of alloreactive T cells via ex vivo blockade of the CD40, CD40L costimulatory

pathway results in the generation of a potent immune regulatory cell. *Blood.,* 2002, 99, 4601-4609.

[114] Kitazawa, Y; Fujino, M; Li, XK; et al. Superagonist CD28 antibody preferentially expanded Foxp3-expressing nTreg cells and prevented graft-versus-host diseases. *Cell Transplant.*, 2009, 18, 627-637.

[115] Suntharalingam, G; Perry, MR; Ward, S; et al. Cytokine storm in a phase 1 trial of the anti-CD28 monoclonal antibody TGN1412. *N.Engl.J.Med.*, 2006, 355, 1018-1028.

[116] Yamazaki, S; Patel, M; Harper, A; et al. Effective expansion of alloantigen-specific Foxp3+ CD25+ CD4+ regulatory T cells by dendritic cells during the mixed leukocyte reaction. *Proc.Natl.Acad.Sci.*, U.S.A 2006, 103, 2758-2763.

[117] Chen, LC; Delgado, JC; Jensen, PE; Chen, X. Direct expansion of human allospecific FoxP3+CD4+ regulatory T cells with allogeneic B cells for therapeutic application. *J.Immunol.,* 2009, 183, 4094-4102.

[118] Tu, W; Lau, YL; Zheng, J; et al. Efficient generation of human alloantigen-specific CD4+ regulatory T cells from naive precursors by CD40-activated B cells. *Blood.,* 2008, 112, 2554-2562.

[119] Zheng, J; Liu, Y; Lau, YL; Tu, W. CD40-activated B cells are more potent than immature dendritic cells to induce and expand CD4(+) regulatory T cells. *Cell Mol.Immunol.*, 2010, 7, 44-50.

[120] Trzonkowski, P; Bieniaszewska, M; Juscinska, J; et al. First-in-man clinical results of the treatment of patients with graft versus host disease with human ex vivo expanded CD4+CD25+C. *Clin.Immunol.*, 2009, 133, 22-26.

[121] Riley, JL; June, CH; Blazar, BR. Human T regulatory cell therapy: take a billion or so and call me in the morning. *Immunity.*, 2009, 30, 656-665.

[122] Claudio Brunstein, et al. Adoptive Transfer of Umbilical Cord *Blood.,* (UCB)-Derived Regulatory T Cells (Tregs) to Recipients of Nonmyeloablative Unrelated Double UCB *Transplantation.,. Blood.,* 2009, 114 Abstract 513,

[123] Mauro Di Ianni and Mauro Di Ianni*; 1 Franca Falzetti1 Alessandra Carotti* 1 Adelmo Terenzi* 1 Elisabetta Bonifacio* 1 Yair Reisner2 Andrea Velardi1 Franco Aversa* 1 and Massimo F Martelli* 1. Adoptive Immunotherapy with Tregs Prevents GvHD and Favours Immune Reconstitution After HLA Haploidentical Transplants for Hematological Malignancies. *Blood.,* . 2009.

[124] Rezvani, K; Barrett, AJ. Characterizing and optimizing immune responses to leukaemia antigens after allogeneic stem cell *Transplantation.,. Best.Pract.Res.Clin.Haematol.,* 2008, 21, 437-453.

[125] Wolf, D; Wolf, AM; Fong, D; et al. Regulatory T-cells in the graft and the risk of acute graft-versus-host disease after allogeneic stem cell transplantation. *Transplantation.,* 2007, 83, 1107-1113.

In: Regulatory T Cells
Editor: Ren S. Hayashi, pp. 137-152

ISBN: 978-1-61761-554-2

Chapter 7

Regulatory T Cells and Hepatocellular Carcinoma: Implication of the Regulatory T Lymphocytes in the Control of the Immune Response

Nadira Delhem**, *Céline Miroux** and Thibaut Vausselin
Laboratoire d'Immunopathologie des Cancers Viro-Induits,
Institut de Biologie de Lille, Lille, France

Abstract

Hepatocellular Carcinoma (HCC) is the fifth most common cancer worldwide and also the third most common cause of cancer-related death. HCC arises most frequently in males with cirrhosis, which is most often a consequence of chronic hepatitis infection (HBV and HCV) or alcohol abuse. To date, the only effective approaches for patients with HCC are resection or liver transplantation.

Immunological mechanisms are important in the surveillance of malignancy and control of tumor progression. Tumor-Infiltrating Lymphocytes (TILs) have been described in HCC, and extensive infiltration has been associated with reduced tumor recurrence following resection. However continued tumor-growth, despite the presence of a lymphocytic infiltration, including tumor-specific T cells within and surrounding tumors, suggests a failure of immune control. Although, many mechanisms have been proposed for this attenuated immune response, it becomes evident that direct suppression of effector cells, supported by regulatory T cells could play a pivotal role in the suppression of immune response to tumors. Initially described in context of immune disorders such as inflammatory autoimmune pathologies, regulatory T lymphocytes are characterized by their capacity to inhibit T helper response. To date, several regulatory T

* Corresponding author: Laboratoire d'Immunorégulation des Cancers Viro-induits :, Institut de Biologie de Lille, UMR 8161, Lille, France, Tel : 0033 320871253, Fax : 0033 320871019, E-mails : Nadira.Delhem@ibl.fr

cells are described, however CD4+CD25+ regulatory T cells and Tr1 sub-populations remain best characterized.

Currently, there is no evidence for direct implication of CD4+CD25+ regulatory T cells in the malignancy and control of HCC progress. However, recent studies showed that both regulatory T cells subpopulations and particularly Tr1 have been implicated in the modulation of the immune response during HCV chronic infection, supporting HCC progress.

Keywords: Regulatory T Cells, Hepatocellular Carcinoma, HBV, HCV, Transplantation.

1. Introduction

Hepatocellular Carcinoma (HCC) is the fifth common cancer world wild and is the third cancer mortality cause, results in an estimated 564 000 new cases annually [1]. Prognosis is usually poor and survival is limited in the majority of patients. The HCC occurs frequently in males who progress to cirrhosis, must often following a chronic viral infection (HCV, HBV) or alcoholic disease. Recent reports in different states suggest that HCC incidence increased, probably because of Hepatitis C Virus (HCV) infection, in spite of a significant increase in alcohol use.

Currently, the survival alternative for patients with end-stage liver disease like HCC is resection or Orthotopic Liver Transplantation (OLTx). However, 83% of patients are confronted to liver pathology recurrence within the five years after OLTx (tumor <5cm or at least 3 tumors<3cm diameter). Other treatment are currently used, like percutaneous alcohol injection, thermic ablation and intra-arterial chemo-embolization, but their efficiency is only observed on localized tumors [2] and less than 10% of the patients with a moderated disease (okuda stade 2) live through 3 years [3].

Immunological mechanisms are important for malignancy surveillance and tumor progression control. Cytotoxic T Leukocytes CD8+ (CTLs) and NK cells (Natural Killer) are potential effector cells in the control of tumor growth, however the activation of a T helper CD4+ immune response is required for the optimal activity of CD8+ and NK cells. The Tumor Infiltrating Leukocytes (TILs) have been described in the HCC, and a large infiltration is associated to a less severe recurrence after resection [4].

The continuous growth of the tumor, despite of the presence of tumor specific T cells both inside and around the tumor, suggests a breakdown of the immunological control. Many mechanisms has been proposed in order to understand this failure of anti-tumor immunological response ; such as partial masking of tumor antigen, the defective antigen presentation by immune cells, a low costimulation, and a direct suppression of the anti-tumor effectors cells [5]. It appears that regulatory T cells could play a role in the suppression of the anti-tumoral immune response.

Initially identified in immunological disorder, like some inflammatory autoimmune pathologies, regulatory T lymphocytes are characterised by their capacity to inhibit the T helper response. These cells have a cytokine profile secretion distinct from those of Th1 and Th2 cells. Even though regulatory T cells are considered to play a role in the self tolerance, recent studies suggest that these cells could be induced *in vivo* against bacterial, viral or parasitic antigens and could also favour the pathogen persistence by inhibiting the protective

Th1 response. [6]. Although many regulatory T cells sub-population have been described, the T CD4+CD25+ and Tr1 cells remain the best characterised.

CD4+CD25+ Regulatory T Cells

CD4+CD25+ regulatory T cells are characterised by the concomitant expression of surface CD4, CD25 (IL2-Rα) markers [7] and the absence of CD127 (IL7-Rα) expression [8]. These cells are induced in the thymus and constitute the central tolerance towards self antigens expressed in the thymus. Some molecules have been suggested as regulatory T cells markers: GITR, CTLA-4 (CD152), OX40 (CD134), ICAM 1 (CD54), and P-Sélectin (CD62P), CD62L, CD103, CD27, CD39, CD70, ... The transcription factor FoxP3 [9] and secondarily the LAG-3 (CD223) molecule, have been identified as strategic markers of regulatory T cells, in murin and human models. More recently, the absence of CD49d (α-chain of the integrin VLA-4 (α4β1)) and CD127 defines functional FoxP3+ Treg cells free of contaminating effector cells [10].

Il-2 is essential for regulatory T cells generation in the thymus, and also for their survival, expansion and suppressive function in the periphery [11]. Regulatory T cells are less responsive to TCR than effector T cells, but exogene IL-2, a strong costimulation with anti-CD28 or a mitogenic stimulation go over the TCR and allow the regulatory T cells to overcome their anergic state [12]. The activation of regulatory T cell is necessary to suppress CD4+ and CD8+ effector T cell proliferation and cytokine secretion, notably by inhibiting IL-2 and IL-2Rα transcription [13, 14, 15]. Regulatory T cells also inhibit proliferation and cytokine secretion of B lymphocytes [16]. Their immunosuppressive mechanism was described dependent of the cellular contact *in vitro*, but could depend of immunosuppressive cytokine secretion *in vivo*, like IL-10 and particularly TGFβ. T CD4+CD25+ are able to anergized effectors T cells by inhibiting IL-2 and IL-2Rα expression by the target cell. The suppressive function induction in regulatory T cells is specific of the antigen and these cells seem to have a stronger avidity, for their specific antigen, in comparison to the CD4+CD25- cells (effectors T cells). The regulatory T cell activity is stimulated by a 10 to 1000 fold lower antigen concentration than those for conventional T lymphocytes [14]. In addition, T CD4+CD25+ depletion induce a more efficient anti-tumoral immune response, confirming the role of regulatory T cells in the control and the impairment of the anti-tumoral immune response [17, 18, 19].

Tr1

Tr1 cells are described by their capacity to produce large amount of IL-10 and TGFβ cytokines, after TCR stimulation, allowing them to inhibit the type 1 and 2 helper response. These cells display an immunosuppressive mechanism dependent on cytokines, mainly through IL-10 secretion. They have an antigenic specificity but their suppression is described « collateral » because they affect all the cells nearby, unlike regulatory T cells CD4+CD25+ who are more specific, thanks to their contact dependent mechanism. Tr1 cells are induced in the periphery by immature dendritic cells $CD11c^{low}$ $CD45RB^{high}$ and in the presence of autocrine IL-10 secretion [20] or from naïve precursor of CD4+CD25+. They directly migrate on the inflammatory site where they inhibit immune responses directed towards antigens which are not expressed in the thymus, and they contribute to the peripheral tolerance. They have a particular cytokine secretion, different of other immune cells, as they produce large amount of IL-10 and TGF-β, moderate secretion of IFN-γ and IL-5 and at a very low level a secretion of IL-2 and IL-4 [21]. Unlike effectors T cells and CD4+CD25+ regulatory T cells, their principal proliferative factor is the IL-15 cytokine, and not IL-2. The IL-10 cytokine play a major role in their differentiation, in association with IFNα. Moreover, Hervé Groux's team has showed that Tr1 cells could also inhibit pathogenic immune response in different inflammatory disease model, in particular intestinal and allergic diseases, or in organ transplantations [22]. Tr1 cells have a particular phenotype, like T1/ST2 expression which discriminate them to the Th1 cells, and CCR5 expression which differentiate them to the Th2 cells [23]. Groux *and al.* have more recently described that the concomitant expression of surface integrin α2 (CD49b) and β2 (CD18) on a CD4+ cell allows the specific characterisation of a Tr1 lymphocyte [24].

An increasing number of studies analyses the potential role of regulatory T cells in the control of anti-viral T specific immune responses, notably during acute and chronic viral infection. For example, it has been described that regulatory T cells alter the T CD8+ IFN-γ secretion during persistant retroviral infection, in a murin model. While, the proliferative capacity of viral specific T CD8+ lymphocytes is not altered. Recent studies have also suggested that regulatory T cells could suppress effector function of viral specific CD8+ T cells during chronic viral infection in humans, like HBV, HCV, HIV or cytomegalovirus infection [25, 26, 27].

2. Regulatory T Cells and HBV Infection

The Hepatitis B Virus (HBV) is a non cytopathic and hepatotrope DNA virus, which infect more than 300 million of people world wide. 10% of patients with HBV infection have a chronic infection. This viral infection causes between 1 and 2 million of death, particularly following liver cirrhosis or hepatocellular carcinoma (HCC), each year.

It has been described that CD4+ T helper and cytotoxic T CD8+ lymphocytes, specific of the virus, are important cells in the control of infection, whereas patients with chronic infection have an inefficient anti-viral immune response. The implication of T lymphocytes with immunosuppressive properties is one of the mechanisms which could explain this T tolerance.

In the 70 and 80 years, scientists begin to interest in a sub-population of T lymphocytes called T suppressive lymphocytes, which expressed surface OKT8 (CD8). The use of antibody against OKT3 (T cells), OKT4 (T helper cells) and OKT8 (cytotoxic / suppressive cells), have showed that the OKT4/OKT8 ratio was significantly lower in patients with chronic HBV infection than in healthy donors, with both a decrease of OKT4 cells frequency and an increase of suppressive OKT8 cells [28]. It has been described *in vitro,* in reactivity test to the HBV surface antigen, that the antigen specific proliferative response is better in the absence of OKT8+ cells. Barnaba *and al.* have thus demonstrated the existence of a regulatory T cells sub-population OKT8+ (CD8+) which is able to inhibit the anti-viral immune response. Though, other studies have showed that the suppressive OKT8+ cell activity was decreased in patients with chronic infection, suggesting the implication of other regulatory T cell sub-population [29]. This information remains unconfirmed, in particular due to the absence of specific isolation or characterisation techniques of the different sub-population.

While since 1993, none studies have been published, in 2003 and 2004, two studies provide new indication in the implication of regulatory T cells in Hepatitis B Virus infection. The first have been carried out on HBV transgenic mice, which express a HBV genome able to replicate and cumulate viral antigen in their endoplasmic reticulum, so these mice are tolerant for viral antigen [30]. They can also develop an HCC and thereby form a model of choice for the study of the human pathology [31]. These mice have a variable anti-viral immunity depending on their age. Thus, 5 months year's old mice develop an antiviral humoral and cellular response more efficient than mice aged of 2 months. The injection of T CD4+ cells from these tolerant mice, in immunocompetent mice, is sufficient to transfer this tolerance. This study brings out, for the first time, a proof of the implication of T CD4+ cells with regular properties in the Hepatitis B virus infection.

The second study [32] was interested in the relationship between healing and different haplotype of the CTLA-4 molecule. The Cytotoxic T Lymphocyte Antigen 4 (CTLA4) is an inhibitory receptor, expressed by activated T cells and also CD4+CD25+ regulatory T cells, where it might be crucial for their activity. The distribution of the various CTLA-4 haplotypes is different between patients with chronic infection or patients which have controlled the infection. While, wild haplotype is associated with the viral persistence, two haplotypes (Alleles +49G and -1722C), whose mutations could interfere with the CTLA-4 function, are associated with healing. This study suggest that the CTLA-4 affect the disease progress, this is consistent with the role of CD4+CD25+ regulatory T cells in the disease pathogenesis.

The study of regulatory T cell implication in HBV infection has grown since 2005, thanks to the work of three teams [33, 34, 35], which have isolated the regulatory T cell population from peripheral blood of HBV infected patients. Their phenotypic characterisation shows that they are CD4+ CD25+/high [33, 34, 35], CD45RO+, CTLA4+ [33][35] and HLADR+ CD95+ [35]. This sub-population is highly enriched in FoxP3 expression, according Q-RT-PCR mRNA analysis [33][35].

The different studies on the proportion of regulatory T cells among the peripheral blood CD4+ T cells give contradictive results. The Stoop *et al.* study [33] reveal a significant increase of the regulatory T cells frequency in patients chronically infected (5,4%), in comparison to healthy control (3,23%) and to patients who have overcame the infection (2,96%). In contrast, Franseze *and al.* [34] show any significant difference between healthy donors (2.5%) and chronically infected patients, whatever the stage of the pathology (active chronic Hepatitis B: 2.8%; Asymptomatic: 1.9%; Immunotolerant: 2.4%). Lastly, Xu *et al.* [35] shows a significant increase of regulatory T cells frequency, only among patients with a severe chronic hepatitis B (6.6%), but not among chronic (3%) or acute patients (3,7%), in comparison to healthy control (2%). These various responses could be explained by the differences in the patients' selection or in the technique of isolation or characterisation of regulatory T cells used.

Functional analysis of regulatory T cells have been realised by the depletion of CD4+CD25+ cells. Results of three teams show that this depletion improve significantly the peripheral blood T cell proliferation, in particular the proliferation of cytotoxic CD8+ T lymphocytes specific of the HBV, and their IFNγ production, after stimulation with HBV peptides, such as capsid [33][35], envelop and DNA polymerase [34] peptide. The regulatory T cells sub-population lead to a drastic decrease in effectors T cell proliferation and their IFNγ secretion [33, 34, 35], after epitopic stimulation and with a dose dependent effect [33]. However, the data concerning regulatory T cells epitopic specificity are contradictive. While Stoop [33] and Xu [35] seem to show an inhibitory effect dependent of the HBV antigens (tetanus toxin effect [33] and anti-CD3 stimulation [34]), Franzese's study [34] show that stimulation of regulatory T cells with a control peptide (Melan A from human melanoma) gives the same results than with HBV peptides.

To date, any functional published assays could allow us to describe the mechanism of action of regulatory T cells. However, we know that HBV antigens from the capsid are able to induce IL-10 secretion by peripheral blood T cells of HBV chronically infected patients [36]. The use of neutralizing antibody could allow to confirm or to infirm the implication of immunoregulatory cytokines IL-10 and TGF-β in the regulatory T cell mechanism. Moreover, culture experiments in Boyden chamber might allow to conclude on the importance of a cellular contact in the regulation by these cells, through specific receptor (not yet identified, CTLA-4, GITR ?) or membranous TGFβ.

Correlation assays with clinical data reveal any association between regulatory T cells frequency and ALAT level in the three studies [33, 34, 35]. Moreover, Stoop [33] and Franzese's [34] works show any correlation between regulatory T cells and viral charge. The Xu study just shows a significant elevation of the regulatory T cell proportion among patients with a viral charge superior to 10^7 copy /ml [35]. During acute infection, the regulatory T cell level is initially low and its increases during the convalescence phases, and then returns to a normal level during the healing [34]. In patients who have resolved the infection, while regulatory T cells are present at a low frequency, they are since able to diminish the cytotoxic T CD8+ proliferation *in vitro* [34]. However Peng *et al.* [37] have demonstrated that in chronic hepatitis B patients, the frequency of CD4+CD25+ regulatory T cells positively correlated with serum viral load, and these cells were capable of suppressing the proliferation and IFN-gamma production of autologous peripheral blood mononuclear cells (PBMC) mediated by HBV antigen stimulation *in vitro*. Another study showed that the frequency of CD4+CD25+Foxp3+ T cells correlated positively with hepatitis B envelope (HBe) antigen status and serum HBV DNA copy numbers and had a converse relationship with HBe antibody status in patients with chronic active hepatitis B patients [38]. These results were confirmed by a study realised in 2010, where Nan *et al.* showed that circulating CD4+CD25+ cell frequency correlated positively with serum HBV DNA load in chronic hepatitis B patients [39].

More recently, a study on 79 patients with chronic hepatitis B have showed that the frequency of Treg cells in peripheral blood was significantly higher than that in normal volunteers. Moreover, the higher level of serum transaminase was associated with higher frequency of Treg cells, which both had a linear correlation relationship. Furthermore, in patients with higher serum transaminase levels, the expansion of HBV-specific CD8+ T cells was higher after removal of Treg cells when compared with patients with lower serum transaminase levels. Based on this observation, it is proposed that liver-injury enhances Treg cell frequency/activity in chronic hepatitis B patients [40]. Moreover, studies have checked the evolution of the CD4+CD25+ regulatory T cells population during exacerbation episodes of chronic hepatitis B, induced by certain treatment like Lamivudin and Prednisolon. These exacerbations are generally clues of the anti-HBV immune response recovery and of the patient's healing. We observe through these phases, a weak decrease of the regulatory T cells percentage, at the same time of an increase of the ALAT level, but it's not seem to be in correlation with the CD8 immune response [34]. These findings provide new insights into the role of regulatory T cells in chronic activity and viral clearance in chronic hepatitis B.

These results are important but don't clarify the role of regulatory T cells on HCC progress. Recently, the Fu *and al.* [41] have enlighten this aspect by demonstrating that circulating CD4+CD25+FoxP3+ Treg frequency was increased significantly and correlated with disease progress in HCC patients. This accumulation of Treg concurrent with significantly reduced infiltration of CD8+ T cells was found in tumor regions compared with nontumor regions. Furthermore, Treg of HCC patients inhibited proliferation, activation, degranulation, and production of granzyme A, granzyme B, and perforin of CD8+ T cells. This increased regulatory T cells impair the effector function of CD8+ T cells and may thus promote disease progress. As, an increased quantity of circulating regulatory T cells was associated with high mortality and reduced survival time of HCC patients, these cells represent both a potential prognostic marker and a therapeutic target for HBV-related HCC individuals.

These results, while being important, are all be obtained in the periphery and so not let know the mechanism which take place in the liver. To date, only two studies were realised on intra-hepatic samples. The Xu and Wang team [35] have thus showed an important infiltration of CD4+ and CD8+ cells in the liver of patients with a severe chronic hepatitis B, in comparison with healthy control or stable chronic hepatitis patients. These liver infiltrating lymphocytes are highly enriched in CD4+CD25+FoxP3+ regulatory T cells, tend to show that regulatory T cells are associated with the chronicity and the severity of the intra-hepatic inflammation, which lead to hepatic lesions apparition and HCC. And the Stoop work [42] has shown that intrahepatic Treg are phenotypically distinct from peripheral blood Treg. Indeed, they showed that the liver contained a population of CD4+FoxP3+ cells that did not express CD25 and lower express HLA-DR and CTLA-4, while these cells were absent from peripheral blood. Moreover, patients with a high viral load have a higher proportion of regulatory T cells in the liver, but not in blood, compared to patients with a low viral load. This study suggests that the higher proportion of intrahepatic Treg observed in patients with a high viral load may explain the lack of control of viral replication.

To date, it is important to outline that, any article have anew studied the suppressive T CD8+ sub-population in HBV infection, in contrast of, for example, the HCV infection [43]. The immunosuppression observed among HBV infected patients must be modulated by some different mechanism, like the implication of different regulatory T cells sub-population, implicating them crucially in the development oh the HBV induced HCC.

3. Regulatory T Cells and HCV Infection

Hepatitis C Virus (HCV) is a major cause of acute hepatitis and of chronic hepatic disease, particularly cirrhosis and liver carcinoma [44]. HCV infection is a real public health problem. It is considered that about 170 millions of people chronically carry the HCV world wide, with 3 to 4 million of new cases each year [45]. In France, the prevalence of infection is important : 500 000 to 650 000 people chronically infected, representing 1,3 % of the population. Hepatitis C is a viral disease, with a parenteral transmission, which was described « non A, non B » until the etiologic agent was discovered in 1989. The HCV is a parenterally transmitted virus by direct contact with human blood.

The HCV virus is a RNA virus, which belong to the Flaviviridae family and the only one to belong to the hepacivirus type. Contrarily to the HBV, the HCV genome does not insert into cellular DNA, its replication is thus necessarily for its persistence. Viral tropism is not limited to hepatocytes, until HCV RNA sequences have been detected, by PCR and *in situ* hybridization, in peripheral blood mononuclear cells [46]

The current treatment is based on pegylated IFNα coupled to the Ribavirin. The efficacity remain low (less than 50% of cases) and very variable. Indeed, analysis of the HCV genetic variability impact on the IFNα-Ribavirin response showed that the treatment response is weaker among genotype 1 infected patients (31%), than in the patients infected with the 2 and 3 types (64%) [47]. Studies have also shown than genotype 1b infection could progress more severely and promote cirrhosis or HCC evolution [48]. This genotype, the most virulent and less responder to the treatment, is unfortunately the most frequent in France.

It is admitted that the HCV infection is characterised by its persistence in spite of an important immune response, both humoral and cellular. Viral eradication is generally associated to a multi-specific and persistent cellular anti-viral response of CD4+ and CD8+ T lymphocytes [44]. CD4+ lymphocytes play a central role in the early control of viral replication. It seems that the effector T cell population, IL-7Ralpha (high) CD25+ CD45RO+ CD4+ FoxP3-, is expanded after liver transplantation of viral or non-viral origin, but that a preexisting HCV infection negatively influences the expansion of this population in liver transplant recipients [49]. In fact, accumulated data suggest that HCV proteins, notably the Core protein, suppress host immune responses through the suppression of functions of immune cells, such as cytotoxic T lymphocytes, natural killer cells, and dendritic cells [50]. A heavy activity of T helper 1 (Th1) lymphocytes; against Core, NS3, NS4 and NS5 proteins; have notably been associated to the spontaneous healing process [51]. On the contrary, a weak or no persistent CD4 amplitude response is rather associated to a poor prognosis of the HCV infection evolution. This immune response is not detectable among chronically infected subjects which have initially controlled the infection but have a reappearance of the viremia [51].

Besides, studies have clearly demonstrated that during the chronic phases of the infection, HCV specific T CD8+ lymphocytes have a functional defect, resulting in an alteration of their *in vitro* proliferation and of their IFNγ production [52]. This altered immune response might spark off the viral tolerance and persistence and possibly its evolution to hepatocellular carcinoma. Although the mechanism leading to HCV specific T cell alteration among chronically infected patients are poorly understood, recent studies have suggested a major contribution of CD4+ and CD8+ regulatory T cells [25] [43] [53] [54].

Role of the Regulatory T Cells in the Chronic Infection

CD4+CD25+ regulatory T cells seem to play a role in the chronic infection by the hepatitis C virus, notably by inhibiting the cellular response of CD4+ T lymphocytes [55].

CD4+CD25+ T cells represent 13,5% (+/-1,8) of the peripheral CD4+ T cells in the blood of HCV chronically infected patients, which is significantly higher than among healthy control (5,3% +/-0,8). These cells have thus an important capacity to suppress the Th1 response against HCV [56]. The persistent HCV infection could be explained by the alteration of the Treg ratio in peripheral blood mononuclear cells (PBMCs) from chronic hepatitis C patients during IFN-α/Ribavirin combination therapy compared with the Treg ratio in liver-infiltrating lymphocytes (LILs) before therapy. The evaluation of Treg is a potentially significant factor for persistent HCV infection, in LILs prior to treatment and in PBMCs during treatment, and could predict the result of combination therapy [57].

Peripheral CD4+CD25+ T lymphocytes of HCV chronically infected patients with normal or abnormal ALAT (elevated), have been analysed for their specificity and their effect on HCV specific CD4+ T cell reactivity [55-58]. Thereby, it has been shown that HCV specific CD4+CD25+/high T cells produce important amount of the growth factor TGF-β but only limited quantity of IL-10 and neither IL-2, nor IFN-γ. Moreover, the production of TGF-β by the HCV specific regulatory T cells is sensitively higher among patients with normal ALAT in comparison to patients with elevated ALAT [55]. In addition, an inverse significant correlation has been found between TGF-β secreted by the HCV specific regulatory T cells and liver inflammation. This study also shows that, among PBMC, HCV specific CD4+ T lymphocytes proliferation are more important in patients with elevated ALAT comparatively to patients with normal ALAT [58]. The CD4+CD25+ depletion in PBMC fraction results in an increase of the HCV specific CD4+ T cells in patients with normal ALAT [55-58]. Thus, CD4 + CD25+ cells in these patients seem to be more efficient in the suppression of the CD4+ T cell activity [55-58]. These data argue the hypothesis that CD4+CD25+ regulatory T cells could play a role by controlling the chronic inflammatory response and hepatic lesions in HCV chronically infected patients [55-58].

In the laboratory, we have realised real time quantitative PCR to analysed the intra-hepatic expression of genes implicated in the immunosuppression, among hepatic biopsy of HCV-1b chronically infected patients divided in 3 groups (healthy liver, cirrhotic liver and HCC). We have outlined a significant increase of CD4+CD25+ and Tr1 regulatory T cells markers, proportionally hepatic fibrosis, among patients with cirrhosis or HCC in comparison to patients without hepatic lesions. We have also observed an intra-hepatic recruitment of regulatory T cells specific of the HCV, because any regulatory T cells have been found in hepatic biopsy of patients with an alcoholic liver carcinoma [59-60]. These results have been recently confirmed by Wang *et al.* , that have shown a significant increase of the Treg number in chronic hepatitis C patients and this increased Treg was associated with HCV genotype 1b [61]. It has also been shown that large numbers of highly activated and differentiated CD4+FoxP3+ Treg localize to the infiltrated chronic HCV-infected liver and may result in limiting the extent of fibrosis. This suggests that CD4+FoxP3+ Treg play a pivotal role in limiting collateral damage by suppressing excessive HCV-induced immune activation [62]. Yoshizawa *et al.* have shown an increase in Treg cells in HCV-related chronic liver disease, particularly in HCC, and these cells were shown to be memory-type Treg cells [63]. Moreover, some data indicate that patients with acute HCV infection, who develop chronicity versus spontaneous resolution exhibit temporal changes in Treg cell function. It is possible that repetitive viral antigenic stimulation alters the function of Treg cells over time [64].

However, it was not described to date, a direct implication of CD4+CD25+ cells in the malignity and the control of the liver carcinoma progress.

Regulatory T Cells in Viral Recurrence after Liver Transplantation

According to OMS statistics, number of cirrhosis and HCC associated to viral infection is expected to double in the next 20 years. The only alternative at the final stage of chronic hepatitis, HCC, is the orthotopic liver transplantation. However, the hepatitis C virus persists among almost all transplanted patients for HCV disease. Thus, graft reinfection by HCV is considered to be systematic and the disease recurrence is accelerated, with a reappearance of cirrhosis in 5 years after transplantation. The factors associated with accelerated disease recurrence; including high viral load pre- (>10(6) IU / mL) and / or early post-transplantation (>10(7) IU / mL at 4 months), donor older than 40-50 years and donor way of life, prolonged ischaemic time, cytomegalovirus coinfection, over immunosuppression and / or abrupt changes in immunosuppression, HIV coinfection, infection by genotype 1b, anti-rejection treatment; give a general survey of the liver damage mechanism and could give indication to prevent or minimize these damages [65-66]. The impact of the viral persistence on long term survival results after orthotopic liver transplantation (OLTx), is variable and the necessity of a re-transplantation is currently debated. To 1993 to 2003, 131 OLTx with a verified histologic HCV recurrence have been retrospectively analysed. Global survival at 1 and 5 years after transplantation are respectively 90,7 and 81,3%. Aged donors (> 70 years) have a higher risk of early HCV reemergence and their life expectancy is reduced if the histological recurrence is observed in the 6 months [67]. 5 years after liver transplantation, patients and graft survival rates are lower among HCV transplanted patients than in patients transplanted for another chronic liver disease [68]. Unfortunately, the viral infection remains persistent in peripheral compartment among all transplanted patients, and is responsible of the systematic reinfection of the graft. This reinfection is characterised by an accelerated fibrosis progress and HCC development in the 5 years after the OLTx, among more than 80% of the transplanted patients. Feray *et al* have shown that the incidence and severity of the graft reinfection are higher among genotype 1b infected patients [69]. Moreover, the fibrosis rate increases proportionally to donor age and creatinin rate of transplanted patient (linked to its renal function) [69]. Immunologic mechanisms responsible for severe fibrosis reappearance after liver transplantation are little-known.

In this context, the regulatory T cell involvement during HCV primo infection and particularly during the tumoral phasis could be in correlation with severe viral recurrence acceleration after liver transplantation. In order to validate this hypothesis, we have realised intra-hepatic (biopsy) and peripheral (sera) analysis of the immunological status, 1 and 5 years after transplantation. 23 genes associated to regulatory T cells sub-populations have been analysed in 3 groups of transplanted patients : stable without viral infection (n=8), patients with minor recurrence (METAVIR<A1F2, n=10) and patients with severe recurrence (METAVIR>A1F2, n=10). The CD4+CD25+ associated markers expression, particularly the CD4, CD25, OX40, GITR and ICAM1, are significantly increased in the liver with hepatic recurrence than in the healthy liver ($p<0.05$). The markers co-expressed by Tr1 cells (CD49b, CD18) are also over-expressed among patients with a severe recurrence in comparison to patients with minor recurrence or stable patients ($p<0.05$) [70]. Moreover, significant correlations were observed between : liver versus serum; liver versus PBMCs; and serum versus PBMCs (r= 0.37, 0.54 and 0.49, respectively). We have concluded that in patients infected with HCV, serum can be used to evaluate expression of liver genes, and thus this alternative, non-invasive technique is sufficient to monitor the progress of liver disease during chronic hepatitis C [60]. Although many factors have been associated to the severity of the hepatic recurrence; as the viremia rate before transplantation, the donor way of life and its age, the cytomegalovirus or HIV infection, the immunosuppressive treatment; the role of regulatory T cells had not been described to date.

All of these results suggest for the first time that regulatory T cells could be implicated in the hepatitis C virus recurrence after liver transplantation. Identification of the regulatory T cells sub-population able to modulate hepatic lesions, like Tr1 which are enhanced during severe HCV recurrence after liver transplantation, could give significant advances in order to develop new therapeutic strategy and could be predictive of HCV recurrence. High levels of IL-10 at 1 year could also be predictive of severe recurrence. It has also been shown that patients developing HCV recurrence after liver transplantation have the same immunoregulatory network as patients with acute hepatitis C evolving to persistent infection likely suggesting that CD4+CD25+ numbers may be a marker to predict recurrence of HCV after liver transplantation [71].

Moreover, we have shown that cyclosporin A, immunosuppressive drug used after liver transplantation, is able to inhibit regulatory T cell activity [72]. This inhibition of regulatory T cells, important for the graft tolerance, could explain, in part, the graft rejection more frequently observed in the presence of cyclosporin A than tacrolimus or rapamycin [73, 74]. Besides, it was known that there is less recurrence in the presence of cyclosporin A than in the presence of other immunosuppressive drugs [73, 74], suggesting the implication of regulatory T cells in the HCV infection recurrence.

These results are in favour of a specific implication of different regulatory T cells implication in the viral pathology evolution to hepatocellular carcinoma. The presence of regulatory T cells in the liver of HCV chronically infected patients and the increase of their frequency during cirrhosis or HCC, suggest a central role of these cells in the hepatic pathology aggravation. Regulatory T cells could, in fact, be implicated in the chronicity status of the pathology.

Acknowledgments

Results from our laboratory and Céline Miroux received support from the “Agence Nationale de Recherche sur le Sida et les Hépatites” (ANRS) and the “Comité du Nord de la Ligue contre le Cancer »

References

[1] Parkin, DM; Bray, F; Ferlay, J; Pisani, P. Global cancer statistics. *Cancer J Clin,* 2002, 55, 74-108.

[2] Dick, EA; Taylor-Robinson, SD; Thomas, HC; Gedroyc, WM. Ablative therapy for liver tumours. *Gut,* 2002, 50, 733-9.

[3] Levy, I; Sherman, M; Liver Cancer Study Group of the University of Toronto. Staging of hepatocellular carcinoma: assessment of the CLIP; Okuda; and Child-Pugh staging systems in a cohort of 257 patients in Toronto. *Gut,* 2002, 50, 881-5.

[4] Wada, Y; Nakashima, O; Kutami, R; Yamamoto, O; Kojiro, M. Clinicopathological study on hepatocellular carcinoma with lymphocytic infiltration. *Hepatology,* 1998, 27, 407-414

[5] Chouaib, S; Asselin-Paturel, C; Mami-Chouaib, F; Caignard, A; Blay, JY. The host-tumor immune conflict: from immunosuppression to resistance and destruction. *Immunol Today,* 1997, 18, 493-7.

[6] Belkaid, Y; Rouse, BT. Natural regulatory, T. cells in infectious disease. *Nature Immunology,* 2005, 6, *353-360.*

[7] Sakaguchi, S; Sakaguchi, N; Asano, M; Itoh, M; Toda, M. Immunologic self-tolerance maintained by activated T cells expressing IL-2 receptor a-chains (CD25). *The Journal of Immunology,* 1995, 155, 1151-1164.

[8] Liu, W; Putnam, AL; Xu-Yu, Z; et al. CD127 expression inversely correlates with FoxP3 and suppressive function of human CD4+ T reg cells; *The Journal of Experimental Medicine,* 2006, 7, 1701.

[9] Hori, S; Nomura, T; Sakagushi, S. Control of regulatory T cells development by the transcription factor FoxP3. *Science,* 2003, 299, 1057-61.

[10] Kleinewietfeld, M; Starke, M; Di Mitri, D; Borsellino, G; Battistini, L; Rotzschke, O; Falk, K. CD49d provides access to "untouched" human FoxP3+ Treg free of contamining effector cells. *Blood,* 2009, 113(4), 827.

[11] Burchill, MA; Yang, J; Vang, KB; Farrar, MA. Interleukin-2 receptor signaling in regulatory T cell development and homeostasis. *Immunol Lett.,* 2007, 30, 114(1), 1-8.

[12] Takahata, Y; Nomura, A; Takada, H; Ohga, S; Furuno, K; Hikino, S; Nakayama, H; Sakaguchi, S; Hara, T. CD25+CD4+ T cells in human cord blood: an immunoregulatory subset with naive phenotype and specific expression of forkhead box p3 (Foxp3) gene. *Exp Hematol.,* 2004, 32(7), 622-9.

[13] Thornton, AM; Shevach, EM. J Exp Med. CD4+CD25+ immunoregulatory T cells suppress polyclonal T cell activation in vitro by inhibiting interleukin 2 production. *J Exp Med.*, 1998, 188(2), 287-96.

[14] Thornton, AM; Shevach, EM. Suppressor effector function of CD4+CD25+ immunoregulatory T cells is antigen nonspecific. *J Immunol.,* 2000, 164(1), 183-90.

[15] Piccirillo, CA; Shevach, EM. Cutting edge: control of CD8+ T cell activation by CD4+CD25+ immunoregulatory cells. *J Immunol.,* 2001, 167(3), 1137-40.

[16] Bystry, RS; Aluvihare, V; Welch, KA; Kallikourdis, M; Betz, AG. B cells and professional APCs recruit regulatory T cells via CCL4. *Nat Immunol.,* 2001, 2(12), 1126-32.

[17] Onizuka, S; Tawara, I; Shimizu, J; Sakaguchi, S; Fujita, T; Nakayama, E. Tumor rejection by in vivo administration of anti-CD25 (interleukin-2 receptor alpha) monoclonal antibody. *Cancer Res.,* 1999, 59(13), 3128-33.

[18] Shimizu, J; Yamazaki S; Sakaguchi, S. Induction of tumor immunity by removing CD25+CD4+ T cells: a common basis between tumor immunity and autoimmunity. *Immunology.,* 2002, 107(1), 5-9.

[19] Gallimore, A; Sakaguchi, S. Regulation of tumour immunity by CD25+ T cells. 2002, *Immunology.,* 107(1), 5.

[20] Levings, MK; Gregori, S; Tresoldi, E; Cazzaniga, S; Bonini, C; Roncarolo, MG. Differentiation of Tr1 cells by immature dendritic cells requires IL-10 but not CD25+CD4+ Treg cells. *Blood,* 2005, 105, 1162-9.

[21] Groux, H; O'Garra, A; Bigler, M; Rouleau, M; Antonenko, S; de Vries, JE; *et al.* A CD4+ T-cell subset inhibits antigen-specific T-cell responses and prevent colitis. *Nature,* 1997, 389, 737-742.

[22] Groux, H. Type 1 T-regulatory cells: their role in the control of immune responses. *Transplantation,* 2003, 75(9 Suppl) : 8S-12S.

[23] McGuirk, P; McCann, C; Mills, KH. Pathogen-specific T regulatory 1 cells induced in the respiratory tract by a bacterial molecule that stimulates interleukin 10 production by dendritic cells: a novel strategy for evasion of protective T helper type 1 responses by Bordetella pertussis. *J Exp Med,* 195(2), 221-31.

[24] Rahmoun, M; Foussat, A; Groux, H; Pene, J; Yssel, H; Chanez, P. Enhanced frequency of CD18- and CD49b-expressing T cells in peripheral blood of asthmatic patients correlates with disease severity. *Int Arch Allergy Immunol,* 2006, 140, 139-49.

[25] Aandahl, EM; Michaelsson, J; Moretto, WJ; Hecht, FM; Nixon, DF. Human CD4+ CD25+ regulatory T cells control T-cell responses to human immunodeficiency virus and cytomegalovirus antigens; *J Virol*, 2004, 78, 2454-2459.

[26] Cabrera, R; Tu, Z; Xu, Y; Firpi, RJ; Rosen, HR; Liu, C; Nelson, DR. An immunomodulatory role for CD4 (+) CD25 (+) regulatory T lymphocytes in hepatitis C virus infection. *Hepatology,* 2004, 40, 1062-1071.

[27] Kinter, AL; Hennessey, M; Bell, A; Kern, S; Lin, Y; Daucher, M; *et al.* CD25(+)CD4(+) regulatory T cells from the peripheral blood of asymptomatic HIV-infected individuals regulate CD4(+) and CD8(+) HIV-specific T cell immune responses in vitro and are associated with favorable clinical markers of disease status. *J Exp Med,* 2004, 200, 331-343.

[28] Barnaba, V; Zaccari, C; Levrero, M; Ruocco, G; Balsano, F. Immunoregulatory T cells in HBV-induced chronic liver disease as defined by monoclonal antibodies. *Clin Immunol Immunopathol,* 1983, 26, 83-90.

[29] Lemm, G; Salzer, K; Warnatz, H. Studies on immunoregulatory mechanisms in acute and chronic hepatitis B. *Clin Exp Immunol,* 1983, 52, 250-258.

[30] Roh, S; Kim, K. Overcoming tolerance in hepatitis B virus transgenic mice: a possible involvement of regulatory T cells. *Microbiol Immunol,* 2003, 47, 453-460.

[31] Chisari, FV; Filippi, P; Buras, J; McLachlan, A; Popper, H; Pinkert, CA; *et al.* Structural and pathological effects of synthesis of hepatitis B virus large envelope polypeptide in transgenic mice; *Proc Natl Acad Sci, U S A* 1987, 84, 6909-6913.

[32] Thio, CL; Mosbruger, TL; Kaslow, RA; Karp, CL; Strathdee, SA; Vlahov, D; *et al.* Cytotoxic T-lymphocyte antigen 4 gene and recovery from hepatitis B virus infection. *J Virol,* 2004, 78, 11258-11262.

[33] Stoop, JN; van der Molen, RG; Baan, CC; van der Laan, LJ; Kuipers, EJ; Kusters, JG; *et al.* Regulatory T cells contribute to the impaired immune response in patients with chronic hepatitis B virus infection. *Hepatology,* 2005, 41, 771-778.

[34] Franzese, O; Kennedy, PT; Gehring, AJ; Gotto, J; Williams, R; Maini, MK; *et al*. Modulation of the CD8+-T-cell response by CD4+ CD25+ regulatory T cells in patients with hepatitis B virus infection; *J Virol,* 2005, 79, 3322-3328.

[35] Xu, D; Fu, J; Jin, L; Zhang, H; Zhou, C; Zou, Z; Zhao, JM; Zhang, B; Shi, M; Ding, X; Tang, Z; Fu, YX; Wang, FS. Circulating and liver resident CD4+CD25+ regulatory T cells actively influence the antiviral immune response and disease progression in patients with hepatitis B. *J Immunol,* 2006, 177, 739-747.

[36] Hyodo, N; Nakamura, I; Imawari, M. Hepatitis B core antigen stimulates interleukin-10 secretion by both T cells and monocytes from peripheral blood of patients with chronic hepatitis B virus infection. *Clin Exp Immunol,* 2004, 135, 462-466.

[37] Peng, G; Li, S; Wu, W; Sun, Z; Chen, Y; Chen, Z. Circulating CD4+ CD25+ regulatory T cells correlate with chronic hepatitis B infection. *Immunology.,* 2008, 123(1), 57-65.

[38] Yang, G; Liu, A; Xie, Q; Guo, TB; Wan, B; Zhou, B; Zhang, JZ. Association of CD4+CD25+Foxp3+ regulatory T cells with chronic activity and viral clearance in patients with hepatitis B. *Int Immunol.,* 2007, 19(2), 133-40.

[39] Nan, XP; Zhang, Y; Yu, HT; Li, Y; Sun, RL; Wang, JP; Bai, XF. Circulating CD4+CD25high regulatory T cells and expression of PD-1 and BTLA on CD4+ T cells in patients with chronic hepatitis B virus infection. *Viral Immunol.*, 2010, 23(1), 63-70.

[40] Lin, CY; Tsai, MC; Huang, CT; Hsu, CW; Tseng, SC; Tsai, IF; Chen, YC; Yeh, CT; Sheen, IS; Chien, RN. Liver injury is associated with enhanced regulatory T-cell activity in patients with chronic hepatitis B. *J Viral Hepat.*, 2007, 14(7), 503-11.

[41] Fu, J; Xu, D; Liu, Z; Shi, M; Zhao, P; Fu, B; Zhang, Z; Yang, H; Zhang, H; Zhou, C; Yao, J; Jin, L; Wang, H; Yang, Y; Fu, YX; Wang, FS. Increased regulatory T cells correlate with CD8 T-cell impairment and poor survival in hepatocellular carcinoma patients. *Gastroenterology*., 2007, 132(7), 2328-39.

[42] Stoop, JN; Claassen, MA; Woltman, AM; Binda, RS; Kuipers, EJ; Janssen, HL; van der Molen, RG; Boonstra, A. Intrahepatic regulatory T cells are phenotypically distinct from their peripheral counterparts in chronic HBV patients. *Clin Immunol.*, 2008, 129(3), 419-27.

[43] Accapezzato, D; Francavilla, V; Paroli, M; Casciaro, M; Chircu, LV; Cividini, A; *et al.* Hepatic expansion of a virus-specific regulatory CD8+ T cell population in chronic HCV infection. *The Journal of Clinical Investigation,* 2004, 113, 963-97.

[44] Ward, J; Robinson, PJ. How to detect hepatocellular carcinoma in cirrhosis. *European Radiology,* 2002, 9, 2258-2272.

[45] Lauer, GM; Walker, BD. Hepatitis C virus infection. *N.Engl.J.Med,* 2001, 345, 41-52.

[46] Lamelin, JP; Zoulim, F; Trepo, C. Lymphotropism oh hepatitis B and C viruses: an update and a newcomer. *Int J Clin Lab Res,* 1995, 25, 1.

[47] Muir, AJ; Bornstein, JD; Killenberg, PG. Atlantic Coast Hepatitis Treatment Group. Peginterferon alfa-2b and ribavirin for the treatment of chronic hepatitis C in blacks and non-Hispanic whites. *N Engl J Med,* 2004, 350, 2265-2271.

[48] Takahashi, K; Iwata, K; Matsumoto, M; Matsumoto, H; Nakao, K; Hatahara, T; *et al.* Hepatitis C virus (HCV) genotype 1b sequences from fifteen patients with hepatocellular carcinoma: the 'progression score' revisited. *Hepatology Research,* 2001, 2, 161-171.

[49] Ciuffreda, D; Codarri, L; Buhler, L; Vallotton, L; Giostra, E; Mentha, G; Morel, P; Pantaleo, G; Pascual, M. Hepatitis C virus infection after liver transplantation is

associated with lower levels of activated CD4(+)CD25(+)CD45RO(+)IL-7ralpha(high) T cells. *Liver Transpl.*, 2010, Jan;16(1), 49-55.

[50] Langhans, B; Braunschweiger, I; Arndt, S; Schulte, W; Satoguina, J; Layland, LE; Vidovic, N; Hoerauf, A; Oldenburg, J; Sauerbruch, T; Spengler, U. Core-specific adaptive regulatory T-cells in different outcomes of hepatitis C; *Clinical Science,* 2010, 119(2), 97-109.

[51] Gerlach; JT; Diepolder, HM; Jung, MC; Gruener, NH; Schraut, WW; Zachoval, R; *et al.* Recurrence of hepatitis C virus after loss of virus-specific CD4(+) T-cell response in acute hepatitis C. *Gastroenterology,* 1999, 117, 933-941.

[52] Wedemeyer, H; He, XS; Nascimbeni, M; Davis, AR; Greenberg, HB; Hoofnagle, JH; *et al.* Impaired effector function of hepatitis C virus-specific CD8+ T cells in chronic hepatitis C virus infection. *J Immunol,* 2002, 169, 3447-3458.

[53] Dittmer, U; He, H; Messer, RJ; Schimmer, S; Olbrich, AR; Ohlen, C; *et al.* Functional impairment of CD8(+) T cells by regulatory T cells during persistent retroviral infection. *Immunity,* 2004, 20, 293-303.

[54] Sugimoto, K; Ikeda, F; Stadanlick, J; Nunes, FA; Alter, HJ; Chang, KM. Suppression of HCV-specific T cells without differential hierarchy demonstrated ex vivo in persistent HCV infection. *Hepatology,* 2003, 38, 1437-1448.

[55] Bolacchi, F; Sinistro, A; Ciaprini, C; Demin, F; Capozzi, M; Carducci, FC; *et al.* Increased hepatitis C virus (HCV)-specific CD4+CD25+ regulatory T lymphocytes and reduced HCV-specific CD4+ T cell response in HCV-infected patients with normal versus abnormal alanine aminotransferase levels. *Clin Exp Immuno,* 2006, 144, 188-196.

[56] Yang, JH; Zhang, YX; Yu, RB; Su, C; Sun, NX. CD4+ CD25+ regulatory T cells suppress CD4+ T cell responses in patients with persistent hepatitis C virus infection. *Zhonghua Nei Ke Za Zhi,* 2006, 45, 29-33.

[57] Akiyama, M; Ichikawa, T; Miyaaki, H; Motoyoshi, Y; Takeshita, S; Ozawa, E; Miuma, S; Shibata, H; Taura, N; Nakao, K. Relationship between regulatory T cells and the combination of pegylated interferon and ribavirin for the treatment of chronic hepatitis type C. *Intervirology.,* 2010, 53(3), 154-60.

[58] Itose, I; Kanto, T; Kakita, N; Takebe, S; Inoue, M; Higashitani, K; Miyazaki, M; Miyatake, H; Sakakibara, M; Hiramatsu, N; Takehara, T; Kasahara, A; Hayashi, N. Enhanced ability of regulatory T cells in chronic hepatitis C patients with persistently normal alanine aminotransferase levels than those with active hepatitis. *J Viral Hepat.,* 2009, 16(12), 844-52.

[59] Delhem, N; Cottrez, F; Carpentier, A; Miroux, C; Moralès, O; François, V; Groux, H; Auriault, C; Pancré, V. Role of the Regulatory T lymphocytes in hepatitis C fibrosis progression. *Bull Cancer.,* 2008, 95(11), 1029-38.

[60] Carpentier, A; Conti, F; Carrière, M; Aoudjehane, L; Miroux, C; Moralès, O; Calmus, Y; Groux, H; Auriault, C; Pancré, V; Delhem, N; Podevin, P. Analysis of gene transcription in sera during chronic hepatitis C infection. *J Med Virol.*, 2009, 81(3), 473-80.

[61] Wang, JP; Zhang, Y; Wei, X; Li, J; Nan, XP; Yu, HT; Li, Y; Wang, PZ; Bai, XF. Circulating Toll-like receptor (TLR) 2, TLR4, and regulatory T cells in patients with chronic hepatitis C. *APMIS,* 2010, 118(4), 261-70.

[62] Claassen, MA; de Knegt, RJ; Tilanus, HW; Janssen, HL; Boonstra, A. Abundant numbers of regulatory T cells localize to the liver of chronic hepatitis C infected patients and limit the extent of fibrosis. *Journal of Hepatoogyl.*, 2010, 52(3), 315-21.

[63] Yoshizawa, K; Abe, H; Kubo, Y; Kitahara, T; Aizawa, R; Matsuoka, M; Aizawa, Y. Expansion of CD4(+)CD25(+)FoxP3(+) regulatory T cells in hepatitis C virus-related chronic hepatitis; cirrhosis and hepatocellular carcinoma. *Hepatol Res.*, 2010, 40(2), 179-87.

[64] Smyk-Pearson, S; Golden-Mason, L; Klarquist, J; Burton, JR; Jr; Tester, IA; Wang, CC; Culbertson, N; Vandenbark, AA; Rosen, HR. Functional suppression by FoxP3+CD4+CD25(high) regulatory T cells during acute hepatitis C virus infection. *Journal of Infectious Diseases.*, 2008, 197(1), 46-57

[65] Schlenker, C; Trotter, J. Recurrent hepatitis C following liver transplantation: a review of factors associated with graft injury. *Minerva Gastroenterol Dietol,* 2006, 52, 175-186.

[66] Roche, B; Samuel, D. Risk factors for hepatitis C recurrence after liver transplantation. *Journal of Viral Hepatitis,* 2007, 14 Suppl 1:89-96.

[67] Ercolani, G; Grazi, GL; Ravaioli, M; Del Gaudio, M; Cescon, M; Varotti, G; *et al.* Histological recurrent hepatitis C after liver transplantation: Outcome and role of retransplantation. *Liver Transpl,* 2006, 12, 1104-1111.

[68] Velidedeoglu, E; Mange, KC; Frank, A; Abt, P; Desai, N; Markmann, JW; *et al.* Factors differentially correlated with the outcome of liver transplantation on HCV+ and HCV- recipients. *Transplantation,* 2004, 77, 1834-42.

[69] Feray, C; Gigou, M; Samuel, D; Paradis, V; Mishiro, S; Maertens, G; *et al.* Influence of the genotypes of hepatitis C virus on the severity of recurrent liver disease after transplantation. *Gastroenterology,* 1995, 108, 1088-1096.

[70] Carpentier, A; Conti, F; Stenard, F; Aoudjehane, L; Miroux, C; Podevin, P; Morales, O; Chouzenoux, S; Scatton, O; Groux, H; Auriault, C; Calmus, Y; Pancre, V; Delhem, N. Increased expression of regulatory Tr1 cells in recurrent hepatitis C after liver transplantation. *Am J Transplant.*, 2009, 9(9), 2102-12.

[71] Perrella, A; Arenga, G; Pisaniello, D; Rampone, B; Di Costanzo, GG; Atripaldi, L; Esposito, C; Di Florio, E; Perrella, O; Cuomo, O. Elevated CD4+/CD25+ T-cell frequency and function during hepatitis C virus recurrence after liver transplantation. *Transplantation Proceedings,* 2009, 41(5), 1761-6.

[72] Miroux, C; Moralès, O; Carpentier, A; Dharancy, S; Conti, F; Boleslowski, E; Podevin, P; Auriault, C; Pancré, V; Delhem, N. Inhibitory effects of cyclosporine on human regulatory T cells in vitro. *Transplant Proc.*, 2009, 41(8), 3371-4.

[73] Rayhill, SC; Barbeito, R; Katz, D; et al. A cyclosporine-based immunosuppression regimen may be better than tacrolimus for long-term liver allograft survival in recipients transplanted for hepatitis C; *Transplantation Proceedings,* 2006, 38, 3625.

[74] Villamil, F; Levy, G; Grazi, GL; et al. Long-term outcomes in liver transplant patients with hepatic C infection receiving tacrolimus or cyclosporine; *Transplantation Proceedings,* 2006, 38, 2964.

In: Regulatory T Cells
Editor: Ren S. Hayashi, pp. 153-168

ISBN: 978-1-61761-554-2

Chapter 8

Regulatory T Cells and Tumor Immunity in the Tumor Microenvironment

Lisong Shen[1], Matthew R. Thompson[2] and Dakang Xu[2]
[1]Department of Clinical Laboratory, Xinhua Hospital Affiliated to Shanghai Jiaotong University School of Medicine, Shanghai, 200092, China
[2]Monash Institute of Medical Research, Monash University, Melbourne, Australia

Abstract

A strong relationship exists between regulatory T (Treg) cells and the development and progression of cancer. Increasingly, evidence suggests that Treg cells can migrate into tumors and suppress effective antitumor responses in the tumor microenvironment. The critical question that needs to be addressed is the mechanism of action of these Treg cells *in vivo*. In this chapter, we consider the nature and characteristics of Treg cells in the tumor microenvironment, including dynamic processes of Treg recruitment, activation and induction during tumor pathogenesis and their potential mechanism for multiple suppressive effects. Enhancing our understanding of the underlying molecular mechanisms will offer an insight into how Treg cell function can be manipulated more effectively for antitumor therapy.

Introduction

The immune system interacts with tumors throughout their development, and these host responses may inhibit tumor growth and progression. However, a failure of immunity can predispose to tumor development [1; 2]. T cell mediated immunosuppression was studied for decades without clarification as to which factors were responsible for these effects. The identification of regulatory T (Treg) cells represents a milestone in the field of immunology, and recently a number of publications have reported that progressive tumor growth evokes generation of immunosuppressor cells and suppressor T cells [3]. Treg cells expressing Foxp3 (forkhead winged-helix transcription factor) are a distinct lineage of lymphocytes developed by the immune system to control self-tolerance and normal immune homeostasis [4]. Malignant tumors evade immune attack apparently through this potent regulatory mechanism. There are a growing number of reports describing accumulations of Treg cells inside of tumors, adjacent neoplastic tissues, draining lymph nodes and also in peripheral blood of patients suffering with solid tumors and hematologic malignancies [5; 6; 7]. Furthermore, the expansion and accumulation of these immunosuppressive cells *in situ* correlates with advanced tumor growth and predicts poor disease prognosis as shown by both laboratory and clinical investigations. Elucidating the mechanisms by which these cells traffic and accumulate in the tumor microenvironment has provided us with potential immunotherapeutic strategies for the treatment of cancer. Several mechanisms have been established to illuminate how Treg cells are recruited or converted in the tumor microenvironment to create a kind of suppressive milieu that restricts the immune response of $CD4^+$ effector cells and the release of cytotoxic granules from Cytotoxic T Lymphocytes (CTL), protecting tumor cells from being killed. Treg cells have been shown to be inversely correlated to the outcome of several human malignancies [8]. Depletion of Treg cells can enhance rejection of tumors which highlights the impact of Treg cells in antitumor immunity [9]. However, the molecular and cellular features responsible for the increase and maintenance of Treg cell levels in cancer remain elusive. In this chapter, we provide insight into recent advances in our understanding of the properties, subsets, antigen specificity and immune regulation of Treg cells. We consider the nature and characteristics of Treg cells in the tumor microenvironment, including dynamic processes of Treg recruitment, activation and induction during tumor pathogenesis and their potential mechanism for multiple suppressive effects.

Tumor Microenvironment

The microenvironment of a developing tumor is a crucial regulator of carcinogenesis was originally proposed by Paget in his famous 'seed and soil' hypothesis. Current data indicates that carcinogenesis and tumor angiogenesis result not only from the interaction of cancer cells with endothelial cells of various origin such as those of the vasculature or lymphatics, but also that surrounding 'normal' stromal and inflammatory cells and tissue also have a crucial role in directing the formation of the blood vessels that nourish a developing tumor. So the new hypothesis now enables us to consider stromal cells and their associated matrix, as well as cells of immune system for their roles in cancer progression. For the latter, we must also consider the role of immune mediators including chemokines, cytokines and chemicals [10; 11; 12]. The dialogue between the tumor cells and the immune cells and their associated cytokines, is an important interplay that drives the process of tumor progression. There is a correlation between survival and the frequency of tumor-infiltrating T lymphocytes (TILs) in cancer patients [13; 14]. However, most studies have not characterized TILs, which are known to comprise distinct subsets, each with differing roles in the complex tumor microenvironment. TILs include populations of $CD4^+$ effectors and Treg cells, and the involvement of the latter in tumor progression is widely accepted. Accumulations of TILs have been linked to improved patient outcome and TILs are recognized as effectors of antitumor immune responses [15; 16]. Recent studies have revealed that a subset of $CD4^+$ T cells, referred to as $CD4^+CD25^+Foxp3^+$ Treg cells, may accumulate in the tumor microenvironment and suppress tumor-specific T-cell responses, thereby hindering tumor rejection [17]. So the mechanism of action of these Treg cells *in vivo* needs to be further characterized in the tumor microenvironment, including dynamic processes of Treg recruitment, activation, induction and multiple suppressive effects during tumor pathogenesis **(Figure).**

Nature and Property of Tregs in Tumors

Treg cell populations are now widely considered to be an important component of the immune system as professional suppressors of immune responses. The concept of regulatory T cells was first described in the late sixties, however they were not definitively defined by a phenotypic marker until Sakaguchi defined $CD4^+CD25^{hi}$ (interleukin-2 receptor alpha-chain) as the constitutional marker in 1995 [18]. It was recently shown that transcription factor Foxp3 is specifically expressed by $CD4^+CD25^+$ Treg cells in rodents and humans. Also there are several other molecules associated with Treg cells to better identify and characterize them, e.g. CTL antigen 4 (CTLA4) and glucocorticoid induced tumor necrosis factor receptor related protein (GITR) [19; 20]. Anomalies of Tregs are responsible for the development of autoimmune responses, inflammatory diseases, tumors and allergic diseases in humans and rodents, characterized by a deficiency or hyperactivity in the development and function of $CD4^+CD25^+$ Treg cells. Therefore, a clear understanding of the characteristics of Treg cells would help us develop new clinical tools to control pathological immune responses. In the last 10 years, high levels of Treg cells were found in many human tumors including ovarian cancer [15; 21], lung cancer [21; 22], gastric cancer [7; 23], breast cancer [24], pancreas [24], melanoma [25], Hodgkin's [22] and non-Hodgkin's [26] lymphomas [25; 27; 28; 29]. Furthermore, Foxp3 expression of Treg cells in the tumor bearing host are inversely associated with patient survival [15; 23]. In general, tumor growth leads to an increase in $CD25^+Foxp3^+$ Tregs, but there is a clear pattern of Treg accumulation. The highest frequencies of Treg have always been found in the tumor tissue itself. In contrast, Treg frequencies in the peripheral compartment, such as draining lymph nodes, spleen and peripheral blood may vary, depending on tumor types and stage of cancer progression [5; 30]. One theory is that the functional inactivation of Treg cells will maintain them at high numbers in tumors and avoid their replenishment from the peripheral lymphocyte pool, which has the capacity to further suppress the effector lymphocyte antitumor response. Furthermore, Tregs at tumor sites are found to be highly activated and appear to undergo extensive cellular proliferation. In addition to CD25, CTLA4 and GITR, several other functionally important molecules such as transforming growth factor ß (TGFß) are found to be highly expressed on $Foxp3^+$ Tregs inside tumor tissues. Treg cells are currently classified into two main subsets, according to their origin and suppressive activity [31]. Natural Treg (nTreg) cells are developed from the thymus through strong T cell antigen receptor (TCR) signaling during the course of positive and negative selection [32]. They have been classically described to suppress the proliferation of responder T cells *in vitro* in a contact-dependent, cytokine-independent manner. nTreg cells have high-level expression of the transcription factor Foxp3 which is the most distinctive marker for the regulatory lineage. They also constitutively express CD25, CTLA4, GITR and OX40, which are molecules that characterize activated lymphocytes [33; 34; 35]. In addition to TCR stimulation, common γ-chain cytokine signaling has been found to be essential for nTreg development. Recent work provides new evidence suggesting that IL-2 and TGFß appear to be most important for driving nTreg development, based on studies of IL-2 and TGFß double knockouts [36]. On the contrary, adaptive Treg cells (aTreg) or inducible Tregs (iTreg) are antigen-specific suppressive cells that are generated from peripheral naïve T cells in the presence of IL-10 or TGFß, in case of T regulatory 1 or T helper 3 subsets, respectively [37; 38]. So far there is no phenotypic or functional marker available to distinguish aTregs from nTregs *in vitro* or *in vivo*. In contrast to nTreg cells, aTreg cells exert suppression mostly through an IL-10 and/or TGFß dependent manner, and their suppressive function is not strictly associated to high-level Foxp3 expression. When the number of Treg cells increases, the immune response is unbalanced and favors unresponsiveness. The number and function of Treg cells should be tightly controlled to maximize immunity against foreign pathogens and preserve self tolerance. It may be possible to inhibit Treg conversion and function through innate activation, inflammation and strong antigenic stimulation. This is evident in many murine and human cancers, at a critical point where tumors are no longer attacked by immune cells. Recent data suggests that Treg cell expansion can occur through the proliferation of pre-existing Treg cells [39; 40], and from the conversion of peripheral naive precursors into newly derived Treg cells [41; 42] **(Figure)**. Based on adoptive cell transfer and antigen-specific TCR transgenic mice studies, it has been shown that pre-existing Tregs induced from non-Treg precursor cells contribute to the total pool of tumor-infiltrating $CD25^{+}Foxp3^{+}$Tregs [41; 42]. Thus, through mechanisms, which still need to be examined in detail, tumors turn the host Treg cells from helpful guardians, against exacerbated immunity, into detrimental suppressors of the antitumor immune response.

Treg Recruitment, Activation and Induction during Tumor Pathogenesis

The development of malignant tumors includes several distinct stages: 1) the initial hyperplasia; 2) neoplastic growth; 3) angiogenesis and local invasion; and 4) distant metastasis. Several human and mouse models have revealed the dynamic nature of the Treg response to tumor onset. In human gastric cancer, higher expression of Foxp3 Treg cells in tumor tissue, ascites and peripheral blood was observed even at the TNM stage I, and persisted at the TNM II, TNM III and TNM IV stages (TNM- Classification of Malignant Tumors), indicating that Treg cell accumulation in tumor tissue is an early event in tumor development and progression [9; 30]. It also suggests that these T cells might suppress immune activation at the tumor site. A similar observation found that $CD25^{+}Foxp3^{+}$ Tregs begin to infiltrate into the tumor lesion at very early stage in KRAS mutation induced spontaneous pancreatic carcinoma [43]. This study also demonstrated that partial depletion of Tregs by CD25 antibody treatment reduced the tumor incidence. Conversely, an increase of Tregs leads to accelerated growth of the tumor mass [44]. It remains to be answered whether increased Treg cells in the tumor microenvironment can be observed directly though expansion, conversion and induction at onset of disease as an early event in the context of tumor progression.

1) Recruitment of Regulatory T Cells

It has been found that donor Foxp3+ Tregs rapidly home to the newly formed tumor lesion directly from the blood circulation within a few hours. There are several mechanisms that may explain this recruitment of Treg cells. Treg cells express a number of chemokine receptors such as CCR2, CCR4, CCR5, CCR7, CCR8 and CXCR4 and can respond to a variety of chemokines such as CCL22, CCL17, CCL1 and CCL4. Functional Treg cells are found in the tumor microenvironment; these include thymus-derived 'natural' Treg cells and locally induced T-regulatory 1 (T_R1) cells. Treg cells differentiate in the thymus. The mouse thymus produces Treg cells as a separate lineage that expresses Foxp3 [45; 46]. Because tumor-associated Treg cells express Foxp3 mRNA and protein [15; 47], it is possible that these cells traffic to tumors from the thymus, bone marrow [48], lymph nodes and peripheral blood under the influence of tumor microenvironmental CC-chemokine ligand 22 (CCL22) [9] (Figure 1). It has been found that such chemokine receptors and chemokines play an important role in recruiting Tregs in malignancies such as Hodgkin's lymphoma, B-cell NHL and gastric cancer [49; 50; 51]. In addition, Treg trafficking to tumor lesions appears to be controlled by myeloid cell infiltration. A human study showed that macrophages promote Treg recruitment to the tumor mass in ovarian cancer patients [15]. It is known that myeloid cells have the capacity to attract Tregs. Oncogene activity can trigger intracellular signaling pathways leading to the production of a range of pro-inflammatory chemokines and cytokines from tumor precursor cells. In lung cancer patients, it was found that high expression of CXCR2 ligands was induced in early alveolar neoplastic lesions by oncogenic KRAS. In a mouse model, the murine ligands for CXCR2, KC and MIP2, were functionally responsible for recruiting immune cells to the incipient tumor lesion [52]. Whether these chemokines were in part derived from myeloid cells remains to be identified. Most tumor tissues express Toll-like receptors (TLR) which may initiate inflammatory signals that attract myeloid cells, which will lead to further production of chemokines and cytokines, trafficking circulating Treg cells to the tumor site.

2) Expansion of Regulatory T cells

Tregs accumulate above the reference range in patients with carcinoma. Although in human, the increase of $CD4^+CD25^+$ T cells has been reported to be associated with the reduction of $CD4^+CD25^-$ T cells[23], the origin and mechanism governing Treg increase in cancer patients remains elusive. Tumors convert DCs into TGFß expressing immature myeloid DCs that are capable of promoting Treg cell proliferation. In mouse model, mature DCs stimulated the *in vivo* expansion of self-antigen-specific Treg. Thymus-derived Treg cells in the tumor microenvironment might clonally expand following stimulation by tumor associated DCs [39]. However, proliferation alone cannot account for their large expansion and, moreover, does not explain the corresponding reduction of $CD4^+CD25^-$ T cells. How Treg cell expansion occurs in tumors was reported by Valzasina et al. Tumor injection in thymectomized and CD25-depleted mice shows that $CD4^{+}CD25^{+}$ T-cell expansion occurs even in the absence of the thymus and independently from proliferation of pre-existing $CD25^{+}$ cells [41]. This suggested that tumor development in mice led to a *de novo* generation of Treg cells. In transplantable tumor models, an increased number of Treg cells was observed early during tumor development. This short time span of tumor development is sufficient to generate a highly suppressive Treg cell population, which can be effectively transferred to other animals suppressing the antitumor immune response [53]. In addition to DCs stimulated the expansion of Treg, *de novo* generation of Treg is another important mechanism for Treg expansion. How Treg cell expansion occurs in tumor bearing Treg cell expansion is not restricted to the tumor microenvironment, as increased Treg cell numbers are observed in tumor-draining lymph nodes and peripheral blood, suggesting that heightening of Treg frequencies is a general mechanism contributing to the overall suppressed immune response in tumor patients. It has been clearly established over the last few years that Treg populations are expanded in most tumors and can be correlated to overall survival [54] [6].

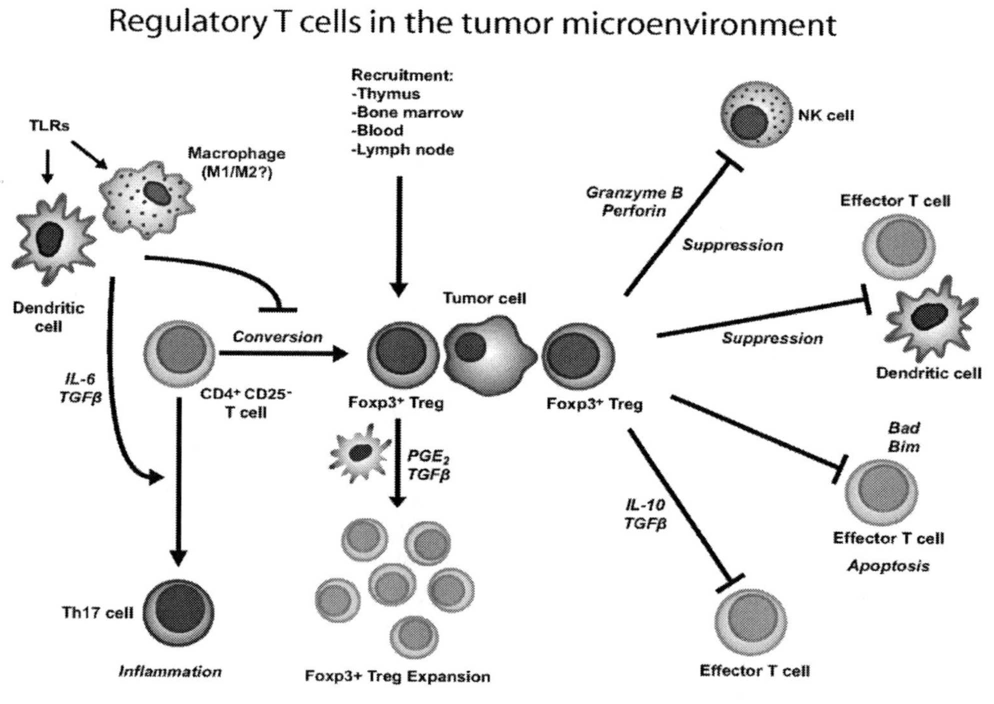

Figure 1. Regulatory T cells and tumor immunity in the tumor microenvironment. Characteristics of Treg cells in the tumor microenvironment, including the dynamic processes of Treg recruitment, expansion, conversion and differentiation during tumor pathogenesis and the potential mechanism for their multiple suppressive effects through cytokines, cell-cell contact, apoptosis etc. mediated immune suppression

3) Conversion of Conventional T Cells into Regulatory T Cells

The conversion of conventional T cells into Treg cells might be an important mechanism through which myeloid-derived suppressor cells (MDSC) exert their immunosuppressive function. Thymic function is largely reduced after adolescence, whereas Treg cells persist throughout the human lifespan. Presumably, Treg cells can be induced to differentiate in the periphery, such as in the tumor microenvironment. Tumor environmental factors, such as vascular endothelial growth factor (VEGF), IL-10 and TGFß suppressed DC differentiation and function, resulting in immature and/or partially differentiated DCs [55]. Plasmacytoid DCs associated with ovarian cancer induced $CD8^+FOXP3^-IL\text{-}10^+$ Treg cells [56; 57], and dysfunctional myeloid DCs also induced $IL\text{-}10^+$ Treg cells *in vitro [58]* and *in vivo* in patients with cancer [59; 60]. Tumors convert DCs into TGFß-expressing immature myeloid DCs that are capable of promoting Treg cell proliferation [39]. Normal mature DCs stimulated the *in vivo* expansion of self-antigen-specific Treg cells in mice [61; 62]. Analogously, thymus-derived Treg cells in the tumor microenvironment might clonally expand following stimulation by tumor-associated DCs. MDSC and other immune cells might be important for the induction or conversion of Treg cell [63]. Macrophages are the most abundant antigen presenting cells (APC) in the periphery and can be found in virtually all organs and tissues. Macrophages are polarized into M1 (with antitumor function) and M2 (with pro-tumor function) forms. It has been recently reported that the M2 subset of human macrophages are able to tune inflammatory responses and adaptive Th1 immunity, scavenge debris, and promote angiogenesis, tissue remodeling and repair leading to tumor progression. M2 macrophages induce Treg cells with a strong suppressive capability in a cell-cell contact dependent manner, which is mediated by membrane-bound TGFß expressed on the Treg population, as inhibition of TGFß signaling in target cells abrogated their suppressive function. The finding that a subset of macrophages has the capability to induce Treg cells in the periphery has important implications for our understanding how Treg cell frequencies are influenced in the tumor microenvironment. Macrophages are in abundance in the microenvironment of a number of human tumors and are believed to contribute to disease control or exacerbation. Monocytes can also contribute to the induction of Treg cells, as LPS treatment resulted in the suppression of T cell immune responses by a COX-PGE2 dependent mechanism resulting in Foxp3 expression in resting conventional $CD4^+CD25^-$ T cells. This mechanism was inhibited by COX-2 inhibitors, a PGE2 neutralizing antibody, or cyclic adenosine monophosphate (cAMP) antagonists [30; 64]. In addition to trafficking, differentiation and expansion of Treg cells, Treg cells can also be present in a tumor as a result of conversion from $CD4^+CD25^-$ T cells [65; 66]. TGFß, which is present at high levels in the tumor microenvironment, might mediate this conversion [67; 68]. Similarly, IL-10 is often found in the tumor microenvironment [68], where it supports the differentiation of $CD4^+IL\text{-}10^+TGF\beta^+$ Treg cells [69; 70] and induces T_R1 cells [71] . However, it is unknown whether IL-10 is implicated in the differentiation and expansion of Treg cells.

Tregs and Immune Suppression

The suppressive functions of Tregs have been addressed using many *in vitro* and *in vivo* mouse models. Tregs employ multiple suppressive mechanisms including those dependant on cell–cell contact or the release of soluble factors [4; 72]. Various molecules may be secreted or expressed on the cell surface of Tregs directly contributing to their suppressive functions.

These mechanisms have not been exclusively investigated in tumor settings. Here we summarize the relevant information in the context of tumor immunity **(Figure)**.

1) Cytokine Mediated Immune Suppression

IL-10 and TGFß were first suspected to mediate Treg suppression. In human tumors, investigation into the role of IL-10 and TGFß in the suppressive activity of Tregs has been performed by experiments separating Tregs from responder cells, utilizing transwell inserts, followed by antibody neutralization of IL-10 or TGFß, to evaluate the mechanisms used by Treg cells to suppress responder cell proliferation [33; 73]. IL-10 or TGFß antibody abrogated Treg mediated suppression, indicating that in the tumor microenvironment suppression is dependent on Tregs which secrete IL-10 and TGFß, and independent of cell-cell contact between Tregs and responder cells. Other models suggest that Treg cells mediate suppression by both TGFß dependent and independent mechanisms. *In vitro* suppression of proliferation can be achieved in T cells insensitive to TGFß [74], indicating the dispensability of TGFß. However, there is data from both *in vitro* and *in vivo* experiments, which show that T cells insensitive to TGFß cannot be suppressed by Treg cells [75]. Another study, using $CD4^+$ and $CD8^+$ cells from mice possessing impaired TGFß receptors, demonstrated that blocking TGFß allowed immune-mediated tumor rejection [76]. However, it is unclear whether TGFß is used by Treg cells to mediate suppression and/or is important for the generation of tumor-sensitized Treg cells that mediate suppression [77]. Antisense oligonucleotides are already being employed in clinical trials to block TGFß either at the tumor vaccine site [78] or in the tumor microenvironment [79]. More recent results document that TGFß is critical for the generation of tumor-sensitized adaptive Treg cells that are efficient suppressors of effective T cell priming. The experiment used tumor-sensitized Treg cells which were separated from the primed tumor-specific effector T cells and evaluated for their therapeutic activity *in vivo*. The effect of TGFß was modeled by using transgenic mice that express dominant-negative TGF-ß receptor type II (dnTGFßRII) in $CD8^+$ and $CD4^+$ cells. In these transgenic mice, $CD4^+$ and $CD8^+$ cells have only a minimal sensitivity to TGF-β, because the receptor complex formed by the TGFßRI and the dnTGFßRII impairs TGFß signaling due to the truncated intracellular signaling domain of the dnTGFßRII. Furthermore, they show that tumor-sensitized Treg cells can mediate effective suppression of a therapeutic immune response through a mechanism that is independent of the effect of TGFß on T cells. Secretion of TGFß is not limited to the mouse model, as multiple studies have demonstrated secretion of TGFß by human tumors, including tumors of the breast [80], colon [81] , and pancreas [82]. These studies identified that TGFß plays an essential role in the generation of tumor-sensitized Treg cells, but is not required as a suppressive molecule that blocks tumor-specific priming. Furthermore, the suppressive mechanism used by tumor-sensitized Treg cells was independent of TGFß acting directly on the responding T cells. It was found that lack of immune suppression in impaired TGFß signaling mouse model, due to a deficiency of tumor-sensitized adaptive Treg cells or the inability of tumor-sensitized natural Treg cells to be activated by TGFß in the tumor bearing host. These findings have lead to further investigations in clinical trials with cancer patients [83]. Their current protocol for the reconstitution of patients made lymphopenic, using chemotherapy, is to reinfuse an apheresis product depleted of Treg cells. Given the observation that TGFß is necessary for the *de novo* induction of tumor-sensitized Treg cells, the blockade of TGFß could be used to prevent the generation of new tumor-sensitized Treg cells in reconstituted patients.

PGE2 itself not only activates Treg cells, but also has been indicated as an effector molecule released by Treg cells to exert their suppressive function. Adaptive Treg cells produce PGE2 via induction of COX-2 and suppress effector T cell responses in a manner that is reversed by COX inhibitors and PGE2 receptor-specific antagonists. In gastric cancer patients, the suppression of effector T cell responses by Treg cells was observed to be reversed *in vitro* by the COX inhibitor indomethacin [84]. Apart from cytokines involved in the putative suppressive mechanisms described above, mouse strains deficient in specific genes such as IL-2 and CTLA4 have developed anomalies in Treg suppressive activity, resulting in an onset of systemic autoimmunity, similar to that observed in Foxp3 deficient mice [85]. It will be interesting to identify a possible core contribution of CTLA4 or IL-2 dependent suppression in tumor immunity.

2) Cell-Cell Contact Mediated Immune Suppression

Blockade of CTLA4 abrogates the inhibitory activity of Treg cells while CD28 deficient Treg cells are able to suppress T cells, indicating that CD28 can be replaced during activation of Treg cells. It was also found that CTLA4 played an additional role in blocking the activity of TGFß signaling for Treg to target cell interactions. Treg cells express high levels of cAMP, which as a second messenger has potent immunosuppressive activity on effector T cell proliferation and IL-2 synthesis, and mediates suppression via its cell-cell contact dependent transfer from Treg cells to effector T cells through gap junctions [86].

3) Apoptosis Mediated Immune Suppression

Treg induced apoptosis of responder T cells is associated with increased levels of the pro-apoptotic gene Bim, activation of the pro-apoptotic gene Bad, as well as a decrease of the pro-survival kinase Akt supporting that cytokine deprivation-induced apoptosis is a mechanism through which Treg cells inhibit the responder T cell response. Another mechanism introduced recently is the direct killing of responder T cell by Treg cells. Cytotoxic T lymphocytes and natural killer cells use the perforin/granzyme pathway to kill virally infected cells and tumor cells. Grossman showed that high expression of granzyme B was observed in a subset of Treg cells in the tumor microenvironment. Treg cells derived from the tumor microenvironment could induce NK and $CD8^{+}$ T cell death in a granzyme B or perforin-dependent manner. These data also indicated that tumor granzyme B and perforin might be relevant for Treg cell mediated suppression of tumor clearance [87]. Thus, Treg cells in certain contexts can differentiate to become “cytotoxic-suppressor” cells.

All of the above mentioned mechanisms either act in concert to induce suppression of effector cells, or certain mechanisms might be selectively used under certain conditions. The inhibitory effect of Treg cells might not be limited to a single suppressive mechanism. Additional uncharacterized mechanisms might also be in place, which are responsible for Treg cell mediated suppression.

Conclusion

Treg cells are a major factor contributing to an immunosuppressive tumor microenvironment, as demonstrated by a large collection of experimental and clinical studies. Significant interest has recently focused on the premise that tumors may take advantage of the imbalanced immune system, which may result from a deregulation of cytokines or immune cells in inflammation. Importantly, tumors may evade immunity by promoting the expansion, recruitment and conversion of Tregs resulting in subsequent suppression of effector cells including $CD4^{+}$, $CD8^{+}$, NK cells etc. in the tumor microenvironment **(Figure)**. It remains to be answered whether the increase in Treg cells in the tumor microenvironment is the result of expansion, conversion or induction at onset of disease. Tumor cells may also contribute the development of inflammatory immune cells such as M1/M2 macrophages and TH17 cells along with elevated number of Treg cells. A new challenge will be to understand how the interplay between these cells may combine towards antitumor immunity. Another challenge will be to determine the precise mechanisms employed by Tregs to suppress other lymphocytes at the molecular level *in vitro* and *in vivo*.

References

[1] Dougan, M. & Dranoff, G. (2009). Immune therapy for cancer. *Annu Rev Immunol*, *27*, 83-117.

[2] Fujimoto, S., Greene, M. I. & Sehon, A. H. (1976). Regulation of the immune response to tumor antigens. I. Immunosuppressor cells in tumor-bearing hosts. *J Immunol*, *116*, 791-9.

[3] Sakaguchi, S. (2004). Naturally arising CD4+ regulatory t cells for immunologic self-tolerance and negative control of immune responses. *Annu Rev Immunol*, *22*, 531-62.

[4] Sakaguchi, S. (2005). Naturally arising Foxp3-expressing CD25+CD4+ regulatory T cells in immunological tolerance to self and non-self. *Nat Immunol*, *6*, 345-52.

[5] Zou, W. (2006). Regulatory T cells, tumor immunity and immunotherapy. *Nat Rev Immunol*, *6*, 295-307.

[6] Beyer, M. & Schultze, J. L. (2006). Regulatory T cells in cancer. *Blood*, *108*, 804-11.

[7] Shen, L. S., Wang, J., Shen, D. F., Yuan, X. L., Dong, P., Li, M. X., Xue, J., Zhang, F. M., Ge, H. L. & Xu, D. (2009). CD4(+)CD25(+)CD127(low/-) regulatory T cells express Foxp3 and suppress effector T cell proliferation and contribute to gastric cancers progression. *Clin Immunol*, *131*, 109-18.

[8] Menard, C., Ghiringhelli, F., Roux, S., Chaput, N., Mateus, C., Grohmann, U., Caillat-Zucman, S., Zitvogel, L. & Robert, C. (2008). Ctla-4 blockade confers lymphocyte resistance to regulatory T-cells in advanced melanoma: surrogate marker of efficacy of tremelimumab? *Clin Cancer Res*, *14*, 5242-9.

[9] Curiel, T. J. (2007). Tregs and rethinking cancer immunotherapy. *J Clin Invest*, *117*, 1167-74.

[10] Bissell, M. J. & Labarge, M. A. (2005). Context, tissue plasticity, & cancer: are tumor stem cells also regulated by the microenvironment? *Cancer Cell*, *7*, 17-23.

[11] Bhowmick, N. A., Neilson, E. G. & Moses, H. L. (2004). Stromal fibroblasts in cancer initiation and progression. *Nature*, *432*, 332-7.

[12] Lewis, C. E. & Pollard, J. W. (2006). Distinct role of macrophages in different tumor microenvironments. *Cancer Res*, *66*, 605-12.

[13] Oble, D. A., Loewe, R., Yu, P. & Mihm, M. C. (2009). Jr., Focus on TILs: prognostic significance of tumor infiltrating lymphocytes in human melanoma. *Cancer Immun., 9,* 3.

[14] Ohtani, H. (2007). Focus on TILs: prognostic significance of tumor infiltrating lymphocytes in human colorectal cancer. *Cancer Immun., 7,* 4.

[15] Curiel, T. J., Coukos, G., Zou, L., Alvarez, X., Cheng, P., Mottram, P., Evdemon-Hogan, M., Conejo-Garcia, J. R., Zhang, L., Burow, M., Zhu, Y., Wei, S., Kryczek, I., Daniel, B., Gordon, A., Myers, L., Lackner, A., Disis, M. L., Knutson, K. L., Chen, L. & Zou, W. (2004). Specific recruitment of regulatory T cells in ovarian carcinoma fosters immune privilege and predicts reduced survival. *Nat Med., 10,* 942-9.

[16] Strauss, L., Bergmann, C., Gooding, W., Johnson, J. T. & Whiteside, T. L. (2007). The frequency and suppressor function of CD4+CD25highFoxp3+ T cells in the circulation of patients with squamous cell carcinoma of the head and neck. *Clin Cancer Res., 13,* 6301-11.

[17] Viguier, M., Lemaitre, F., Verola, O., Cho, M. S., Gorochov, G., Dubertret, L., Bachelez, H., Kourilsky, P. & Ferradini, L. (2004). Foxp3 expressing CD4+CD25(high) regulatory T cells are overrepresented in human metastatic melanoma lymph nodes and inhibit the function of infiltrating T cells. *J Immunol., 173,* 1444-53.

[18] Sakaguchi, S., Sakaguchi, N., Asano, M., Itoh, M. & Toda, M. (1995). Immunologic self-tolerance maintained by activated T cells expressing IL-2 receptor alpha-chains (CD25). Breakdown of a single mechanism of self-tolerance causes various autoimmune diseases. *J Immunol., 155,* 1151-64.

[19] McHugh, R. S., Whitters, M. J., Piccirillo, C. A., Young, D. A., Shevach, E. M., Collins, M. & Byrne, M. C. (2002). CD4(+)CD25(+) immunoregulatory T cells: gene expression analysis reveals a functional role for the glucocorticoid-induced TNF receptor. *Immunity., 16,* 311-23.

[20] Takahashi, T., Tagami, T., Yamazaki, S., Uede, T., Shimizu, J., Sakaguchi, N., Mak, T. W. & Sakaguchi, S. (2000). Immunologic self-tolerance maintained by CD25(+)CD4(+) regulatory T cells constitutively expressing cytotoxic T lymphocyte-associated antigen 4. *J Exp Med., 192,* 303-10.

[21] Woo, E. Y., Chu, C. S., Goletz, T. J., Schlienger, K., Yeh, H., Coukos, G., Rubin, S. C., Kaiser, L. R. & June, C. H. (2001). Regulatory CD4(+)CD25(+) T cells in tumors from patients with early-stage non-small cell lung cancer and late-stage ovarian cancer. *Cancer Res., 61,* 4766-72.

[22] Ishibashi, Y., Tanaka, S., Tajima, K., Yoshida, T. & Kuwano, H. (2006). Expression of Foxp3 in non-small cell lung cancer patients is significantly higher in tumor tissues than in normal tissues, especially in tumors smaller than 30 mm. *Oncol Rep., 15,* 1315-9.

[23] Sasada, T., Kimura, M., Yoshida, Y., Kanai, M. & Takabayashi, A. (2003). CD4+CD25+ regulatory T cells in patients with gastrointestinal malignancies: possible involvement of regulatory T cells in disease progression. *Cancer., 98,* 1089-99.

[24] Liyanage, U. K., Moore, T. T., Joo, H. G., Tanaka, Y., Herrmann, V., Doherty, G., Drebin, J. A., Strasberg, S. M., Eberlein, T. J., Goedegebuure, P. S. & Linehan, D. C.

(2002). Prevalence of regulatory T cells is increased in peripheral blood and tumor microenvironment of patients with pancreas or breast adenocarcinoma. *J Immunol., 169,* 2756-61.

[25] Appay, V., Jandus, C., Voelter, V., Reynard, S., Coupland, S. E., Rimoldi, D., Lienard, D., Guillaume, P., Krieg, A. M., Cerottini, J. C., Romero, P., Leyvraz, S., Rufer, N. & Speiser, D. E. (2006). New generation vaccine induces effective melanoma-specific CD8+ T cells in the circulation but not in the tumor site. *J Immunol., 177,* 1670-8.

[26] Yang, Z. Z., Novak, A. J., Stenson, M. J., Witzig, T. E. & Ansell, S. M. (2006). Intratumoral CD4+CD25+ regulatory T-cell-mediated suppression of infiltrating CD4+ T cells in B-cell non-Hodgkin lymphoma. Blood *107,* 3639-46.

[27] Meloni, F., Morosini, M., Solari, N., Passadore, I., Nascimbene, C., Novo, M., Ferrari, M., Cosentino, M., Marino, F., Pozzi, E. & Fietta, A. M. (2006). Foxp3 expressing CD4+ CD25+ and CD8+CD28- T regulatory cells in the peripheral blood of patients with lung cancer and pleural mesothelioma. *Hum Immunol., 67,* 1-12.

[28] Ormandy, L. A., Hillemann, T., Wedemeyer, H., Manns, M. P., Greten, T. F. & Korangy, F. (2005). Increased populations of regulatory T cells in peripheral blood of patients with hepatocellular carcinoma. *Cancer Res., 65,* 2457-64.

[29] Lau, K. M., Cheng, S. H., Lo, K. W., Lee, S. A., Woo, J. K., van Hasselt, C. A., Lee, S. P., Rickinson, A. B. & Ng, M. H. (2007). Increase in circulating Foxp3+CD4+CD25(high) regulatory T cells in nasopharyngeal carcinoma patients. *Br J Cancer., 96,* 617-22.

[30] Yuan, X. L., Chen, L., Li, M. X., Dong, P., Xue, J., Wang, J., Zhang, T. T., Wang, X. A., Zhang, F. M., Ge, H. L., Shen, L. S. & Xu, D. Elevated expression of Foxp3 in tumor-infiltrating Treg cells suppresses T-cell proliferation and contributes to gastric cancer progression in a COX-2-dependent manner. *Clin Immunol*, *134*, 277-88.

[31] Bluestone, J. A. & Abbas, A. K. (2003). Natural versus adaptive regulatory T cells. *Nat Rev Immunol., 3,* 253-7.

[32] Aschenbrenner, K., D'Cruz, L. M., Vollmann, E. H., Hinterberger, M., Emmerich, J., Swee, L. K., Rolink, A. & Klein, L. (2007). Selection of Foxp3+ regulatory T cells specific for self antigen expressed and presented by Aire+ medullary thymic epithelial cells. Nat Immunol *8,* 351-8.

[33] Takahashi, T., Kuniyasu, Y., Toda, M., Sakaguchi, N., Itoh, M., Iwata, M., Shimizu, J. & Sakaguchi, S. (1998). Immunologic self-tolerance maintained by CD25+CD4+ naturally anergic and suppressive T cells: induction of autoimmune disease by breaking their anergic/suppressive state. *Int Immunol., 10,* 1969-80.

[34] Shimizu, J., Yamazaki, S., Takahashi, T., Ishida, Y. & Sakaguchi, S. (2002). Stimulation of CD25(+)CD4(+) regulatory T cells through GITR breaks immunological self-tolerance. *Nat Immunol*, *3,* 135-42.

[35] Takeda, I., Ine, S., Killeen, N., Ndhlovu, L. C., Murata, K., Satomi, S., Sugamura, K. & Ishii, N. (2004). Distinct roles for the OX40-OX40 ligand interaction in regulatory and nonregulatory T cells. *J Immunol., 172,* 3580-9.

[36] Liu, Y., Zhang, P., Li, J., Kulkarni, A. B., Perruche, S. & Chen, W. (2008). A critical function for TGF-beta signaling in the development of natural CD4+CD25+Foxp3+ regulatory T cells. *Nat Immunol, 9,* 632-40.

[37] Faria, A. M. & Weiner, H. L. (2005). Oral tolerance. *Immunol Rev*, *206,* 232-59.

[38] Roncarolo, M. G., Gregori, S., Battaglia, M., Bacchetta, R., Fleischhauer, K. & Levings, M. K. (2006). Interleukin-10-secreting type 1 regulatory T cells in rodents and humans. *Immunol Rev*, *212,* 28-50.

[39] Ghiringhelli, F., Puig, P. E., Roux, S., Parcellier, A., Schmitt, E., Solary, E., Kroemer, G., Martin, F., Chauffert, B. & Zitvogel, L. (2005). Tumor cells convert immature myeloid dendritic cells into TGF-beta-secreting cells inducing CD4+CD25+ regulatory T cell proliferation. *J Exp Med., 202,* 919-29.

[40] Zhou, G., Drake, C. G. & Levitsky, H. I. (2006). Amplification of tumor-specific regulatory T cells following therapeutic cancer vaccines. Blood *107,* 628-36.

[41] Valzasina, B., Piconese, S., Guiducci, C. & Colombo, M. P. (2006). Tumor-induced expansion of regulatory T cells by conversion of CD4+CD25- lymphocytes is thymus and proliferation independent. *Cancer Res., 66,* 4488-95.

[42] Zhou, G. & Levitsky, H. I. (2007). Natural regulatory T cells and de novo-induced regulatory T cells contribute independently to tumor-specific tolerance. *J Immunol., 178,* 2155-62.

[43] Clark, C. E., Hingorani, S. R., Mick, R., Combs, C., Tuveson, D. A. & Vonderheide, R. H. (2007). Dynamics of the immune reaction to pancreatic cancer from inception to invasion. *Cancer Res., 67,* 9518-27.

[44] Teng, M. W., Swann, J. B., von Scheidt, B., Sharkey, J., Zerafa, N., McLaughlin, N., Yamaguchi, T., Sakaguchi, S., Darcy, P. K. & Smyth, M. J. (2010). Multiple antitumor mechanisms downstream of prophylactic regulatory T-cell depletion. *Cancer Res., 70,* 2665-74.

[45] Hori, S., Nomura, T. & Sakaguchi, S. (2003). Control of regulatory T cell development by the transcription factor Foxp3. *Science*, *299,* 1057-61.

[46] Fontenot, J. D., Rasmussen, J. P., Williams, L. M., Dooley, J. L., Farr, A. G. & Rudensky, A. Y. (2005). Regulatory T cell lineage specification by the forkhead transcription factor foxp3. *Immunity., 22,* 329-41.

[47] Ko, K., Yamazaki, S., Nakamura, K., Nishioka, T., Hirota, K., Yamaguchi, T., Shimizu, J., Nomura, T., Chiba, T. & Sakaguchi, S. (2005). Treatment of advanced tumors with agonistic anti-GITR mAb and its effects on tumor-infiltrating Foxp3+CD25+CD4+ regulatory T cells. *J Exp Med., 202,* 885-91.

[48] Zou, L., Barnett, B., Safah, H., Larussa, V. F., Evdemon-Hogan, M., Mottram, P., Wei, S., David, O., Curiel, T. J. & Zou, W. (2004). Bone marrow is a reservoir for CD4+CD25+ regulatory T cells that traffic through CXCL12/CXCR4 signals. *Cancer Res., 64,* 8451-5.

[49] Burchill, M. A., Yang, J., Vogtenhuber, C., Blazar, B. R. & Farrar, M. A. (2007). IL-2 receptor beta-dependent STAT5 activation is required for the development of Foxp3+ regulatory T cells. *J Immunol., 178,* 280-90.

[50] Ishida, T., Ishii, T., Inagaki, A., Yano, H., Komatsu, H., Iida, S., Inagaki, H. & Ueda, R. (2006). Specific recruitment of CC chemokine receptor 4-positive regulatory T cells in Hodgkin lymphoma fosters immune privilege. *Cancer Res., 66,* 5716-22.

[51] Mizukami, Y., Kono, K., Kawaguchi, Y., Akaike, H., Kamimura, K., Sugai, H. & Fujii, H. (2008). CCL17 and CCL22 chemokines within tumor microenvironment are related to accumulation of Foxp3+ regulatory T cells in gastric cancer. *Int J Cancer., 122,* 2286-93.

[52] Wislez, M., Fujimoto, N., Izzo, J. G., Hanna, A. E., Cody, D. D., Langley, R. R., Tang, H., Burdick, M. D., Sato, M., Minna, J. D., Mao, L., Wistuba, I., Strieter, R. M. & Kurie, J. M. (2006). High expression of ligands for chemokine receptor CXCR2 in alveolar epithelial neoplasia induced by oncogenic KRAS. *Cancer Res., 66,* 4198-207.

[53] Elpek, K. G., Yolcu, E. S., Franke, D. D., Lacelle, C., Schabowsky, R. H. & Shirwan, H. (2007). Ex vivo expansion of CD4+CD25+FoxP3+ T regulatory cells based on synergy between IL-2 and 4-1BB signaling. *J Immunol., 179,* 7295-304.

[54] Griffiths, R. W., Elkord, E., Gilham, D. E., Ramani, V., Clarke, N., Stern, P. L. & Hawkins, R. E. (2007). Frequency of regulatory T cells in renal cell carcinoma patients and investigation of correlation with survival. *Cancer Immunol Immunother., 56,* 1743-53.

[55] Gabrilovich, D. (2004). Mechanisms and functional significance of tumor-induced dendritic-cell defects. *Nat Rev Immunol., 4,* 941-52.

[56] Zou, W., Machelon, V., Coulomb-L'Hermin, A., Borvak, J., Nome, F., Isaeva, T., Wei, S., Krzysiek, R., Durand-Gasselin, I., Gordon, A., Pustilnik, T., Curiel, D. T., Galanaud, P., Capron, F., Emilie, D. & Curiel, T. J. (2001). Stromal-derived factor-1 in human tumors recruits and alters the function of plasmacytoid precursor dendritic cells. *Nat Med., 7,* 1339-46.

[57] Wei, S., Kryczek, I., Zou, L., Daniel, B., Cheng, P., Mottram, P., Curiel, T., Lange, A. & Zou, W. (2005). Plasmacytoid dendritic cells induce CD8+ regulatory T cells in human ovarian carcinoma. *Cancer Res., 65,* 5020-6.

[58] Jonuleit, H., Schmitt, E., Schuler, G., Knop, J. & Enk, A. H. (2000). Induction of interleukin 10-producing, nonproliferating CD4(+) T cells with regulatory properties by repetitive stimulation with allogenic immature human dendritic cells. *J Exp Med., 192,* 1213-22.

[59] Dhodapkar, M. V., Steinman, R. M., Krasovsky, J., Munz, C. & Bhardwaj, N. (2001). Antigen-specific inhibition of effector T cell function in humans after injection of immature dendritic cells. *J Exp Med., 193,* 233-8.

[60] Chakraborty, N. G., Chattopadhyay, S., Mehrotra, S., Chhabra, A. & Mukherji, B. (2004). Regulatory T-cell response and tumor vaccine-induced cytotoxic T lymphocytes in human melanoma. *Hum Immunol., 65,* 794-802.

[61] Yamazaki, S., Iyoda, T., Tarbell, K., Olson, K., Velinzon, K., Inaba, K. & Steinman, R. M. (2003). Direct expansion of functional CD25+ CD4+ regulatory T cells by antigen-processing dendritic cells. *J Exp Med., 198,* 235-47.

[62] Tarbell, K. V., Yamazaki, S., Olson, K., Toy, P. & Steinman, R. M. (2004). CD25+ CD4+ T cells, expanded with dendritic cells presenting a single autoantigenic peptide, suppress autoimmune diabetes. *J Exp Med., 199,* 1467-77.

[63] Denning, T. L., Granger, S. W., Mucida, D., Graddy, R., Leclercq, G., Zhang, W., Honey, K., Rasmussen, J. P., Cheroutre, H., Rudensky, A. Y. & Kronenberg, M. (2007). Mouse TCRalphabeta+CD8alphaalpha intraepithelial lymphocytes express genes that down-regulate their antigen reactivity and suppress immune responses. *J Immunol., 178,* 4230-9.

[64] Bryn, T., Yaqub, S., Mahic, M., Henjum, K., Aandahl, E. M. & Tasken, K. (2008). LPS-activated monocytes suppress T-cell immune responses and induce FOXP3+ T cells through a COX-2-PGE2-dependent mechanism. *Int Immunol., 20,* 235-45.

[65] Curotto de Lafaille, M. A., Lino, A. C., Kutchukhidze, N. & Lafaille, J. J. (2004). CD25- T cells generate CD25+Foxp3+ regulatory T cells by peripheral expansion. *J Immunol., 173,* 7259-68.

[66] Liang, S., Alard, P., Zhao, Y., Parnell, S., Clark, S. L. & Kosiewicz, M. M. (2005). Conversion of CD4+ CD25- cells into CD4+ CD25+ regulatory T cells in vivo requires B7 costimulation, but not the thymus. *J Exp Med., 201,* 127-37.

[67] Fantini, M. C., Becker, C., Monteleone, G., Pallone, F., Galle, P. R. & Neurath, M. F. (2004). Cutting edge: TGF-beta induces a regulatory phenotype in CD4+CD25- T cells through Foxp3 induction and down-regulation of Smad7. *J Immunol., 172,* 5149-53.

[68] Wan, Y. Y. & Flavell, R. A. (2005). Identifying Foxp3-expressing suppressor T cells with a bicistronic reporter. *Proc Natl Acad Sci.,* U S A *102,* 5126-31.

[69] Yu, P., Lee, Y., Liu, W., Krausz, T., Chong, A., Schreiber, H. & Fu, Y. X. (2005). Intratumor depletion of CD4+ cells unmasks tumor immunogenicity leading to the rejection of late-stage tumors. *J Exp Med., 201,* 779-91.

[70] Seo, N., Hayakawa, S., Takigawa, M. & Tokura, Y. (2001). Interleukin-10 expressed at early tumor sites induces subsequent generation of CD4(+) T-regulatory cells and systemic collapse of antitumor immunity. *Immunology., 103,* 449-57.

[71] Groux, H., O'Garra, A., Bigler, M., Rouleau, M., Antonenko, S., de Vries, J. E. & Roncarolo, M. G. (1997). A CD4+ T-cell subset inhibits antigen-specific T-cell responses and prevents colitis. Nature *389,* 737-42.

[72] Wang, H. Y., Lee, D. A., Peng, G., Guo, Z., Li, Y., Kiniwa, Y., Shevach, E. M. & Wang, R. F. (2004). Tumor-specific human CD4+ regulatory T cells and their ligands: implications for immunotherapy. *Immunity., 20,* 107-18.

[73] Thornton, A. M. & Shevach, E. M. (1998). CD4+CD25+ immunoregulatory T cells suppress polyclonal T cell activation in vitro by inhibiting interleukin 2 production. *J Exp Med., 188,* 287-96.

[74] Piccirillo, C. A., Letterio, J. J., Thornton, A. M., McHugh, R. S., Mamura, M., Mizuhara, H. & Shevach, E. M. (2002). CD4(+)CD25(+) regulatory T cells can mediate suppressor function in the absence of transforming growth factor beta1 production and responsiveness. *J Exp Med., 196,* 237-46.

[75] Li, M. O., Sanjabi, S. & Flavell, R. A. (2006). Transforming growth factor-beta controls development, homeostasis, & tolerance of T cells by regulatory T cell-dependent and -independent mechanisms. *Immunity., 25,* 455-71.

[76] Gorelik, L. & Flavell, R. A. (2001). Immune-mediated eradication of tumors through the blockade of transforming growth factor-beta signaling in T cells. *Nat Med., 7,* 1118-22.

[77] Shevach, E. M., Tran, D. Q., Davidson, T. S. & Andersson, J. (2008). The critical contribution of TGF-beta to the induction of Foxp3 expression and regulatory T cell function. Eur *J Immunol., 38,* 915-7.

[78] Nemunaitis, J., Dillman, R. O., Schwarzenberger, P. O., Senzer, N., Cunningham, C., Cutler, J., Tong, A., Kumar, P., Pappen, B., Hamilton, C., DeVol, E., Maples, P. B., Liu, L., Chamberlin, T., Shawler, D. L. & Fakhrai, H. (2006). Phase II study of belagenpumatucel-L, a transforming growth factor beta-2 antisense gene-modified allogeneic tumor cell vaccine in non-small-cell lung cancer. *J Clin Oncol, 24,* 4721-30.

[79] Hau, P., Jachimczak, P., Schlingensiepen, R., Schulmeyer, F., Jauch, T., Steinbrecher, A., Brawanski, A., Proescholdt, M., Schlaier, J., Buchroithner, J., Pichler, J., Wurm, G.,

Mehdorn, M., Strege, R., Schuierer, G., Villarrubia, V., Fellner, F., Jansen, O., Straube, T., Nohria, V., Goldbrunner, M., Kunst, M., Schmaus, S., Stauder, G., Bogdahn, U., & Schlingensiepen, K. H. (2007). Inhibition of TGF-beta2 with AP 12009 in recurrent malignant gliomas: from preclinical to phase I/II studies. Oligonucleotides *17,* 201-12.

[80] Walker, R. A. & Dearing, S. J. (1992). Transforming growth factor beta 1 in ductal carcinoma in situ and invasive carcinomas of the breast. *Eur J Cancer, 28,* 641-4.

[81] Shim, K. S., Kim, K. H., Han, W. S. & Park, E. B. (1999). Elevated serum levels of transforming growth factor-beta1 in patients with colorectal carcinoma: its association with tumor progression and its significant decrease after curative surgical resection. *Cancer, 85,* 554-61.

[82] Green, E. A., Gorelik, L., McGregor, C. M., Tran, E. H. & Flavell, R. A. (2003). CD4+CD25+ T regulatory cells control anti-islet CD8+ T cells through TGF-beta-TGF-beta receptor interactions in type 1 diabetes. *Proc Natl Acad Sci.,* U S A *100,* 10878-83.

[83] Ruttinger, D., van den Engel, N. K., Winter, H., Schlemmer, M., Pohla, H., Grutzner, S., Wagner, B., Schendel, D. J., Fox, B. A., Jauch, K. W. & Hatz, R. A. (2007). Adjuvant therapeutic vaccination in patients with non-small cell lung cancer made lymphopenic and reconstituted with autologous PBMC: first clinical experience and evidence of an immune response. *J Transl Med, 5,* 43.

[84] Yuan, X. L., Chen, L., Li, M. X., Dong, P., Xue, J., Wang, J., Zhang, T. T., Wang, X. A., Zhang, F. M., Ge, H. L., Shen, L. S. & Xu, D. (2010). Elevated expression of Foxp3 in tumor-infiltrating Treg cells suppresses T-cell proliferation and contributes to gastric cancer progression in a COX-2-dependent manner. Clin Immunol *134,* 277-88.

[85] Schmidt, E. M., Wang, C. J., Ryan, G. A., Clough, L. E., Qureshi, O. S., Goodall, M., Abbas, A. K., Sharpe, A. H., Sansom, D. M. & Walker, L. S. (2009). Ctla-4 controls regulatory T cell peripheral homeostasis and is required for suppression of pancreatic islet autoimmunity. *J Immunol., 182,* 274-82.

[86] Sojka, D. K., Hughson, A. & Fowell, D. J. (2009). CTLA-4 is required by CD4+CD25+ Treg to control CD4+ T-cell lymphopenia-induced proliferation. Eur *J Immunol., 39,* 1544-51.

[87] Grossman, W. J., Verbsky, J. W., Barchet, W., Colonna, M., Atkinson, J. P. & Ley, T. J. (2004). Human T regulatory cells can use the perforin pathway to cause autologous target cell death. *Immunity., 21,* 589-601.

In: Regulatory T Cells
Editor: Ren S. Hayashi, pp. 169-183

ISBN: 978-1-61761-554-2

Chapter 9

Advantages and Disadvantages of and Cooperation between $CD8^+$ and $CD4^+$ Regulatory T Cells

***Haruhiko Suzuki*[1*], *Yusuke Okuno*[1], *and Agustina Tri Endharti*[1, 2, 3]**

[1]Department of Immunology, Nagoya University Graduate School of Medicine, Nagoya, Japan

[2]Japanese Society for Promotion of Science, Tokyo, Japan

[3]Faculty of Medicine, Bravijava University, East Java, Indonesia

Abstract

Adaptive immune responses are based on the cooperative harmony between $CD4^+$ and $CD8^+$ T cells. This principle applies not only to typical effector T cells but also to regulatory T cells. Studies of $CD4^+$ regulatory T cells have greatly advanced after the discovery of $CD4^+CD25^+Foxp3^+$ regulatory T cells (Treg), whereas progress in understanding the dynamics of $CD8^+$ regulatory T cells has been much slower. Although several $CD8^+$ regulatory T cells with different markers have been reported, these cells have not been studied as extensively as $CD4^+$ Treg. Among $CD8^+$ regulatory T cells, the $CD8^+CD122^+$ regulatory T cells in mice and their human counterparts, $CD8^+CXCR3^+$ cells, appear to be the most promising for further study. Following the initial discovery of $CD8^+CD122^+$ regulatory T cells, the same research group has clarified part of the molecular mechanism of these regulatory T cells and determined the important roles of these cells in certain autoimmune diseases by using experimental mouse models. In this review, we summarize the advantages and disadvantages of each $CD8^+$ regulatory T cell population in comparison with $CD4^+$ Treg and other $CD8^+$ regulatory T cells. Furthermore, we present experimental results showing the synergistic effect of

* Corresponding author: Department of Immunology, Nagoya University Graduate School of Medicine, 65 Tsurumai-cho, Showa-ku, Nagoya, 466-8550, Japan., E-mail: harusuzu@med.nagoya-u.ac.jp

$CD8^+CD122^+$ regulatory T cells and $CD4^+$ Treg and suggest that both $CD4^+$ and $CD8^+$ regulatory T cells be used for the treatment of immune diseases.

Introduction

T cells are a central part of the adaptive immune system and control other immune-related cells, including B cells, NK cells, macrophages, and dendritic cells. T cells are further divided into subpopulations according to various criteria. For example, there are different types of T cell receptors (TCRs), including αβTCR-type T cells and γδTCR-type T cells. Based on the accessory marker molecules on the cell surface, αβTCR-type T cells are further classified as $CD4^+$ T cells and $CD8^+$ T cells. T cells can also be functionally classified into helper T cells, cytotoxic T cells, and regulatory (suppressor) T cells. Their CD4 and CD8 surface markers play important roles in their functions. For example, $CD4^+$ T cells are related to helper T cells and $CD8^+$ T cells are mostly thought to be cytotoxic T cells. Regulatory T cells, which were previously known as suppressor T cells, are currently thought to be included in $CD4^+$ T cell populations, but are also known to exist in $CD8^+$ T cell populations. There is no doubt that the knowledge that regulatory T cells are predominantly found in $CD4^+$ T cell populations depends on the discovery of $CD4^+CD25^+$ regulatory T cells [1]. Moreover, discovery of Foxp3 as a master gene for $CD4^+CD25^+$ regulatory T cells further accelerated the study of $CD4^+CD25^+$ regulatory T cells and confirmed the importance of $CD4^+$ regulatory T cells [2]. However, progress in understanding the functions of $CD4^+CD25^+Foxp3^+$ regulatory T cells appears to have slowed recently. It seems that the intense study of $CD4^+CD25^+$ regulatory T cells reached a peak and is now in a plateau phase. There may be many reasons for this, but we should keep in mind that the enormous advances in the study of $CD4^+CD25^+$ regulatory T cells have come at the expense of studies of other regulatory cells. It may be a good time to re-focus attention on other regulatory cells.

Based on the hypothesis described above, the main theme of this review article is to discuss regulatory cells other than $CD4^+CD25^+$ regulatory T cells, with a particular focus on $CD8^+$ regulatory T cells. The aim of this review is not to downplay the importance of $CD4^+CD25^+$ regulatory T cells but rather to reconsider the importance of $CD8^+$ regulatory T cells in comparison and in combination with $CD4^+CD25^+$ regulatory T cells. Therefore, we begin this review article with a discussion of the principles of $CD4^+CD25^+$ regulatory T cells and then discuss other regulatory T cells in comparison with $CD4^+CD25^+$ regulatory T cells.

Essential Summary of $CD4^+CD25^+$ Regulatory T Cells

The existence of $CD4^+CD25^+$ regulatory T cells was first reported by Sakaguchi et al in 1995 [1]. After extensive debate regarding the existence of suppressor T cells, it was considered appropriate to describe this new class of suppressor T cells as "regulatory" T cells instead of "suppressor" T cells. One important aspect of the argument with regard to the $CD4^+CD25^+$ regulatory T cells was that they had been identified as "naturally arising" or "naturally occurring" regulatory T cells. However, it is now known that the same regulatory T cells can be induced to arise from other non-regulatory T cells [3-6]. Importantly, the study of these $CD4^+CD25^+$ regulatory T cells expanded quickly because of the development of an easy and reliable assay. $CD4^+CD25^+$ regulatory T cells are anergic to *in vitro* stimulation with soluble anti-CD3 antibody plus antigen presenting cells (APCs). Therefore, $CD4^+CD25^+$ regulatory T cells themselves never proliferate but suppress the proliferation of other non-regulatory $CD4^+$ T cells in co-cultures. This facilitates evaluation of their proliferation suppressive activity by a simple ^{3}H-thymidine uptake assay. The next driving force that encouraged further study of these $CD4^+CD25^+$ regulatory T cells was the discovery of Foxp3 [2]. Foxp3 is often called the "master gene" of $CD4^+CD25^+$ regulatory T cells and determines the fate of $CD4^+$ T cells with respect to differentiation into regulatory cells.

Because there have been numerous original papers and review articles on the subject of $CD4^+CD25^+$ regulatory T cells, further discussion of these cells is not necessary at this point.

General Cautions with Respect to Discussions of $CD8^+$ Regulatory T Cells

Before the discovery of $CD4^+CD25^+$ regulatory T cells, T cells with suppressive activity were known as suppressor T cells. These cells were the subject of intense investigation. In these studies, the majority of suppressor T cells were $CD8^+$ cells. After gene analysis led to the abandonment of the term "suppressor T cells," the existence of T cells with suppressive activity could not be neglected and many reports describing such T cells continued to be published. The term "T cells with regulatory activity" or similar terms [7-9] were employed to describe these cells. Therefore, it was never doubted that the $CD8^+$ T cell population includes cells with regulatory functions. However, there does not appear to be an accepted method for identifying regulatory T cells. To develop such a method, it is important to select "marker" molecules whose expression is indicative of regulatory T cells.

One more very important point to consider is whether the cells are "naturally arising" (naturally occurring) regulatory cells or "induced" regulatory T cells. This used to be an exclusively important point in the process of expanding the study of $CD4^+CD25^+Foxp3^+$ regulatory T cells that were exclusively described as naturally arising regulatory T cells. Although it was found that induced-type $CD4^+Foxp3^+$ regulatory T cells also exist [3-6], the importance of naturally arising regulatory T cells should not be underestimated. It may be difficult to evaluate the biological relevance of induced-type regulatory T cells *in vivo*. On the other hand, it is easy to evaluate the biological function of the naturally arising regulatory T cells. Conversely, it is very difficult to determine the biological function of induced type regulatory T cells *in vivo*. In this context, we provide a discussion of the different classes of $CD8^+$ regulatory T cells.

$CD8^+$ Regulatory T Cells without Distinct Markers

Before describing $CD8^+$ T cells with distinct markers that distinguish regulatory T cells from other non-regulatory T cells, we will discuss examples of representative $CD8^+$ regulatory cells without special markers. Most of these regulatory cells were reported before the importance of $CD4^+CD25^+Foxp3^+$ cells became recognized and before arriving at the concept that regulatory T cells could be distinguished from other cells by special markers. In 1986, Elmasry et al. first used the term "regulatory T cells" although most investigators, including Elmasry et al, still mainly used the term "suppressor T cells" [10]. In 1993, Kemeny et al published three papers describing rat $CD8^+$ regulatory T cells [11-13]. In the 1990s, there appeared several reports that used the term "regulatory T cells" in context of $CD8^+$ T cell populations [14-17]. These cells were thought to be independent of the $CD4^+CD25^+$ regulatory T cells identified by Sakaguchi et al. After 2000, additional reports of $CD8^+$ regulatory T cells were published, possibly as a result of the increasing interest in $CD4^+CD25^+$ "regulatory T cells" [18-22]. In any case, it was not possible to judge whether those $CD8^+$ regulatory T cells were of the "naturally arising" type or the "induced" type because there was no information about the existence of the regulatory cells or their precursors at that time. Although these reports might have been evaluated with respect to the findings that regulatory T cells appear in various situations, significant advances in understanding the function of $CD8^+$ regulatory T cells were not made.

$CD8^+CD28^-$ Regulatory T Cells

When referring to studies on $CD8^+CD28^-$ regulatory T cells, consider whether human or rodent cells were used. The majority of reports describing $CD8^+CD28^-$ regulatory T cells used human cells, and many of these reports have been published by Suciu-Foca et al. The earliest of these reports were published in 1998 [23, 24]. In these papers, it was shown that $CD8^+CD28^-$ T cells suppress the MHC-reactive T cell allotype and the MHC-reactive T cell xenotype. Although it was not indicated whether the T cells were naturally arising or induced, it is a fact that human $CD8^+$ cells are clearly separated into $CD28^+$ and $CD28^-$ cells. It is unknown whether these preexisting $CD8^+CD28^-$ cells themselves are regulatory cells or the precursors of regulatory cells; however, it is possible that in humans, $CD8^+CD28^-$ cell populations contain the naturally arising type of regulatory T cells [25-31].

The problematic observations that lead to confusion with respect to $CD8^+CD28^-$ regulatory T cells in humans appear in rodent $CD8^+CD28^-$ cells [32-34]. Studies of the regulatory activity of rodent $CD8^+CD28^-$ cells were mainly performed using the experimental autoimmune encephalomyelitis (EAE) model and the inflammatory bowel disease (IBD) model. The results of these studies revealed significant effects of $CD8^+CD28^-$ regulatory T cells *in vivo*. First, however, the study of CD28-knockout mice should be more carefully interpreted. Although the authors of these papers were aware that $CD8^+$ cells obtained from CD28-knockout mice were not equivalent to $CD8^+CD28^-$ cells obtained from wild-type mice, their primary motivation for these experiments appears to have been based on the consensus that human $CD8^+CD28^-$ cells were regulatory T cells. Second, rodent $CD8^+$ cells are not clearly distinguishable as $CD28^+$ cells and $CD28^-$ cells. This is obviously different from the human $CD8^+$ cells. The levels of expression of CD28 continuously change in $CD28^+$ cells and $CD28^-$ cells, and this makes it quite challenging to distinguish between $CD28^+$ cells and $CD28^-$ cells. In our opinion, rodent $CD8^+CD28^-$ cells (if separated by force) should not be considered equivalent to human $CD8^+CD28^-$ cells. It appears that the authors were of the belief that human $CD8^+CD28^-$ cells represent regulatory T cells and it is also suggestive that no subsequent reports of investigations of rodent $CD8^+CD28^-$ regulatory T cells have been published.

CD8αα Regulatory T Cells with MHC-Class Ib Restriction

Reports of the existence of $CD8^+$ regulatory T cells in humans and in mice were first published in 1995 [35, 36]. The response of $CD8^+$ regulatory T cells is restricted by Qa-1 (mouse) or HLA-E (human). They are one of the molecules of MHC class Ib. After the first appearance of such Qa-1-restricted $CD8^+$ regulatory T cells, only a few reports regarding the Qa-1-restricted regulatory T cells were published in the following decades [37-39]. In 2006, Kumar et al found that Qa-1-restricted regulatory T cells were characterized by the unique expression of CD8αα homodimer [40-42]. An additional unique feature of these regulatory T cells was that they possessed cytotoxic activity and thus, they were considered to regulate T cells by killing the activated T cells.

It appears that the authors of these reports on CD8αα regulatory T cells were not concerned about whether $CD8\alpha\alpha^+$ regulatory T cells are of the naturally arising-type or the induced-type. However, simple analyses of mouse spleen cells or lymph node cells revealed that a small percentage of cells (less than 1% of total $CD8^+$ cells) are of the $CD8\alpha\alpha^+$ phenotype. This suggests that the existence of the naturally arising type of $CD8\alpha\alpha^+$ regulatory T cells was unlikely. Comparisons of these cells with the later described $CD8^+CD122^+$ regulatory T cells indicate that these $CD8\alpha\alpha^+$ regulatory T cells may represent a further differentiated subpopulation of $CD8^+CD122^+$ regulatory T cells.

$CD8^+CD103^+$ Regulatory T Cells

This new type of regulatory T cells was first identified by ten Berge et al in human peripheral blood lymphocytes (PBLs) in 2006 [43]. The corresponding regulatory T cells in rodents have not yet been reported. As described in the review article, the cells were categorized as adaptive (induced) type regulatory T cells [44]. However, it was also indicated that approximately 4% of the total $CD8^+$ cells in human PBLs express CD103. This leaves the possibility that these cells may be a naturally arising type of regulatory T cells [43, 45]. Identification of the rodent counterparts of these cells is awaited.

$CD8^+$ Regulatory T Cells That Express Markers of $CD4^+$ Regulatory T Cells

There have been several reports that $CD8^+$ regulatory cells express markers, which are also expressed by $CD4^+$ regulatory T cells. These findings may be based on the idea that regulatory T cells share common markers even if the $CD4^+$ cells and $CD8^+$ cells are fundamentally different. As $CD4^+$ regulatory T cells are characterized by the expression of CD25, $CD8^+CD25^+$, T cells were first examined to determine whether they possess regulatory activity [46, 47]. In one case, the regulatory activity of $CD8^+CD25^+$ cells was examined in comparison with that of $CD4^+CD25^+$ cells, and the $CD8^+CD25^+$ cells showed significant regulatory activity in both *ex vivo* and *in vivo* experiments although the regulatory activity was much weaker than that of $CD4^+CD25^+$ regulatory T cells [46]. In spleen and lymph node tissues of mice, $CD25^+$ cells represent less than 1% of the total of $CD8^+$ cells. It is therefore quite unlikely that $CD8^+CD25^+$ cells function as naturally arising-type regulatory T cells.

Because CD25 (IL-2 receptor α chain) expression is induced in activated T cells, this finding might merely indicate that $CD8^+$ cells became activated to express CD25 under the unusual conditions which promote GVH (graft vs. host) reaction, and some of the activated cells therefore showed weak regulatory activity. Another report described $CD8^+CD25^+$ regulatory T cells in MHC class II-deficient mice [47]. However, MHC class II-deficient mice lack the majority of $CD4^+$ cells and therefore production of $CD8^+CD25^+$ regulatory T cells that function in a manner similar to $CD4^+CD25^+$ regulatory T cells could compensate for the lack of certain critical activities of $CD4^+CD25^+$ regulatory T cells.

As CD25 is not a specific marker of $CD4^+$ regulatory T cells, another possible approach for identification of $CD8^+$ regulatory T cells is to search for $CD8^+$ cells expressing Foxp3, which is known to be highly and specifically expressed in $CD4^+$ regulatory T cells. One disadvantage to using Foxp3 as a specific marker for $CD4^+$ regulatory T cells is that Foxp3 is found mainly in the nucleus and we cannot detect Foxp3 in live cells with the usual methods. There have been several reports describing $CD8^+Foxp3^+$ regulatory T cells, most of which are studies using human samples [48-52]. These reports have suggested that human $CD8^+$ regulatory T cells may also express Foxp3. In the case of mice, total $CD8^+$ cells do not show expression of this protein as examined by both real time RT-PCR analysis methods and by intra-cellular staining with anti-Foxp3 antibody (our unpublished observation). This suggests that there are differences between Foxp3 expression in human $CD8^+$ regulatory T cells and mouse $CD8^+$ regulatory T cells. In any case, the possibility of $CD8^+Foxp3^+$ regulatory T cells of the naturally arising-type existing in mice is considered to be very low. Further studies of $CD8^+Foxp3^+$ regulatory cells in mice are needed.

Other $CD8^+$ Regulatory T Cells Identified by Different Marker Molecules

There are other $CD8^+$ regulatory T cells defined by other markers, with the exception of CD122, which will be described in the following paragraphs. Examples of such regulatory T cells are $CD8^+CD11c^+$ cells, $CD8^+CD45RC^{low}$ cells, $CD8^+PD\text{-}1^+$ cells, and $CD8^+CD75s^+$ cells.

$CD8^+CD75s^+$ regulatory T cells in mice were first reported in 1995 [53]. These cells were classified as induced-type regulatory T cells. Thus, their history is somehow longer than that of the other $CD8^+$ regulatory T cells. However, there have been few reports on these regulatory T cells. This may suggest that there have been difficulties in the analyses or fundamental problems encountered while working with these cells.

$CD8^+CD45RC^{low}$ cells in rats were first reported in 2004. These cells were classified as naturally arising-type regulatory T cells [54, 55]. These cells are only found in rats. Counterpart cells in mice and humans have not been reported yet. This presents limitations with respect to analysis of these cells because experimental systems including antibody panels for rats are not as complete as those for mice and humans.

$CD8^+PD\text{-}1^+$ regulatory T cells were first reported in 2007 [56]. These cells were identified in mice and classified as induced-type regulatory T cells. Accurate evaluation of these regulatory T cells will be dependent upon further analysis and reports.

$CD8^+CD11c^+$ regulatory T cells were also identified in mice and classified as induced-type-regulatory T cells. There is a low content of ~3% $CD11c^+$ cells in the total $CD8^+$ cell population [57]. However, the authors' opinion is that such preexisting $CD8^+CD11c^+$ cells are not the precursor of induced $CD8^+CD11C^+$regulatory T cells. These regulatory T cells should be further analyzed to determine their developmental origin and other characteristics.

$CD8^+CD122^+$ Regulatory T Cells

$CD8^+CD122^+$ regulatory T cells in mice were first reported in 2004 by Suzuki and colleagues [58]. CD122 is the β subunit of IL-2 receptor and IL-15 receptor, indicating a clear contrast with CD25 (the α subunit of IL-2 receptor) which is expressed in $CD4^+CD25^+$ regulatory T cells. These cells are both classified as naturally arising regulatory T cells. In spleen or lymph node tissue of wild-type 6~10-week-old C57BL/6 mice, $CD122^+$ cells represent 10–15% of the total of $CD8^+$ cells. This is in agreement with the observation that ~10% of total $CD4^+$ cells are also $CD25^+$. Because most $CD8^+$ cells express low levels of CD122, it might be better to express this characteristic as "$CD122^{high}$ cells" instead of "$CD122^+$" cells. However, it is clearly seen in the flow cytometric analysis where the distinctive line between "$CD122^{-\text{ or low}}$ cells" and "$CD122^{+\text{ or high}}$ cells." is. We prefer to express this characteristic as "$CD122^+$ cells" rather than "$CD122^{high}$ cells" and will do so throughout this article.

Discovery of $CD8^+CD122^+$ Regulatory T Cells

It has been established that $CD8^+CD122^+$ regulatory T cells were discovered through the analysis of CD122-knockout mice [59]. CD122-knockout mice have various abnormalities in hematopoietic cells and the fundamental cause of the abnormalities is the increased levels of abnormally activated T cells, which are $CD4^+$ and $CD8^+$. By analyzing the mechanism which produces the increased levels of activated T cells in CD122-knockout mice, we concluded that it is caused by a defect in regulatory T cells. Now, because it is clear that CD122-knockout mice lack functional $CD4^+CD25^+$ regulatory T cells, the phenotype of CD122-knockout mice is caused to some extent by the defect of $CD4^+CD25^+$ regulatory T cells. However, it is also true that the overall phenotypes of CD122-knockout mice are not merely explained by the defect of the $CD4^+CD25^+$ regulatory T cells. For example, increased granulocytopoiesis and decreased erythropoiesis are not featured in Foxp3-knockout mice or scurfy mice that only lack $CD4^+CD25^+$ regulatory T cells. Combined with the finding that there are $CD8^+CD122^+$ cells in the spleen and lymph nodes of mice, we hypothesized that these $CD8^+CD122^+$ cells are other regulatory T cells. When deficient in these cells, the overall phenotype of CD122-knockout mice is more complex. The result of experiments showing that transfer of $CD8^+CD122^+$ cells taken from wild-type C57BL/6 mice into CD122-knockout neonates dramatically improved the phenotype of $CD8^+CD122^+$ cell-transferred mice was the first evidence to prove that $CD8^+CD122^+$ cells are real regulatory T cells. Additional evidence was obtained from the experiments involving transfer of $CD8^+CD122^-$ cells taken from wild-type C57BL/6 mice into lymphocyte-deficient *RAG* (recombination activating gene)-2-knockout mice. This caused hematopoietic abnormalities similar to the abnormalities exhibited by CD122-knockout mice and the mice died within 10 weeks of the T cell transfer. However, co-transfer of $CD8^+CD122^+$ cells together with $CD8^+CD122^-$ cells caused no abnormalities.

Suppressive Mechanism of CD8⁺CD122⁺ Regulatory T Cells

The mechanism of suppression by $CD8^+CD122^+$ regulatory T cells was also investigated by Suzuki and colleagues. By using an *ex vivo* culture system to reproduce the suppressive action of $CD8^+CD122^+$ regulatory T cells, it was proven that these regulatory T cells first recognize activated T cells as their target cells for regulation. In this recognition phase, interaction of conventional MHC class I molecules on the target-activated T cells with TCR on the $CD8^+CD122^+$ regulatory T cells is required [60]. Furthermore, in the recognition phase, the interaction of CD80/86 on the target cells with CD28 on the $CD8^+CD122^+$ regulatory T cells was found to be important [61]. Once $CD8^+CD122^+$ regulatory T cells recognize the target-activated T cells, the regulatory T cells convert to the active form of regulatory T cells and start to produce IL-10, which suppresses the action of activated T cells [62].

Application of CD8⁺CD122⁺ Regulatory T Cells to Autoimmune Diseases

The *in vivo* relevance of $CD8^+CD122^+$ regulatory T cells has also been investigated by Suzuki et al who examined the effect of $CD8^+CD122^+$ regulatory T cells in an EAE model [63]. Mice with depleted levels of $CD8^+CD122^+$ regulatory T cells induced by *in vivo* administration of anti-CD122 antibody were tested by inducing EAE. In such $CD8^+CD122^+$ cell-depleted mice, the symptoms of EAE, including paralysis of tail and limbs, continued for more than 20 days after induction of EAE. In contrast, mice which had normal levels of $CD8^+CD122^+$ rapidly recovered from the peak of EAE symptoms 14–16 days after induction of EAE. This experimental result indicates the possibility that $CD8^+CD122^+$ regulatory T cells play an important role in the recovery phase of EAE. To confirm this possibility, $CD8^+CD122^+$ cells collected from other syngeneic C57BL/6 mice were administered to anti-CD122 antibody-treated mice at the peak of EAE symptoms. The result clearly indicated reduction of EAE symptoms in mice treated with the $CD8^+CD122^+$ cells but not in untreated control mice. These experimental results suggested the possibility that individuals suffering from autoimmune diseases might benefit from administration of $CD8^+CD122^+$ regulatory T cells. Another striking effect of $CD8^+CD122^+$ regulatory T cells related to possible prevention and cure of IBD was seen in mouse models. Adoptive transfer of $CD4^+CD45RB^+$ cells into lymphocyte-deficient RAG-2-knockout mice causes IBD. The mechanism of IBD development is not well known, but $CD4^+CD45RB^+$ cells, which lack $CD4^+(CD25^+Foxp3^+)$ regulatory T cells, are thought to become dysregulated and cause IBD. This hypothesis is probably correct because transfer of $CD4^+CD25^+$ cells together with $CD4^+CD45RB^+$ cells does not cause IBD. However, the transfer of $CD8^+CD122^+$ cells together with $CD4^+CD45RB^+$ cells also prevents the development of IBD, suggesting a more complex mechanism of IBD development. Whatever the mechanism, $CD8^+CD122^+$ cells again had regulatory activity, which is almost as strong as that of $CD4^+CD25^+$ cells *in vivo*. This confirms the relevance of these regulatory T cells.

Human Counterpart of Mouse CD8⁺CD122⁺ Regulatory T Cells

Although mouse $CD8^+$ regulatory T cells were distinguished by the expression of CD122, there are no $CD8^+CD122^+$ cells in human PBLs [64]. As $CD8^+CD122^+$ regulatory T cells are essential and irreplaceable in mice, it is quite unlikely that cells with same function as mouse $CD8^+CD122^+$ cells do not exist in humans. DNA microarray analysis comparing gene expression profiles between $CD8^+CD122^+$ cells and $CD8^+CD122^-$ cells revealed that one chemokine receptor, CXCR3, was expressed exclusively in $CD8^+CD122^+$ cells. In mice, $CD8^+CD122^+$ cells and $CD8^+CXCR3^+$ cells were found to largely overlap, whereas $CD8^+CD122^+$ cells were found to be absent while $CD8^+CXCR3^+$ cells are present in human PBLs. Although the *ex vivo* assay for regulatory T cells is only available for human $CD8^+CXCR3^+$ cells, these cells were found to produce IL-10 when they were properly stimulated and were also found to suppress production of IFN-γ from $CD8^+CXCR3^-$ cells, indicating that human $CD8^+CXCR3^+$ cells possess the same function as mouse $CD8^+CD122^+$ regulatory T cells. Taken together, $CD8^+CXCR3^+$ cells are supposed to be the counterpart of mouse $CD8^+CD122^+$ ($CXCR3^+$) regulatory T cells. $CD4^+$ regulatory T cells are commonly distinguished by the expression of CD25 and Foxp3 both in mice and humans [65, 66]. On the other hand, most $CD8^+$ regulatory T cells have only been identified either in humans or in rodents. Although the marker is different between mice and humans, it should be noted that functionally corresponding regulatory T cells are identified in mice and in humans in the case of $CD8^+$ ($CD122^+$) $CXCR3^+$ regulatory T cells [64].

The Search for Additional Specific Markers of $CD8^+CD122^+$ Regulatory T Cells

As it is problematic from the beginning of identification of $CD8^+CD122^+$ regulatory T cells, CD122 is not a specific marker of $CD8^+$ regulatory T cells. CD122 is strongly expressed in activated $CD8^+$ T cells and memory $CD8^+$ T cells. Thus, $CD8^+CD122^+$ cells must contain a fairly large percentage of non-regulatory T cells including memory T cells, even if the cells are obtained from very young mice [67]. At present, the most convincing feature of regulatory T cells may be whether the cells gain the ability to produce IL-10, which is the most reliable effector molecule of $CD8^+CD122^+$ regulatory T cells. However, detection of IL-10 is only available either for average production from bulk cells (ELISA of culture supernatant, etc.), which cannot identify which individual cells are IL-10-producing regulatory cells, or in assays that kill the cells (ELISPOT, intracellular staining, RT-PCR, etc.), which places limitations on subsequent experiments. Therefore, more specific maker molecules are needed, which are expressed predominantly in real regulatory T cells. It is preferable that these molecules are expressed on the cell surface.

Cooperation of $CD4^+CD25^+$ Regulatory T Cells and $CD8^+CD22^+$ Regulatory T Cells

We recently obtained preliminary experimental results that indicate cooperative activity of $CD4^+CD25^+$ regulatory T cells and $CD8^+CD122^+$ regulatory T cells in two different experimental systems. In our opinion, the results are reasonable. There have been no previous reports describing cooperative work between $CD4^+$ regulatory T cells and $CD8^+$ regulatory T cells,. Since the results obtained by the analysis of a single regulatory T cell population is becoming limited, it may be an appropriate time to begin research work to investigate at least two different types of regulatory T cells and clarify their shared roles, cooperative activity, and other characteristics.

Concluding Remarks

In this review article, we have described the essential characteristics of $CD4^+CD25^+$ regulatory T cells and several different $CD8^+$ regulatory T cells. The majority of $CD8^+$ regulatory T cells are the $CD8^+CD122^+$ regulatory T cells that we are presently investigating. We believe that these $CD8^+CD122^+$ regulatory T cells (human counterpart: $CD8^+CXCR3^+$ regulatory T cells) are the only $CD8^+$ regulatory T cell population that could be compared with $CD4^+CD25^+$ regulatory T cells. Our reasons for this are as follows: **(i) The regulatory activity of the cells was clearly demonstrated in both *in vivo* and *ex vivo* experiments. (ii) A regulatory T cell population with the same function was identified in both humans and mice. (iii) Both types of regulatory T cells account for 5~15% of the total $CD4^+$ or $CD8^+$ T cell population, which is considered to be a reasonable percentage range of naturally arising regulatory T cells.** Studies to clarify the relationship between the two different types of regulatory T cells may open new avenues in this research field and guide the formulation of strategies for using these regulatory T cells in medicine.

References

[1] Immunologic self-tolerance maintained by activated T cells expressing IL-2 receptor α-chains (CD25). Breakdown of a single mechanism of self-tolerance causes various autoimmune diseases. Sakaguchi S, Sakaguchi N, Asano M, Itoh M, Toda M. *J Immunol.* 1995, 155, 1151-1164.

[2] Control of regulatory T cell development by the transcription factor Foxp3. Hori S, Nomura T, Sakaguchi S. *Science.* 2003, 299, 1057-1061.

[3] Natural and induced $CD4^+CD25^+$ cells educate $CD4^+CD25^-$ cells to develop suppressive activity: the role of IL-2, TGF-β, and IL-10. Zheng SG, Wang JH, Gray JD, Soucier H, Horwitz DA. 2004. *J Immunol.* 172, 5213-5221.

[4] Conversion of peripheral $CD4^+CD25^-$ naive T cells to $CD4^+CD25^+$ regulatory T cells by TGF-β induction of transcription factor Foxp3. Chen W, Jin W, Hardegen N, Lei KJ, Li L, Marinos N, McGrady G, Wahl SM. 2003. *J Exp Med.,* 198, 1875-1886.

[5] TGF-β induces Foxp3 T-regulatory cells from $CD4^+CD25^-$ precursors. In: S; Fu, N; Zhang, AC; Yopp, D; Chen, M; Mao, D; Chen, H; Zhang, Y; Ding, JS. Bromberg, 2004, *Am J Transplant.,* 4, 1614-1627.

[6] Cutting edge: TGF-β induces a regulatory phenotype in $CD4^+CD25^-$ T cells through Foxp3 induction and down-regulation of Smad7. Fantini MC, Becker C, Monteleone G, Pallone F, Galle PR, Neurath MF. 2004. *J Immunol.,* 172, 5149-5153.

[7] Functional analysis of human T cell subsets defined by monoclonal antibodies. VI. Distinct and opposing immunoregulatory functions within the OKT8+ population.

Thomas Y, Rogozinski L, Rabbani L, Chess A, Goldstein G, Chess L. *J Mol Cell Immunol.,* 1984, 1, 103-113.

[8] Human suppressor T cells: induction, differentiation, and regulatory functions. Rich RR, elMasry MN, Fox EJ. *Hum Immunol.* 1986, 17, 369-387.

[9] Regulatory activity of the human CD8$^+$ cell subset: a comparison of CD8$^+$ cells from the intestinal lamina propria and blood. Lee A, Sugerman H, Elson CO. *Eur J Immunol.,* 1988, 18, 21-27.

[10] Opposing immunoregulatory functions of CD8$^+$ lymphocytes: a requirement for monocytes in suppressor cell induction. Elmasry MN, Fox EJ, Rich RR. *J Immunol.,* 1986, 137, 2468-2477.

[11] Ricin enhances IgE responses by inhibiting a subpopulation of early-activated IgE regulatory CD8$^+$ T cells. Diaz-Sanchez D, Lee TH, Kemeny DM. *Immunology.,* 1993, 78, 226-236.

[12] Elimination of IgE regulatory rat CD8$^+$ T cells in vivo differentially modulates interleukin-4 and interferon-γ but not interleukin-2 production by splenic T cells. Diaz-Sanchez D, Noble A, Staynov DZ, Lee TH, Kemeny DM. *Immunology.* 1993, 78, 513-519.

[13] Elimination of IgE regulatory rat CD8$^+$ T cells in vivo increases the co-ordinate expression of Th2 cytokines IL-4, IL-5 and IL-10. Noble A, Staynov DZ, Diaz-Sanchez D, Lee TH, Kemeny DM. *Immunology,* 1993, 80, 326-329.

[14] Inhibition of a novel model of murine experimental autoimmune orchitis by intravenous administration with a soluble testicular antigen: participation of CD8$^+$ regulatory T cells. In: A; Mukasa, M; Itoh, Y; Tokunaga, C; Hiramine, K. Hojo, *Clin Immunol Immunopathol.,* 1992, 62, 210-219.

[15] Suppression of tumour-specific cytotoxic T-cell responses against the syngeneic BALB/c plasmacytoma ADJ-PC-5 by tumour-induced CD8$^+$ regulatory T cells via IFN-γ. Pauels HG, Specht C, Becker C, Kölsch E. *Scand J Immunol.,* 1996, 43, 421-430.

[16] Intraocular injection of class II-restricted peptide induces an unexpected population of CD8 regulatory cells. Kosiewicz MM, Streilein JW. *J Immunol.,* 1996, 157, 1905-1912.

[17] Antigen presenting cells integrate opposing signals from CD4$^+$ and CD8$^+$ regulatory T lymphocytes to arbitrate the outcomes of immune responses. Lederman S, Suciu-Foca N. *Hum Immunol.,* 1999, 60, 533-561.

[18] In vitro generation of regulatory CD8$^+$ T cells similar to those found in mice with anterior chamber-associated immune deviation. Kezuka T, Streilein JW. *Invest Ophthalmol Vis Sci.,* 2000, 41, 1803-1811.

[19] Blockade of CTLA-4 signals inhibits Th2-mediated murine chronic graft-versus-host disease by an enhanced expansion of regulatory CD8$^+$ T cells. In: J, Sakurai, J; Ohata, K; Saito, H; Miyajima, T; Hirano, T; Kohsaka, S; Enomoto, K; Okumura, M. Azuma, *J Immunol.,* 2000, 164, 664-669.

[20] Oral exposure to alloantigen generates intragraft CD8$^+$ regulatory cells. In: J; Zhou, RI; Carr, RS; Liwski, AW; Stadnyk, TD. Lee, *J Immunol.,* 2001, 167, 107-113.

[21] Antigen-bearing immature dendritic cells induce peptide-specific CD8$^+$ regulatory T cells in vivo in humans. In: MV; Dhodapkar, RM. Steinman, *Blood.,* 2002, 100, 174-177.

[22] Generation of human CD8 T regulatory cells by CD40 ligand-activated plasmacytoid dendritic cells. In: M; Gilliet, YJ. Liu, *J Exp Med.,* 2002, 195, 695-704.

[23] Specific suppression of T helper alloreactivity by allo-MHC class I-restricted $CD8^+CD28^-$ T cells. Liu Z, Tugulea S, Cortesini R, Suciu-Foca N. *Int Immunol.,* 1998, 10, 775-783.

[24] Specific suppression of human $CD4^+$ Th cell responses to pig MHC antigens by $CD8^+CD28^-$ regulatory T cells. In: R; Ciubotariu, AI; Colovai, G; Pennesi, Z; Liu, D; Smith, P; Berlocco, R; Cortesini, N. Suciu-Foca, *J Immunol.,* 1998, 161, 5193-5202.

[25] Overlap between molecular markers expressed by naturally occurring $CD4^+CD25^+$ regulatory T cells and antigen specific $CD4^+CD25^+$ and $CD8^+CD28^-$ T suppressor cells. In: L; Scotto, AJ; Naiyer, S; Galluzzo, P; Rossi, JS; Manavalan, S; Kim-Schulze, J; Fang, RD; Favera, R; Cortesini, N; Suciu-Foca, *Hum Immunol.,* 2004, 65, 1297-1306.

[26] Reduced immunosuppression in pediatric liver-intestine transplant recipients with $CD8^+CD28^-$ T-suppressor cells. In: R; Sindhi, JS; Manavalan, A; Magill, N; Suciu-Foca, A; Zeevi, *Hum Immunol.,* 2005, 66, 252-257.

[27] Foxp3 expressing $CD4^+CD25^+$ and $CD8^+CD28^-$ T regulatory cells in the peripheral blood of patients with lung cancer and pleural mesothelioma. In: F; Meloni, M; Morosini, N; Solari, I; Passadore, C; Nascimbene, M; Novo, M; Ferrari, M; Cosentino, F; Marino, E; Pozzi, AM. Fietta, *Hum Immunol*., 2006, 67, 1-12.

[28] $CD8^+CD28^-$ T regulatory lymphocytes inhibiting T cell proliferative and cytotoxic functions infiltrate human cancers. In: G; Filaci, D; Fenoglio, M; Fravega, G; Ansaldo, G; Borgonovo, P; Traverso, B; Villaggio, A; Ferrera, A; Kunkl, M; Rizzi, F; Ferrera, P; Balestra, M, Ghio, P; Contini, M; Setti, D; Olive, B; Azzarone, G; Carmignani, JL; Ravetti, G; Torre, F. Indiveri, *J Immunol.,* 2007, 179, 4323-4334.

[29] The role of $CD8^+CD28^-$ regulatory cells in suppressing myasthenia gravis-associated responses by a dual altered peptide ligand. In: H; Ben-David, A; Sharabi, M; Dayan, M; Sela, E. Mozes, *Proc Natl Acad Sci U S A.,* 2007, 104, 17459-17464.

[30] Advancements on phenotypic and functional characterization of non-antigen-specific $CD8^+CD28^-$ regulatory T cells. In: D; Fenoglio, F; Ferrera, M; Fravega, P; Balestra, F; Battaglia, M; Proietti, C; Andrei, D; Olive, LC; Antonio, F; Indiveri, G. Filaci, *Hum Immunol.,* 2008, 69, 745-50.

[31] Antigen-specific recognition is critical for the function of regulatory $CD8^+CD28^-$ T cells. In: P; Renner, FC; Popp, E; Eggenhofer, P; Slowik, P; Piso, EK; Geissler, HJ; Schlitt, MH. Dahlke, *Transpl Immunol.,* 2010, 22, 144-149.

[32] Regulatory functions of $CD8^+CD28^-$ T cells in an autoimmune disease model. In: N; Najafian, T; Chitnis, AD; Salama, B; Zhu, C; Benou, X; Yuan, MR; Clarkson, MH; Sayegh, SJ. Khoury, *J Clin Invest.,* 2003, 112, 1037-1048.

[33] $CD8^+CD28^-$ regulatory T lymphocytes prevent experimental inflammatory bowel disease in mice. In: I; Ménager-Marcq, C; Pomié, P; Romagnoli, JP; Van Meerwijk, *Gastroenterology.,* 2006, 131, 1775-1785.

[34] Changes of $CD8^+CD28^-$ T regulatory cells in rat model of colitis induced by 2,4-dinitrofluorobenzene. In: WB; Xiao, YL. Liu, *World J Gastroenterol.,* 2003, 9, 2528-2532.

[35] Human $CD8^+$ T lymphocyte clones specific for T cell receptor V β families expressed on autologous $CD4^+$ T cells. In: R; Ware, H; Jiang, N; Braunstein, J; Kent, E; Wiener, B; Pernis, L. Chess, *Immunity.,* 1995, 2, 177-184.

[36] Murine $CD8^+$ T cells that specifically delete autologous $CD4^+$ T cells expressing V β 8 TCR: a role of the Qa-1 molecule. In: H; Jiang, R; Ware, A; Stall, L; Flaherty, L; Chess, B. Pernis, *Immunity.*, 1995, 2, 185-194.
[37] Analysis of regulatory CD8 T cells in Qa-1-deficient mice. In: D; Hu, K; Ikizawa, L; Lu, ME; Sanchirico, ML; Shinohara, H. Cantor, *Nat Immunol.*, 2004, 5, 516-523.
[38] The Qa-1 dependent $CD8^+$ T cell mediated regulatory pathway. In: H. Jiang, *Cell Mol Immunol.*, 2005, 2, 161-167.
[39] Qa-1/HLA-E-restricted regulatory $CD8^+$ T cells and self-nonself discrimination: an essay on peripheral T-cell regulation. In: H; Jiang, L. Chess, *Hum Immunol.*, 2008, 69, 721-727.
[40] Regulation of immunity by a novel population of Qa-1-restricted $CD8\alpha\alpha^+TCR\alpha\beta^+$ T cells. In: X; Tang, I; Maricic, N; Purohit, B; Bakamjian, LM; Reed-Loisel, T; Beeston, P; Jensen, V. Kumar, *J Immunol.*, 2006, 177, 7645-7655.
[41] Anti-TCR antibody treatment activates a novel population of nonintestinal CD8 $\alpha\alpha^+$ TCR $\alpha\beta^+$ regulatory T cells and prevents experimental autoimmune encephalomyelitis. In: X; Tang, I; Maricic, V. Kumar, *J Immunol.*, 2007, 178, 6043-6050.
[42] Dendritic cells use endocytic pathway for cross-priming class Ib MHC-restricted $CD8\alpha\alpha^+TCR\alpha\beta^+$ T cells with regulatory properties. In: TR; Smith, X; Tang, I; Maricic, Z; Garcia, S; Fanchiang, V. Kumar, *J Immunol.*, 2009, 182, 6959-6568.
[43] CD103 is a marker for alloantigen-induced regulatory $CD8^+$ T cells. Uss In: E; Rowshani, AT; Hooibrink, B; Lardy, NM; van Lier, RA; ten Berge, IJ. *J Immunol.* 2006, 177, 2775-2783.
[44] Alloantigen-induced regulatory $CD8^+CD103^+$ T cells. In: SD; Koch, E; Uss, RA; Van Lier, IJ;.Ten Berge, *Hum Immunol.*, 2008, 69, 737-744.
[45] Rapamycin enhances the number of alloantigen-induced human $CD103^+CD8^+$ regulatory T cells in vitro. In: E; Uss, SL; Yong, B; Hooibrink, RA; Van Lier, IJ. Ten Berge, *Transplantation.*, 2007, 83, 1098-1106.
[46] Tracking ex vivo-expanded $CD4^+CD25^+$ and $CD8^+CD25^+$ regulatory T cells after infusion to prevent donor lymphocyte infusion-induced lethal acute graft-versus-host disease. In: G; Xia, M; Kovochich, RL; Truitt, BD. Johnson, *Biol Blood Marrow Transplant.*, 2004, 10, 748-760.
[47] Peripheral $CD8^+CD25^+$ T lymphocytes from MHC class II-deficient mice exhibit regulatory activity. In: B; Bienvenu, B; Martin, C; Auffray, C; Cordier, C; Bécourt, B. Lucas, *J Immunol.*, 2005, 175, 246-253.
[48] Parallel expansion of human virus-specific $FoxP3^-$ effector memory and de novo-generated $FoxP3^+$ regulatory $CD8^+$ T cells upon antigen recognition in vitro. In: E; Billerbeck, HE; Blum, R; Thimme, *J Immunol.*, 2007, 179, 1039-1048.
[49] $CD8^+$ $Foxp3^+$ regulatory T cells mediate immunosuppression in prostate cancer. In: Y; Kiniwa, Y; Miyahara, HY; Wang, W; Peng, G; Peng, TM; Wheeler, TC; Thompson, LJ; Old, RF. Wang, *Clin Cancer Res.*, 2007, 13, 6947-6958.
[50] Generation of highly suppressive adaptive $CD8^+CD25^+FOXP3^+$ regulatory T cells by continuous antigen stimulation. In: M; Mahic, K; Henjum, S; Yaqub, BA; Bjørnbeth, KM; Torgersen, K; Taskén, EM. Aandahl, *Eur J Immunol.*, 2008, 38, 640-646.

[51] Costimulatory strength influences the differential effects of transforming growth factor beta1 for the generation of $CD8^+$ regulatory T cells. TM; Fan, DM; Kranz, RA; Flavell, EJ. Roy, *Mol Immunol.,* 2008, 45, 2937-2950.

[52] Critical role for $CD8^+FoxP3^+$ regulatory T cells in colon cancer immune response in humans. In: I; Sobhani, S. Le Gouvello, *Gut.,* 2009, 58, 743-734.

[53] CD75s is a marker of murine $CD8^+$ suppressor T cells. In: JC; Zimring, SB;Levery, B; Kniep, LM; Kapp, M; Fuller, Kapp JA. *Int Immunol.,* 2003, 15, 1389-1399.

[54] Identification of a novel natural regulatory CD8 T-cell subset and analysis of its mechanism of regulation. In: E; Xystrakis, AS; Dejean, I; Bernard, P; Druet, R; Liblau, D; Gonzalez-Dunia, A. Saoudi, *Blood.,* 2004, 104, 3294-3301.

[55] Functional and genetic analysis of two CD8 T cell subsets defined by the level of CD45RC expression in the rat. In: E; Xystrakis, P; Cavailles, AS; Dejean, B; Cautain, C; Colacios, D; Lagrange, MJ; van de Gaar, I; Bernard, D; Gonzalez-Dunia, J; Damoiseaux, GJ; Fournié, A. Saoudi, *J Immunol.,* 2004. 173, 3140-3147.

[56] A novel alloantigen-specific $CD8^+PD1^+$ regulatory T cell induced by ICOS-B7h blockade in vivo. In: A; Izawa, K; Yamaura, MJ; Albin, M; Jurewicz, K; Tanaka, MR; Clarkson, T; Ueno, A; Habicht, GJ; Freeman, H; Yagita, R; Abdi, T; Pearson, DL; Greiner, MH; Sayegh, N. Najafian, *J Immunol.,* 2007, 179, 786-796.

[57] Origins and functional basis of regulatory $CD11c^+CD8^+$ T cells. In: DS; Vinay, CH; Kim, BK; Choi, BS. Kwon, *Eur J Immunol.,* 2009. 39, 1552-1563.

[58] Essential roles of $CD8^+CD122^+$ regulatory T cells in the maintenance of T cell homeostasis. In: M; Rifa'i, Y; Kawamoto, I; Nakashima, H. Suzuki, *J Exp Med.,* 2004, 200, 1123-1134.

[59] Deregulated T cell activation and autoimmunity in mice lacking interleukin-2 receptor β. In: H; Suzuki, TM; Kündig, C; Furlonger, A; Wakeham, E; Timms, T; Matsuyama, R; Schmits, JJ; Simard, PS; Ohashi, H; Griesser, T; Taniguchi, CJ; Paige, TW. Mak, *Science.,* 1995, 268, 1472-1476.

[60] $CD8^+CD122^+$ regulatory T cells recognize activated T cells via conventional MHC class I-αβTCR interaction and become IL-10-producing active regulatory cells. In: M; Rifa'i, Z; Shi, SY; Zhang, YH; Lee, H; Shiku, K; Isobe, H. Suzuki, *Int Immunol.,* 2008, 20, 937-947.

[61] Importance of CD80/CD86-CD28 interactions in the recognition of target cells by $CD8^+CD122^+$ regulatory T cells. In: Z; Shi, M; Rifa'i, YH; Lee, H; Shiku, K; Isobe, H. Suzuki, *Immunology.,* 2008, 124, 121-128.

[62] Cutting edge, $CD8^+CD122^+$ regulatory T cells produce IL-10 to suppress IFN-γ production and proliferation of $CD8^+$ T cells. In: AT; Endharti, M; Rifa'i, Z; Shi, Y; Fukuoka, Y; Nakahara, Y; Kawamoto, K; Takeda, K; Isobe, H. Suzuki, *J Immunol.* 2005, 175, 7093-7097.

[63] Essential role of $CD8^+CD122^+$ regulatory T cells in the recovery from experimental autoimmune encephalomyelitis. In: YH; Lee, Y; Ishida, M; Rifa'i, Z; Shi, K; Isobe, H. Suzuki, *J Immunol.,* 2008, 180, 825-832.

[64] Human $CD8^+CXCR3^+$ T cells have the same function as murine $CD8^+CD122^+$ Treg. In: Z; Shi, Y; Okuno, M; Rifa'i, AT; Endharti, K; Akane, K; Isobe, H. Suzuki, *Eur J Immunol.,* 2009, 39, 2106-2119.

[65] Human anergic/suppressive $CD4^+CD25^+$ T cells, a highly differentiated and apoptosis-prone population. In: LS; Taams, J; Smith, MH; Rustin, M; Salmon, LW; Poulter, AN. Akbar, *Eur J Immunol.,* 2001, 31, 1122-1131.
[66] Human $CD4^+CD25^+$ thymocytes and peripheral T cells have immune suppressive activity in vitro. In: LA; Stephens, C; Mottet, D; Mason, F. Powrie, *Eur J Immunol.,* 2001, 31, 1247-1254.
[67] Are $CD8^+CD122^+$ cells regulatory T cells or memory T cells? In: H; Suzuki, Z; Shi, Y; Okuno, K. Isobe, *Hum Immunol.,* 2008, 69, 751-754.

In: Regulatory T Cells
Editor: Ren S. Hayashi, pp. 185-199

ISBN: 978-1-61761-554-2

Chapter 10

Regulatory T Cells in Arthritis: Why Do They Fail?

Halima Moncrieffe*[*] *and Lucy R. Wedderburn
Rheumatology Unit, Institute of Child Health,
University College London, London, United Kingdom

Abstract

Regulatory T cells (Treg) are crucial for suppression of autoimmune responses and employ a number of mechanisms for controlling the immune response. Treg are abundant at the site of inflammation in arthritis yet fail to control the pathological process. Recently the characterisation of surface markers which help define Treg in inflammation have aided investigations into why this failure of immune regulation occurs and has increased our understanding of Treg mechanisms of action. These investigations are complex due to overlap of Treg markers with those found on activated cells. There is increasing evidence that Treg in arthritis have defects in suppressive function. Here we discuss defects in Treg from patients with inflammatory arthritis, with the focus on adult rheumatoid arthritis (RA), and childhood arthritis (juvenile idiopathic arthritis JIA). We review evidence for how these Treg defects might be affected by treatment and possible novel therapeutic strategies.

Introduction

Tight control of the immune system is necessary, in order to mount appropriate responses against invading pathogens while minimising damage to host tissues. A lymphocyte population with a pivotal role in modulation and control of the immune response is the population known as regulatory T cells (Treg). With defects in the mechanisms which allow

[*] Corresponding author: h.moncrieffe@ich.ucl.ac.uk

either the development, maturation or activation of autoreactive lymphocytes, autoimmune diseases can occur. Inflammatory arthritis is one of the most common rheumatological conditions in the United Kingdom with prevalence rates of 1 in 100 for adult rheumatoid arthritis (RA), and 1 in 1000 children up to the age of 16 (juvenile idiopathic arthritis JIA). Despite the lack of known pathogenic triggers characterised for inflammatory arthritis, host joints are affected by inappropriate inflammation and immune responses which lead to damage to synovium, cartilage and bone, and ultimately joint malfunction, damage and destruction. The role of regulatory T cells in inflammatory arthritis, and lines of evidence implicating how and why these cells are unable to control this autoimmune condition are discussed here.

Regulatory T Cells: An Essential Role in Maintaining Tolerance

There are a number of mechanisms which contribute to the maintenance of a balanced immunological repertoire of populations of lymphocytes with vastly varied immune specificities, available to react against a plethora of possible antigens, while also preventing auto-aggressive lymphocytes from being activated. Together such mechanisms are considered to provide immunological ‘tolerance’ [1]. Central tolerance mechanisms are successful in preventing the maturation of the majority of autoreactive T or B lymphocytes: thus for example high-affinity autoreactive T cells, which could cause autoimmune pathology, undergo apoptosis in the thymus. However a proportion of these autoreactive cells escape deletion during development, and enter the peripheral circulation. Peripheral tolerance mechanisms represent a further level of control of such cells by preventing mature autoreactive cells from reacting against the host. Regulatory T cells (Treg) are one of the important modulators maintaining peripheral tolerance, and may arise from the thymus (so called natural or nTreg) or in the periphery (iTreg) Treg are widely thought to modulate the inflammatory response in autoimmune diseases including inflammatory arthritis.

Regulatory Cell Phenotype

A wealth of research into regulatory T cell biology has resulted in greater clarity in our understanding of both the phenotype and mechanisms of action for this important immune cell population. The transcription factor FoxP3 is necessary and sufficient for Treg development and function in mice [2, 3] In humans the potentially fatal condition IPEX (Immunodysregulation, Polyendocrinopathy, and Enteropathy, X-linked) syndrome results from mutations in the FOXP3 gene [4]. Tregs from IPEX patients who do not express FOXP3 protein have severely impaired suppressive capability and these children develop severe and multi organ autoimmune disease. However, there is a highly variable disease course associated with various FOXP3 mutations observed in these patients with defects both Treg and effector cell function observed [5].

The study of Treg has been hampered by the lack of cell surface protein markers with which to purify these cells. Thus while the intranuclear transcription factor FoxP3 has a central role in Treg biology, it cannot be used as a mechanism to purify Treg cells to perform functional studies using due to the need to permeabilise the cell for intranuclear staining to identify FOXP3 protein. It has been known for some years that functional Treg cells express high levels of CD25, the IL-2 receptor α chain on their cell surface [6, 7]. This has allowed many studies of Treg function in both humans and mice. However there are fundamental difficulties in using CD25 and/or FOXP3 as definitive markers for Treg since in human lymphocytes these proteins can also be expressed on activated T cells, such as those T cells found in inflammation. FOXP3 is transiently expressed upon activation and transient expression may not be associated with suppressive activity [8, 9, 10]. Given the continuum of expression of CD25 on T cells, the specific protein expression level separating $CD25^{hi}$ versus CD25 intermediate cells is difficult to define. Thus, distinguishing inflamed cells from regulatory cells using these markers alone may be problematic.

In healthy circulating CD4 T cells, CD25 and FOXP3 protein expression correlate and $CD25^{hi}$ cells include those with potent suppressive activity. Direct evidence also exists that high levels of FOXP3 expression correlate with Treg function. Using a conditionally active form of FOXP3, Levings and colleagues showed that human Treg function is dependent on the amount of active FOXP3 expressed [11].

Recent studies have led to a more detailed characterisation of Treg surface phenotype and these cells are able to identified as $CD4+CD25^{hi}CD127^{lo}$. A member of the TNF-receptor superfamily, CD27, in combination with high CD25 expression has also been suggested to be a marker for Treg in the inflamed joint [12]. Recently Treg have been shown to express the ectonucleotidase CD39 [13], which has led to the elucidation of a novel functional mechanism for Treg activity, which will be discussed below. These findings together allow for isolation of Treg by cell sorting with greater accuracy than has been previously possible.

More recently, further subpopulations of Treg cells, such as memory and effector Treg have been defined, according to the expression of other cell surface proteins including HLA-DR and CD45RA or CD45RO [14, 15, 16]. It is possible that Treg exist within many populations which share FOXP3 expression, with some Treg subpopulations expressing different levels of FOXP3 protein, and have different functional capacities.

Regulatory function is a key measure of Treg phenotype. In vitro measures of Treg function necessarily only model the in vivo actions of these cells. Inhibition of responder cell proliferation and suppression of proinflammatory cytokines are widely used as a standard readout for Treg function. As well as the capacity to suppress the functional activity of both T and B cells [6, 17], Treg themselves are characterised by the fact that they are 'anergic' or less responsive to specific antigenic stimulation and are also unable to produce the cytokine IL-2 which acts to promote T cell proliferation.

Regulatory T Cell Suppression of Inflammation

Treg act via a number of mechanisms to regulate inflammation and there is increasing evidence to suggest defects in Treg function may contribute to disease pathology in arthritis. In vitro, Treg need cell–cell contact for effective suppression and mechanisms proposed for suppression involve IL-2 competition or direct cell death of targets. One proposed mechanism of Treg action that would require close cell contact is the transfer of cAMP through gap junctions from Treg to responder T cells [18]. cAMP inhibits T cell proliferation and differentiation and switches off IL-2 expression by CD4+ T cells (reviewed in [19]). In murine T cells, the mechanism for this has been shown to be via cAMP-mediated induction of the transcription factor inducible cAMP early repressor (ICER) which can mask CREB and Jun DNA binding sites and inhibit IL-2 transcription [19, 20].

There is evidence from mouse models that although in vitro Treg need cell-cell contact to mediate effective suppression, in vivo there may be a role for suppressive cytokines namely IL-10, TGFβ and more recently IL-35 [21]. TGFβ may also play a key role in maintaining Treg, (nTreg and iTreg), both in numbers and in 'fitness for function'. TGFβ upregulates Foxp3 expression in T cells [22] and TGFβ signalling maintains high suppressive activity in peripheral Treg [23]. Recently TGFβ has been shown to maintain and stabilise Foxp3 expression through epigenetic mechanisms [24], and recently reviewed in [25].

The recent characterisation of CD39 on the surface of Treg has been important in revealing a new mechanism of Treg suppression [13]. CD39 is an ectonucleoside triphosphate diphosphohydrolase which is expressed on the cell surface and hydrolyses extracellular ATP. Since extracellular ATP has been shown to have proinflammatory roles via binding of P2X7 receptors on antigen presenting cells and leading to the production of IL-1β [26], hydrolysis of extracellular ATP would be one mechanism of Treg mediated control. The breakdown product of ATP hydrolysis, i.e. AMP is a substrate for downstream ectonucleotidase activity, mediated by CD73 and results in production of immunosuppressive adenosine. The adenosine A2A receptor is expressed on activated T effector cells. Similar to other markers of Treg, CD39 may also be expressed on T cells which are activated. In mice, lack of CD39 results in increased expression of the proinflammatory cytokines IFNγ, IL-1β, IL-6 and TNF [27]. CD39-/- murine CD4 cells which have greatly reduced capacity to hydrolyse ADP and are unable to produce adenosine in vitro [13]. In vivo, Treg from CD39 knockout mice were demonstrated to be ineffective in preventing adenosine-mediated contact hypersensitivity reactions, whereas Treg from wild-type mice were able to [28].

We have recently shown that cells from the joint express higher levels of CD39 and synovial fluid cells have more ATPase activity on a per cell basis. However we have also demonstrated a population of CD4 T cells which express CD39 but not FOXP3, that are more abundant in the joint of patients with JIA compared to peripheral blood; these FOXP3-CD39+ cells can break down ATP yet fail to suppress proliferation of effector cells [29]. We have demonstrated a significantly lower amount of CD73 from cells from the inflamed joint compared to peripheral blood and that these cells fail to produce adenosine [29]. This is new evidence that the production of adenosine by Treg from the inflamed site is defective.

Regulatory T Cell Defects in Arthritis

Treg are abundant in the joint of patients with juvenile idiopathic arthritis (JIA) and other inflammatory arthritis. [30, 31, 32]. These CD25+ CD4+ T cells, when sorted from the inflamed site of patients with either RA or JIA, show potent regulatory function in vitro, indicating that these cells are not all activated cells [31, 33]. Moreover, Treg from the synovial fluid express higher levels of FOXP3 mRNA and protein than those from peripheral blood of the same patients [29, 33]. However Treg from patients with arthritis clearly fail to completely control inflammation. Recognition that these Treg may influence clinical phenotype came from the demonstration that the number of CD4+ Treg found within the CD4 population of synovial T cells directly correlates with clinical disease course in children with arthritis. Thus, the number of synovial CD4+ Foxp3+ Treg in the joint was higher in children with mild, self remitting arthritis called oligoarticular JIA, than those with more severe extended oligo articular JIA [12, 33].

Why Treg fail to suppress inflammation in the arthritis joint is the subject of active investigation. There are at least three possible explanations:

1. The Treg present are intrinsically functional but insufficient in number to prevent disease.
2. The inflammatory milieu at the site of arthritis may switch off Treg function in vivo.
3. There are intrinsic defects in the Treg which render them unable to prevent inflammation.

Treg Numbers

The absolute number of Treg in relation to effector cells, is of functional importance, at least in vitro where suppression observed is more powerful at higher Treg to effector T cell ratios. In arthritis there is a correlation with Treg number in the joint and disease severity in JIA: the number of Treg is higher in the milder form of JIA [33]. We have also recently shown that Treg cell numbers in the joint of these patients correlate inversely with the highly proinflammatory Th17 cells [34]. The number of Treg in the peripheral blood of both JIA and RA patients is similar to that in blood of healthy donors, however a number of defects have been described and are discussed below.

The Inflammatory Milieu in Arthritis

The inflamed arthritic joint has an abundance of proinflammatory cytokines. One recent study used multiple immunoassay to measure and compare a panel of cytokines and chemokines in JIA blood compared synovial fluid and demonstrated that the latter contained significantly higher levels of IL-6, IL-15, CCL2, CCL3, CXCL8, CXCL9 and CXCL10 [35]. In addition increases in IL-6, IL-17, TNFα and IFNγ were observed to be higher in patients with active disease compared to those with inactive disease.

Treg isolated from the joint have been shown to have suppressive capacity. One area of investigation has therefore been to test if the inflammatory milieu may contribute to Treg being insufficient to control inflammation in vivo.

TNFα is a pleiotropic cytokine which may have pro or anti-inflammatory effects (reviewed in [36]) and has a key role in arthritis pathogenesis. Blockade of TNF signalling with anti-TNF therapy is a routine treatment option when standard DMARD treatments fail to work [37]. The role of TNF in Treg biology is gaining increasing prominence with several studies demonstrating that TNF modulates the function of Treg cells [38]. Treg express a receptor for TNF, i.e. TNFR2, and TNF in conjunction with IL-2 activates Tregs and increases their suppressive capacity. While in vitro stimulation of CD4 T cells resulted in upregulated TNFR2 on memory and naïve cells upon activation, Treg still expressed the highest levels of this receptor [39]. Recently a mechanism by which TNF mediates its effects on Treg has been described [40]. In CD45RA- Treg, TNF was shown to induce the NFκB canonical pathway which includes genes such as TNF and OX40 while not changing FOXP3 protein levels. TNF had functional effects on the ability of Treg to suppress effector cell proliferation and IFN γ secretion. This was reversed in vitro by anti-TNF therapy. These data suggest that in the presence of TNF the suppressive functions of Treg are 'overcome;' due to NFκB signalling.

In addition to TNF, the proinflammatory cytokine IL-6 also plays an important role in arthritis pathology, for example both cytokines have been implicated in generating and maintaining joint pain in a rat model of arthritis [41]. IL-6 has also been shown to be able to modulate Treg function and ability to suppress effector cells [42]. Both TNF and IL-6 have been shown to influence the plasticity of Treg cells, further discussed below, and therefore may inhibit Treg function by facilitating 'conversion' of Treg to Th17 cells [43, 44].

Ruprecht and colleagues tested Treg suppression in the presence of a panel of proinflammatory cytokines with cells from the joint of JIA patients [12]. Notably IL-7 and IL-15, which are elevated in the JIA joint, were found to reduce Treg-mediated suppression in JIA synovial Treg.

Several studies have suggested that activated effector cells from the joint are somehow 'resistant' to suppression [31], an effect that has been proposed to be mediated via cytokines including IL-15, IL-21 [45] or IL-6 [42].

Plasticity in Regulatory T Cells

The newly described plasticity between T cell subsets, in particular between Treg and Th17 cells (reviewed in [46]), suggests that these populations are not entirely distinct 'lineages', but rather that these cells may switch from Treg to Teffector cell, under specific conditions such as during inflammation. A double-transgenic reporter murine strain has been generated by Bluestone and colleagues [47] where Foxp3 promoter activation leads to green fluorescence protein expression and, due to Cre recombinase activity, permanent yellow fluorescence protein (YFP) expression. Using this elegant model, cells which had expressed Foxp3 but were no longer doing so were found to be producing effector cytokines such as IFNγ and IL-17. Moreover these 'exFoxp3' cells were autoaggressive.

Given the intrinsic bias of Treg antigen-specific receptors towards self-recognition, such switching of Treg to Th17 cells represents a potential mechanism of generating highly pathogenic T effector cells ('exTreg') from the increased numbers of Treg present at inflamed sites. Treg contain a subset of cells which co-express the master transcription factor of Th17 cells, RORγt (RORC2 in humans). While direct interaction between Foxp3 and RORC2 can suppress downstream target genes of RORC2, including IL-17, a reduction of Foxp3 levels in such cells can lead to a restoration of IL-17 production [48].

Treg and Th17 share a developmental relationship in mice and humans [49] and reviewed in [50]). In both species TGFβ appears to play a role in development or maintenance of stability of both Th17 and Treg cells, while IL-6, IL-1, TNFα may favour Th17 in this balance [51] and reviewed in [52]. In arthritis patients we recently reported a numeric relationship between synovial Treg and Th17 numbers, such that Treg in the joint are inversely correlated with Th17 cells. This suggests that pro- or inflammatory influences within the joint are important to the balance of these two T cell populations and ultimately to clinical disease phenotype in children with arthritis. [34].

Treg Defects in Human Arthritis

Many studies have shown that the number of Treg in the peripheral blood of patients does not differ from healthy controls. However Treg from the peripheral blood of patients with RA have been shown to be defective in some studies. RA Treg were found to be significantly defective in suppressing IFNγ and TNF production from autologous CD4+CD25+ cells, although they could suppress in vitro proliferation [53]. In a contrasting study Treg from RA patients were found to be defective at suppressing both proliferation and cytokine production by effectors [38]. The reasons for the contrast between these studies remain to be clarified.

A recent elegant study of TCR signalling in healthy controls and RA patients has confirmed this defect of RA Treg in suppressing production of inflammatory cytokines (15). It is known that Treg suppressive function, such as inhibition of IL-2 or IFNγ production, requires T cell receptor engagement [54]. The protein kinase C-θ (PKC-θ) pathway is downstream to the TCR signalling molecules Src family kinases and has been implicated in Treg function in the study by Zanin-Zhorov and colleagues [55]. PKC-θ is sequestered away from the immunological synapse in Treg with 11-fold less PKC-θ recruited after TCR triggering compared with effector T cells. The intrinsic defect of Treg from RA patients in IFNγ suppression was largely reversed by treatment with the PKC-θ inhibitor C20. The severity of disease as measured by the DAS score, inversely correlated with the ability of Treg from RA patients to suppress autologous IFNg. C20 was able to increase ex vivo Treg suppressive function irrespective of disease severity in this case.

CTLA-4 (Cytotoxic T Lymphocyte Antigen 4), which is expressed on the T cell surface and constitutively expressed on Treg is a surface receptor for CD80 or CD86 and binds these costimulatory molecules with higher affinity than the alternative binding partner, CD28. CTLA-4 has been widely studied in the role of a negative regulator of responder T cell signalling; mice lacking CTLA-4 die at a young age with massive lymph node enlargement [56]. A CTLA-4 defect has been recently described in RA Treg [57]. The report showed reduced expression of CTLA-4 on the cell surface of Treg from RA patients compared to healthy individuals. In this study, CTLA-4 was able to inhibit T cell receptor (TCR)-mediated signalling in Treg from healthy controls but not from RA patients. By artificially increasing Treg CTLA-4 expression by addition of phorbol ester in vitro, suppression of pro-inflammatory cytokine production was restored. In addition to being a marker of Treg, CTLA-4 is a putative RA susceptibility gene [58].

TNFα has a key role in mediating inflammation in many patients with arthritis and has been shown to inhibit Treg activity [38]. Anti-TNF is now a key biological therapy for inflammatory arthritis [37]. In severe RA disease patients, a significant rise in the number of Treg has been demonstrated in the peripheral blood of patients who respond well to TNFα blockade by Infliximab. These Treg were shown to be functionally and phenotypically different from the pre-existing natural Treg, for example the newly generated Treg were CD62L-, and their induction was dependent upon TGFβ[59]. These peripherally-induced Treg had a distinct surface phenotype (CD62L-) and they suppressed IFNγ and TNFα production through production of IL-10 and TGFβ, in contrast to healthy Treg which suppressed via CTLA-4. Zanin-Zhorov et al also demonstrated a mechanism for these TNFα effects on Treg; they showed that PKC-θ inhibition prevented TNFα-induced down-regulation of Foxp3 in Treg. In addition they showed that TNFα treatment increased PKC-θ recruitment to the immunological synapse in Treg.

These studies of Treg in peripheral blood in arthritis patients may provide only a partial picture, since the biology at the inflamed site will be powerfully affected by inflammatory mediators.

Novel Therapeutic Approaches

Understanding the biology of Treg in arthritis has led to novel therapeutic approaches, which are important given that current therapies have a variable outcome for patients and fail to cure the disease. Several drugs for rheumatoid arthritis have been shown to influence the numbers or function of Treg and drugs such as methotrexate and adenosine agonsists may mimic some of the actions of Treg function via the adenosine pathway.

Adenosine has anti-inflammatory effects and its production is a recently described mechanism of Treg action. There is evidence that methotrexate, the first line treatment for arthritis, results in adenosine release [60, 61]. We have provided evidence that Treg from the joint in JIA maybe defective in adenosine production [29]. Novel therapies which mimic the effects of Treg action or indeed replace a functional defect are emerging. Adenosine and other adenosine receptor agonists are emerging as novel immunosuppressants. There are four characterised adenosine receptors which are a subfamily of G protein-coupled receptors and are expressed on a wide range of cells in the immune system including macrophages, dendritic cells, mast cells and lymphocytes (reviewed in [62]). In patients with early rheumatoid arthritis, a defect in A2A and A3 receptors function has been found [63]. Thus this may result in defective downstream signalling of adenosine produced by Treg. The altered function of A2A and A3 receptors in RA patients was restored by anti-TNF therapy, but not methotrexate [63]. A3AR agonists are currently undergoing clinical trials for the treatment of rheumatoid arthritis [64].

Treg Immunotherapy: Existing Therapies and Novel Approaches

Generation of large number of Treg in vitro for use as therapy is challenging. However with improved characterisation of Treg function and defects in arthritis, immunotherapy to correct these defects or to improve Treg efficacy in vivo is an area of intense investigation. Several challenges may exist to the strategy of giving Treg cells generated in vitro, back to patients, namely the potential instability of Treg function and the capacity of these cells to switch to proinflammatory phenotypes in vivo.

Several agents that are already in clinical use to treat arthritis may have effects on Treg cells. Thus in RA the anti-TNF agent Infliximab has been shown to lead to a rise in Treg number and these are a novel population as described above. IL-6 blockade by Toculizumab, an antibody to the soluble IL-6R is now used in the treatment of RA and JIA [37, 65, 66] and this strategy is likely to have effects upon the Treg: Th17 balance, perhaps by preventing IL-6 from driving CD4+ cells toward a Th17 phenotype. However to date few studies have investigated this aspect of novel biologics. Many other agents may also affect this balance such as retinoic acid [67], and these may prove useful adjuncts in future therapeutic regimes [52].

Treg can be expanded ex vivo following stimulation in IL-2 with anti-CD3 and anti-CD28 antibodies and maintain suppressive function [68]. However there remains a problem in Treg purity due to the difficulties in distinguishing Treg from activated T cells and also subsequent downregulation of FOXP3 in Treg and production of inflammatory cytokines such as IL-2 and IFNγ [69]. Purifying CD45RA+CD25^{hi} CD4 T cells to induce Treg expansion was shown to generate Treg which were able to maintain suppressive capacity [70].

Vasoactive intestinal peptide (VIP) has suppressive effects in collagen induced arthritis (CIA), a murine model. VIP can increase suppressive capacity [71] and has been shown to increase Treg numbers and percentage resulting in reduced CIA disease phenotype [72]. The effects of VIP have recently been characterised on human cells. Upon stimulation of human CD4+ CD25- T cells, VIP was shown to induce CD25^{hi}Foxp3+ Tregs de novo through cAMP induction, but did not expand previously established Treg [73].

Recently transfection of CD4 T cells with FoxP3 and the anti-apoptotic Bcl-2 family member Bcl-xL were used to generate Treg with longer term survival and more effective suppression in a murine model of arthritis [74]. As discussed above anti-TNFα has been shown to induce regulatory T cells in patients with rheumatoid arthritis, a process that involves TGFβ. In vitro, addition of Infliximab to CD25- cells was shown to induce a CD62L- Treg population which were able to suppress TNFα and IFNγ via production of IL-10 and TGFβ. Natural Treg, which were characterised as CD62L+, from the same individuals were still defective.

Treg can suppress immune effector cell functions by transfer of cAMP to target T cells through gap junctions. Molecules which increase cAMP and so mimic this role of Treg would therefore be of therapeutic interest. One such example is vasoactive intestinal peptide which elevates cAMP. Another example are PDE inhibitors which prevent rapid degradation of cAMP and indeed can elevate cAMP in auto-‘aggressive’ murine CD4 T cells [75].

The generation of tolerogenic DC may represent a mechanism for generation of regulatory T cells and restoration of immune tolerance. A recent report described generation of regulatory T cells by interaction with plasmacytoid DC with a tolerogenic phenotype, which expressed high levels of indoleamine 2,3-dioxygenase (IDO) in rheumatoid arthritis patients achieving remission [76]. These newly generated Tr1 cells do not express FOXP3, but secrete IL-10 and suppress in vitro proliferation of autologous CD4 T cells. The generation of ‘semi-mature’ tolerogenic DC, for example in the presence of vitamin D and dexamthasone, has been widely studied in animal models, and recently proposed for use as immunotherapy for RA [77]. Whether these DC act by inducing Treg or by induction of anergy in effector T cells remains unclear.

Conclusions

Our understanding of Treg biology has greatly increased over the last 5 years with a number of key discoveries made explaining new molecular mechanisms of Treg function. In arthritis, although Treg can be readily found in the inflamed site, the inflammatory milieu appears to form an environment that is not conducive to suppressive phenotype. A number of different defects have been identified in Treg of arthritis patients; with this improved characterisation of why Treg are defective novel targets for treatment can be investigated to restore these defective features. Biologic therapies may provide an opportunity to prevent loss of suppressive function, reduce Treg transformation to effector T cells or even generate new Treg.

References

[1] Von Boehmer, H; Melchers, F. Checkpoints in lymphocyte development and autoimmune disease. *Nat Immunol.,* Jan; 11(1), 14-20.

[2] Brunkow, ME; Jeffery, EW; Hjerrild, KA; Paeper, B; Clark, LB; Yasayko, SA; et al. Disruption of a new forkhead/winged-helix protein; scurfin; results in the fatal lymphoproliferative disorder of the scurfy mouse. *Nat Genet.,* 2001, Jan; 27(1), 68-73.

[3] Le Bras, S; Geha, RS. IPEX and the role of Foxp3 in the development and function of human Tregs. *J Clin Invest.,* 2006, Jun; 116(6), 1473-5.

[4] Wildin, RS; Ramsdell, F; Peake, J; Faravelli, F; Casanova, JL; Buist, N; et al. X-linked neonatal diabetes mellitus; enteropathy and endocrinopathy syndrome is the human equivalent of mouse scurfy. *Nat Genet.,* 2001, Jan; 27(1), 18-20.

[5] Bacchetta, R; Passerini, L; Gambineri, E; Dai, M; Allan, SE; Perroni, L; et al. Defective regulatory and effector T cell functions in patients with FOXP3 mutations. *J Clin Invest.,* 2006, Jun; 116(6), 1713-22.

[6] Baecher-Allan, C; Brown, JA; Freeman, GJ; Hafler, DA. CD4+CD25high regulatory cells in human peripheral blood. *J Immunol.,* 2001, Aug 1, 167(3), 1245-53.

[7] Ng, WF; Duggan, PJ; Ponchel, F; Matarese, G; Lombardi, G; Edwards, AD; et al. Human CD4(+)CD25(+) cells: a naturally occurring population of regulatory T cells. *Blood.,* 2001, Nov 1, 98(9), 2736-44.

[8] Wang, J; Ioan-Facsinay, A; van der Voort, EI; Huizinga, TW; Toes, RE. Transient expression of FOXP3 in human activated nonregulatory CD4+ T cells. *Eur J Immunol.,* 2007, Jan; 37(1), 129-38.

[9] Allan, SE; Crome, SQ; Crellin, NK; Passerini, L; Steiner, TS; Bacchetta, R; et al. Activation-induced FOXP3 in human T effector cells does not suppress proliferation or cytokine production. *Int Immunol.,* 2007, Apr; 19(4), 345-54.

[10] Kmieciak, M; Gowda, M; Graham, L; Godder, K; Bear, HD; Marincola, FM; et al. Human T cells express CD25 and Foxp3 upon activation and exhibit effector/memory phenotypes without any regulatory/suppressor function. *J Transl Med.,* 2009, 7, 89.

[11] Allan, SE; Song-Zhao, GX; Abraham, T; McMurchy, AN; Levings, MK. Inducible reprogramming of human T cells into Treg cells by a conditionally active form of FOXP3. *Eur J Immunol.,* 2008, Dec; 38(12), 3282-9.

[12] Ruprecht, CR; Gattorno, M; Ferlito, F; Gregorio, A; Martini, A; Lanzavecchia, A; et al. Coexpression of CD25 and CD27 identifies FoxP3+ regulatory T cells in inflamed synovia. *J Exp Med.,* 2005, Jun 6, 201(11), 1793-803.

[13] Deaglio, S; Dwyer, KM; Gao, W; Friedman, D; Usheva, A; Erat, A; et al. Adenosine generation catalyzed by CD39 and CD73 expressed on regulatory T cells mediates immune suppression. *J Exp Med.,* 2007, Jun 11, 204(6), 1257-65.

[14] Baecher-Allan, C; Wolf, E; Hafler, DA. MHC class II expression identifies functionally distinct human regulatory T cells. *J Immunol.,* 2006, Apr 15, 176(8), 4622-31.

[15] Booth, NJ; McQuaid, AJ; Sobande, T; Kissane, S; Agius, E; Jackson, SE; et al. Different proliferative potential and migratory characteristics of human CD4+ regulatory T cells that express either CD45RA or CD45RO. *J Immunol.,* 2010, Apr 15, 184(8), 4317-26.

[16] Miyara, M; Yoshioka, Y; Kitoh, A; Shima, T; Wing, K; Niwa, A; et al. Functional delineation and differentiation dynamics of human CD4+ T cells expressing the FoxP3 transcription factor. *Immunity.,* 2009, Jun 19, 30(6), 899-911.

[17] Meiler, F; Klunker, S; Zimmermann, M; Akdis, CA; Akdis, M. Distinct regulation of IgE; IgG4 and IgA by T regulatory cells and toll-like receptors. *Allergy.,* 2008, Nov; 63(11), 1455-63.

[18] Bopp, T; Becker, C; Klein, M; Klein-Hessling, S; Palmetshofer, A; Serfling, E; et al. Cyclic adenosine monophosphate is a key component of regulatory T cell-mediated suppression. *J Exp Med.,* 2007, Jun 11, 204(6), 1303-10.

[19] Bodor, J; Fehervari, Z; Diamond, B; Sakaguchi, S. Regulatory T cell-mediated suppression: potential role of ICER. *J Leukoc Biol.,* 2007, Jan; 81(1), 161-7.

[20] Bodor, J; Fehervari, Z; Diamond, B; Sakaguchi, S. ICER/CREM-mediated transcriptional attenuation of IL-2 and its role in suppression by regulatory T cells. *Eur J Immunol.,* 2007, Apr; 37(4), 884-95.

[21] Collison, LW; Workman, CJ; Kuo, TT; Boyd, K; Wang, Y; Vignali, KM; et al. The inhibitory cytokine IL-35 contributes to regulatory T-cell function. *Nature*, 2007, Nov 22, 450(7169), 566-9.

[22] Chen, W; Jin, W; Hardegen, N; Lei, KJ; Li, L; Marinos, N; et al. Conversion of peripheral CD4+CD25- naive T cells to CD4+CD25+ regulatory T cells by TGF-beta induction of transcription factor Foxp3. *J Exp Med.,* 2003, Dec 15, 198(12), 1875-86.

[23] Marie, JC; Letterio, JJ; Gavin, M; Rudensky, AY. TGF-beta1 maintains suppressor function and Foxp3 expression in CD4+CD25+ regulatory T cells. *J Exp Med.,* 2005, Apr 4, 201(7), 1061-7.

[24] Kim, HP; Leonard, WJ. CREB/ATF-dependent T cell receptor-induced FoxP3 gene expression: a role for DNA methylation. *J Exp Med.,* 2007, Jul 9, 204(7), 1543-51.

[25] Huehn, J; Polansky, JK; Hamann, A. Epigenetic control of FOXP3 expression: the key to a stable regulatory T-cell lineage? *Nat Rev Immunol.*, 2009, Feb; 9(2), 83-9.

[26] Ward, JR; West, PW; Ariaans, MP; Parker, LC; Francis, SE; Crossman, DC; et al. Temporal IL-1beta secretion from primary human peripheral blood monocytes by P2X7-independent and P2X7-dependent mechanisms. *J Biol Chem.*, May 21.

[27] Enjyoji, K; Ko, K; Thukral, C; Blumel, B; Sun, X; Wu, Y; et al. Deletion of Cd39/Entpd1 results in hepatic insulin resistance. *Diabetes.*, 2008, Jun 20.

[28] Ring, S; Oliver, SJ; Cronstein, BN; Enk, AH; Mahnke, K. CD4+CD25+ regulatory T cells suppress contact hypersensitivity reactions through a CD39, adenosine-dependent mechanism. *J Allergy Clin Immunol.*, 2009, Jun; 123(6), 1287-96 e2.

[29] Moncrieffe, H; Nistala, K; Kamhieh, Y; Evans, J; Eddaoudi, A; Eaton, S; et al. High Expression of the Ectonucleotidase CD39 on T Cells from the Inflamed Site Identifies

Two Distinct Populations; One Regulatory and One Memory T Cell Population. *J Immunol.,* 2010, May 24.

[30] Cao, D; van Vollenhoven, R; Klareskog, L; Trollmo, C; Malmstrom, V. CD25brightCD4+ regulatory T cells are enriched in inflamed joints of patients with chronic rheumatic disease. *Arthritis Res Ther*, 2004, 6(4), R335-46.

[31] Van Amelsfort, JM; Jacobs, KM; Bijlsma, JW; Lafeber, FP; Taams, LS. CD4(+)CD25(+) regulatory T cells in rheumatoid arthritis: differences in the presence; phenotype; and function between peripheral blood and synovial fluid. *Arthritis Rheum.*, 2004, Sep; 50(9), 2775-85.

[32] Van Amelsfort, JM; Van Roon, JA; Noordegraaf, M; Jacobs, KM; Bijlsma, JW; Lafeber, FP; et al. Proinflammatory mediator-induced reversal of CD4+; CD25+ regulatory T cell-mediated suppression in rheumatoid arthritis. *Arthritis Rheum*, 2007, Mar; 56(3), 732-42.

[33] de Kleer, IM; Wedderburn, LR; Taams, LS; Patel, A; Varsani, H; Klein, M; et al. CD4+CD25bright regulatory T cells actively regulate inflammation in the joints of patients with the remitting form of juvenile idiopathic arthritis. *J Immunol.,* 2004, May 15, 172(10), 6435-43.

[34] Nistala, K; Moncrieffe, H; Newton, KR; Varsani, H; Hunter, P; Wedderburn, LR. Interleukin-17-producing T cells are enriched in the joints of children with arthritis; but have a reciprocal relationship to regulatory T cell numbers. *Arthritis Rheum.*, 2008, Mar; 58(3), 875-87.

[35] de Jager, W; Hoppenreijs, EP; Wulffraat, NM; Wedderburn, LR; Kuis, W; Prakken, BJ. Blood and synovial fluid cytokine signatures in patients with juvenile idiopathic arthritis: a cross-sectional study. *Ann Rheum Dis.*, 2007, May; 66(5), 589-98.

[36] Chen, X; Oppenheim, JJ. TNF-alpha: an activator of CD4+FoxP3+TNFR2+ regulatory T cells. *Curr Dir Autoimmun.*, 2010, 11, 119-34.

[37] Smolen, JS; Landewe, R; Breedveld, FC; Dougados, M; Emery, P; Gaujoux-Viala, C; et al. EULAR recommendations for the management of rheumatoid arthritis with synthetic and biological disease-modifying antirheumatic drugs. *Ann Rheum Dis.*, 2010, Jun; 69(6), 964-75.

[38] Valencia, X; Stephens, G; Goldbach-Mansky, R; Wilson, M; Shevach, EM; Lipsky, PE. TNF downmodulates the function of human CD4+CD25hi T-regulatory cells. *Blood.,* 2006, Jul 1, 108(1), 253-61.

[39] Mercer, F; Kozhaya, L; Unutmaz, D. Expression and function of TNF and IL-1 receptors on human regulatory T cells. *PLoS One*, 2010, 5(1), e8639.

[40] Nagar, M; Jacob-Hirsch, J; Vernitsky, H; Berkun, Y; Ben-Horin, S; Amariglio, N; et al. TNF activates a NF-kappaB-regulated cellular program in human CD45RA- regulatory T cells that modulates their suppressive function. *J Immunol.,* 2010, Apr 1, 184(7), 3570-81.

[41] Schaible, HG; von Banchet, GS; Boettger, MK; Brauer, R; Gajda, M; Richter, F; et al. The role of proinflammatory cytokines in the generation and maintenance of joint pain. *Ann N Y Acad Sci.*, 2010, Apr; 1193(1), 60-9.

[42] Pasare, C; Medzhitov, R. Toll pathway-dependent blockade of CD4+CD25+ T cell-mediated suppression by dendritic cells. *Science.* 2003, Feb 14, 299(5609), 1033-6.

[43] Yang, XO; Nurieva, R; Martinez, GJ; Kang, HS; Chung, Y; Pappu, BP; et al. Molecular antagonism and plasticity of regulatory and inflammatory T cell programs. *Immunity.,* 2008, Jul 18, 29(1), 44-56.

[44] Zhou, L; Chong, MM; Littman, DR. Plasticity of CD4+ T cell lineage differentiation. *Immunity.,* 2009, May; 30(5), 646-55.

[45] Clough, LE; Wang, CJ; Schmidt, EM; Booth, G; Hou, TZ; Ryan, GA; et al. Release from regulatory T cell-mediated suppression during the onset of tissue-specific autoimmunity is associated with elevated IL-21. *J Immunol.,* 2008, Apr 15, 180(8), 5393-401.

[46] Bluestone, JA; Mackay, CR; O'Shea, JJ; Stockinger, B. The functional plasticity of T cell subsets. *Nat Rev Immunol.*, 2009, Nov; 9(11), 811-6.

[47] Zhou, X; Bailey-Bucktrout, SL; Jeker, LT; Penaranda, C; Martinez-Llordella, M; Ashby, M; et al. Instability of the transcription factor Foxp3 leads to the generation of pathogenic memory T cells in vivo. *Nat Immunol.,* 2009, Sep; 10(9), 1000-7.

[48] Ichiyama, K; Yoshida, H; Wakabayashi, Y; Chinen, T; Saeki, K; Nakaya, M; et al. Foxp3 inhibits RORgammat-mediated IL-17A mRNA transcription through direct interaction with RORgammat. *J Biol Chem.*, 2008, Jun 20, 283(25), 17003-8.

[49] Koenen, HJ; Smeets, RL; Vink, PM; van Rijssen, E; Boots, AM; Joosten, I. Human CD25highFoxp3pos regulatory T cells differentiate into IL-17-producing cells. *Blood.,* 2008, Sep 15, 112(6), 2340-52.

[50] Lee, YK; Mukasa, R; Hatton, RD; Weaver, CT. Developmental plasticity of Th17 and Treg cells. *Curr Opin Immunol*, 2009, Jun; 21(3), 274-80.

[51] Zhou, L; Lopes, JE; Chong, MM; Ivanov; II; Min, R; Victora, GD; et al. TGF-beta-induced Foxp3 inhibits T(H)17 cell differentiation by antagonizing RORgammat function. *Nature*, 2008, May 8, 453(7192), 236-40.

[52] Nistala, K; Wedderburn, LR. Th17 and regulatory T cells: rebalancing pro- and anti-inflammatory forces in autoimmune arthritis. *Rheumatology*, (Oxford). 2009, Jun; 48(6), 602-6.

[53] Ehrenstein, MR; Evans, JG; Singh, A; Moore, S; Warnes, G; Isenberg, DA; et al. Compromised function of regulatory T cells in rheumatoid arthritis and reversal by anti-TNFalpha therapy. *J Exp Med.,* 2004, Aug 2, 200(3), 277-85.

[54] Chen, C; Rowell, EA; Thomas, RM; Hancock, WW; Wells, AD. Transcriptional regulation by Foxp3 is associated with direct promoter occupancy and modulation of histone acetylation. *J Biol Chem*, 2006, Dec 1, 281(48), 36828-34.

[55] Zanin-Zhorov, A; Ding, Y; Kumari, S; Attur, M; Hippen, KL; Brown, M; et al. Protein kinase C-theta mediates negative feedback on regulatory T cell function. *Science*, 2010, Apr 16, 328(5976), 372-6.

[56] Tivol, EA; Borriello, F; Schweitzer, AN; Lynch, WP; Bluestone, JA; Sharpe, AH. Loss of CTLA-4 leads to massive lymphoproliferation and fatal multiorgan tissue destruction; revealing a critical negative regulatory role of CTLA-4. *Immunity.,* 1995, Nov; 3(5), 541-7.

[57] Flores-Borja, F; Jury, EC; Mauri, C; Ehrenstein, MR. Defects in CTLA-4 are associated with abnormal regulatory T cell function in rheumatoid arthritis. *Proc Natl Acad Sci*, U S A. 2008, Dec 9, 105(49), 19396-401.

[58] Han, S; Li, Y; Mao, Y; Xie, Y. Meta-analysis of the association of CTLA-4 exon-1 +49A/G polymorphism with rheumatoid arthritis. *Hum Genet.*, 2005, Oct; 118(1), 123-32.
[59] Nadkarni, S; Mauri, C; Ehrenstein, MR. Anti-TNF-alpha therapy induces a distinct regulatory T cell population in patients with rheumatoid arthritis via TGF-beta. *J Exp Med.,* 2007, Jan 22, 204(1), 33-9.
[60] Riksen, NP; Barrera, P; van den Broek, PH; van Riel, PL; Smits, P; Rongen, GA. Methotrexate modulates the kinetics of adenosine in humans in vivo. *Ann Rheum Dis.*, 2006, Apr; 65(4), 465-70.
[61] Cronstein, BN; Eberle, MA; Gruber, HE; Levin, RI. Methotrexate inhibits neutrophil function by stimulating adenosine release from connective tissue cells. Proc *Natl Acad Sci*, U S A. 1991, Mar 15, 88(6), 2441-5.
[62] Hasko, G; Linden, J; Cronstein, B; Pacher, P. Adenosine receptors: therapeutic aspects for inflammatory and immune diseases. *Nat Rev Drug Discov*, 2008, Sep; 7(9), 759-70.
[63] Varani, K; Massara, A; Vincenzi, F; Tosi, A; Padovan, M; Trotta, F; et al. Normalization of A2A and A3 adenosine receptor up-regulation in rheumatoid arthritis patients by treatment with anti-tumor necrosis factor alpha but not methotrexate. *Arthritis Rheum*, 2009, Oct; 60(10), 2880-91.
[64] Gao, ZG; Jacobson, KA. Emerging adenosine receptor agonists. *Expert Opin Emerg Drugs*, 2007, Sep; 12(3), 479-92.
[65] Genovese, MC; McKay, JD; Nasonov, EL; Mysler, EF; da Silva, NA; Alecock, E; et al. Interleukin-6 receptor inhibition with tocilizumab reduces disease activity in rheumatoid arthritis with inadequate response to disease-modifying antirheumatic drugs: the tocilizumab in combination with traditional disease-modifying antirheumatic drug therapy study. *Arthritis Rheum*, 2008, Oct; 58(10), 2968-80.
[66] Woo, P. Systemic juvenile idiopathic arthritis: diagnosis; management; and outcome. *Nat Clin Pract Rheumatol*, 2006, Jan; 2(1), 28-34.
[67] Mucida, D; Park, Y; Kim, G; Turovskaya, O; Scott, I; Kronenberg, M; et al. Reciprocal TH17 and regulatory T cell differentiation mediated by retinoic acid. *Science*, 2007, Jul 13, 317(5835), 256-60.
[68] Levings, MK; Sangregorio, R; Roncarolo, MG. Human CD25(+)CD4(+) t regulatory cells suppress naive and memory T cell proliferation and can be expanded in vitro without loss of function. *J Exp Med.,* 2001, Jun 4, 193(11), 1295-302.
[69] Hoffmann, P; Eder, R; Kunz-Schughart, LA; Andreesen, R; Edinger, M. Large-scale in vitro expansion of polyclonal human CD4(+)CD25high regulatory T cells. *Blood.,* 2004, Aug 1, 104(3), 895-903.
[70] Hoffmann, P; Eder, R; Boeld, TJ; Doser, K; Piseshka, B; Andreesen, R; et al. Only the CD45RA+ subpopulation of CD4+CD25high T cells gives rise to homogeneous regulatory T-cell lines upon in vitro expansion. *Blood.,* 2006, Dec 15, 108(13), 4260-7.
[71] Delgado, M; Chorny, A; Gonzalez-Rey, E; Ganea, D. Vasoactive intestinal peptide generates CD4+CD25+ regulatory T cells in vivo. *J Leukoc Biol.,* 2005, Dec; 78(6), 1327-38.
[72] Gonzalez-Rey, E; Fernandez-Martin, A; Chorny, A; Delgado, M. Vasoactive intestinal peptide induces CD4+ CD25+ T regulatory cells with therapeutic effect in collagen-induced arthritis. *Arthritis Rheum*, 2006, Mar; 54(3), 864-76.

[73] Pozo, D; Anderson, P; Gonzalez-Rey, E. Induction of alloantigen-specific human T regulatory cells by vasoactive intestinal peptide. *J Immunol.,* 2009, Oct 1, 183(7), 4346-59.

[74] Haque, R; Lei, F; Xiong, X; Wu, Y; Song, J. FoxP3 and Bcl-xL cooperatively promote regulatory T cell persistence and prevention of arthritis development. *Arthritis Res Ther*, 2010, Apr 12, 12(2), R66.

[75] Bopp, T; Dehzad, N; Reuter, S; Klein, M; Ullrich, N; Stassen, M; et al. Inhibition of cAMP degradation improves regulatory T cell-mediated suppression. *J Immunol.,* 2009, Apr 1, 182(7), 4017-24.

[76] Kavousanaki, M; Makrigiannakis, A; Boumpas, D; Verginis, P. Novel role of plasmacytoid dendritic cells in humans: induction of interleukin-10-producing Treg cells by plasmacytoid dendritic cells in patients with rheumatoid arthritis responding to therapy. *Arthritis Rheum.*, 2010, Jan; 62(1), 53-63.

[77] Hilkens, CM; Isaacs JD; Thomson AW. Development of dendritic cell-based immunotherapy for autoimmunity. *Int Rev Immunol*. 2010 Apr;29(2):156-83.

In: Regulatory T Cells
Editor: Ren S. Hayashi, pp. 201-223

ISBN: 978-1-61761-554-2

Chapter 11

Immune Homeostasis Maintained by Tregs

Myoung-Ho Jang,[1,2] Yun-Jae Jung[3] Hyung-Ran Kim,[4] Byung-In Moon[5] and Ju-Young Seoh[4*]

[1]Laboratory of Gastrointestinal Immunology, World Premier International Immunology Frontier Research Center, Osaka University, Osaka, Japan
[2]Division of Integrative Bioscience and Biotechnology, Pohang University of Science and Technology, Pohang, Korea
[3]Department of Microbiology, Graduate School of Medicine, Gachon University of Medicine and Science, Incheon, Korea
[4]Department of Microbiology and [5]department of Surgery, Ewha Womans University Medical School, Seoul, Korea

Abstract

Regulatory T cells (Tregs) play a key role in the maintenance of immune homeostasis *in vivo*. The unique characteristics of Tregs endowed by foxp3, inability to produce IL-2 while constitutively expressing IL-2Rα (CD25), make themselves absolutely dependent on the responding cells for IL-2. When IL-2 is produced and secreted by the responder cells, it would bind preferentially to IL-2Rs on nearby Tregs, as the density of IL-2R on Tregs is higher than that on the responding cells. Therefore, Tregs start to proliferate earlier than the responding cells, and suppress the latter, resulting in an elegant loop of negative feedback. Furthermore, Tregs suppress the responder cells by lowering the production of IL-2 as well as by inducing earlier shedding of IL-2Rα from the responding cells. Meanwhile, the fate of secreted IL-2 and the mode of suppression are different according to the concentration of IL-2; if the

* Corresponding author: Department of Microbiology, Ewha Womans University Medical School, Mok-6-Dong 911-1, Yangchon-Gu, Seoul 158-710, South Korea, Tel: + 82 2 2650 5738 (Office), + 82 10 7222 5738 (Cell), Fax: + 82 2 2644 8667, E-mail: jyseoh@ewha.ac.kr

stimulating signal is weak, a small amount of IL-2 is secreted from the responder cells and will bind preferentially to the IL-2Rs on Tregs. Therefore, Tregs will proliferate earlier and suppress the responder cells. If the stimulating signal is strong and the initially secreted IL-2 is enough, it would also bind to the IL-2Rs on the responder cells. In this case, the responder cells proliferate initially, but their late expansion is suppressed by Tregs. Either way, in the steady state, Tregs always win the responder cells in the competition for IL-2, hence the state of tolerance. However, in infection or inflammatory state, the suppressive activities of Tregs are suppressed (contra-suppression) by certain molecules secreted from the innate immune cells (i.e., dendritic cells), such as IL-6 or IL-21, resulting in the proliferation of the responder cells, hence immune response is elicited. In conclusion, regulation of IL-2 and IL-2Rα is an important mechanism of immune homeostasis maintained by Tregs.

Introduction

Regulatory T cells (Tregs) play a key role in the maintenance of immune homeostasis *in vivo* [1]. Tregs were originally identified for their critical role to prevent the development of autoimmune diseases [2, 3]. In addition to the maintenance of self-tolerance, Tregs are also important in the regulation of almost all kinds of immune responses, including infection, graft rejection, hypersensitivity reaction, tumor immunity and so on [4]. They exert their regulatory effects by suppressing the proliferation and function of immune effector cells, including $CD4^+$ helper T cells [5], $CD8^+$ cytotoxic lymphocytes [6], B cells [7] and NK cells [8]. Many investigators have studied the mechanism of suppression by Tregs, and several fascinating mechanisms and associated molecules have been proposed, such as CTLA-4 [9] and IDO [10], but it is more plausible that multiple mechanisms and molecules are involved in the suppressive action of Tregs [11].

Initially, Tregs were identified in the $CD4^+CD25^+$ fraction that constitutively express IL-2Rα [2]. Several investigators have studied on the functional significance of CD25 expression on Tregs [12], but in our opinion, it has not yet been satisfactory. Because, we can learn much more about how immune homeostasis is maintained by Tregs through the modulation of IL-2 and IL-2Rα circuit. In the present study, we will focus on the regulatory mechanism of Tregs through the modulating IL-2 and IL-2Rα circuit.

1. Proliferation of Treg is Dependent on Exogenous IL-2 or the Responding Cells

IL-2 and IL-2R signaling is essential in the survival and function of Tregs. Mice deficient in IL-2, IL-2R or STAT-5 that transmits signals generated from IL-2R, have reduced numbers of Tregs and a lethal phenotype with lymphoproliferative diseases [13, 14]. Now, it is generally accepted that the sole non-redundant function of IL-2 is maintenance of the survival and function of Tregs in the periphery [15, 16]. Nonetheless, Tregs cannot produce IL-2 by themselves while they are always ready to respond to IL-2 as they constitutively express IL-2Rα (CD25) [17, 18]. Therefore, Tregs cannot proliferate in an autocrine fashion, and are dependent on exogenous IL-2 or the cells that secrete IL-2 for their proliferation [19]. *In vivo*, it should be activated T cells that provide Tregs with IL-2. This situation can be reproduced *in vitro*, where proliferation of Tregs is dependent on responder cells (Figure 1).

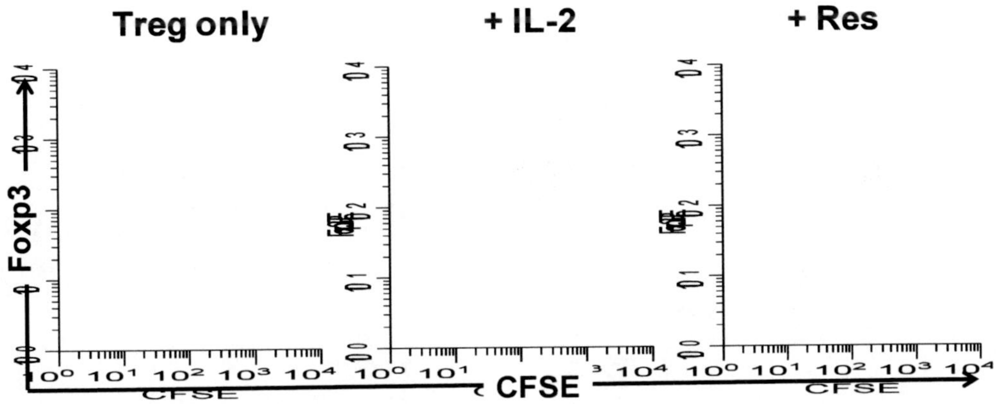

Figure 1. Regulatory T cells (Tregs) depend on exogenous IL-2 or the responder cells (Res) for their proliferation. $CD4^+CD25^+$ cells were isolated from the spleens of BALB/c mice using a regulatory T cells isolation kit (Miltenyi Biotec, Bergisch Gladbach, Germany), and used for Tregs. $CD4^+CD25^-$ cells were also isolated and used for the responder cells (Res). $CD11c^+$ cells were isolated using a spleen dissociation medium (StemCell technologies, Vancouver, BC), density gradient centrifugation over 15.5% Accudenz (Accurate Chemical & Scientific, Westbury, NY), and immunomagnetic selection using anti-CD11c-PE (Pharmingen, San Jose, CA) and microbead anti-PE (Miltenyi Biotec), and used for dendritic cells (DCs). The purity of Tregs and DCs was over 95%. Ten thousands Tregs stained with CFSE (Invitrogen, Carlsbad, CA), were cultured with 2,000 DCs in each well of round bottomed 96-well plates. The cells were stimulated with anti-CD3e (1 ug/mL, Pharmingen) with or without recombinant mouse IL-2 (300 pg/mL, Chemicon, Temecula, CA) or equal numbers of the responder cells. On the 3rd day, the cells were harvested and stained for intranuclear FoxP3 using a FoxP3 staining kit (eBioscience, San Diego, CA).

Inability to produce IL-2 while constitutively expressing IL-2Rα is an inherent characteristic of Tregs, because foxp3, the master gene for the development and function of Tregs, suppresses transcription of IL-2 while activating the IL-2Rα gene through interacting with NFAT and Runx1 [20]. This unique characteristic makes Treg proliferation absolutely dependent on their target of regulation, formulating an elegant negative feedback loop [21]. The negative feedback is so exquisite that Tregs proliferate only when necessary and only to the necessary extent. If the stimulating signal is weak, a small amount of IL-2 is produced from the activated T cells, and a small number of Tregs, that can suppress the activated T cells, will proliferate (Figure 2A). If the stimulating signal is strong, an abundant amount of IL-2 is produced and many activated T cells will proliferate, but many Tregs, which can counterbalance the expanding activated T cells, will also proliferate (Figure 2B). On the other hand, activated T cells cannot escape from the regulation by Tregs, as the growth factor IL-2 necessary for proliferation, is shared by Tregs and the activated T cells. In steady state, Tregs will always win the activated T cells in the competition for IL-2, as they constitutively express a higher level of IL-2Rα [12]. Therefore, Treg proliferation always predominates over and the proliferation of the activated T cells is blocked (Figure 2C). Consequently, immune response cannot occur at all in steady state! Then, how can immune response occur when necessary? This issue will be discussed later.

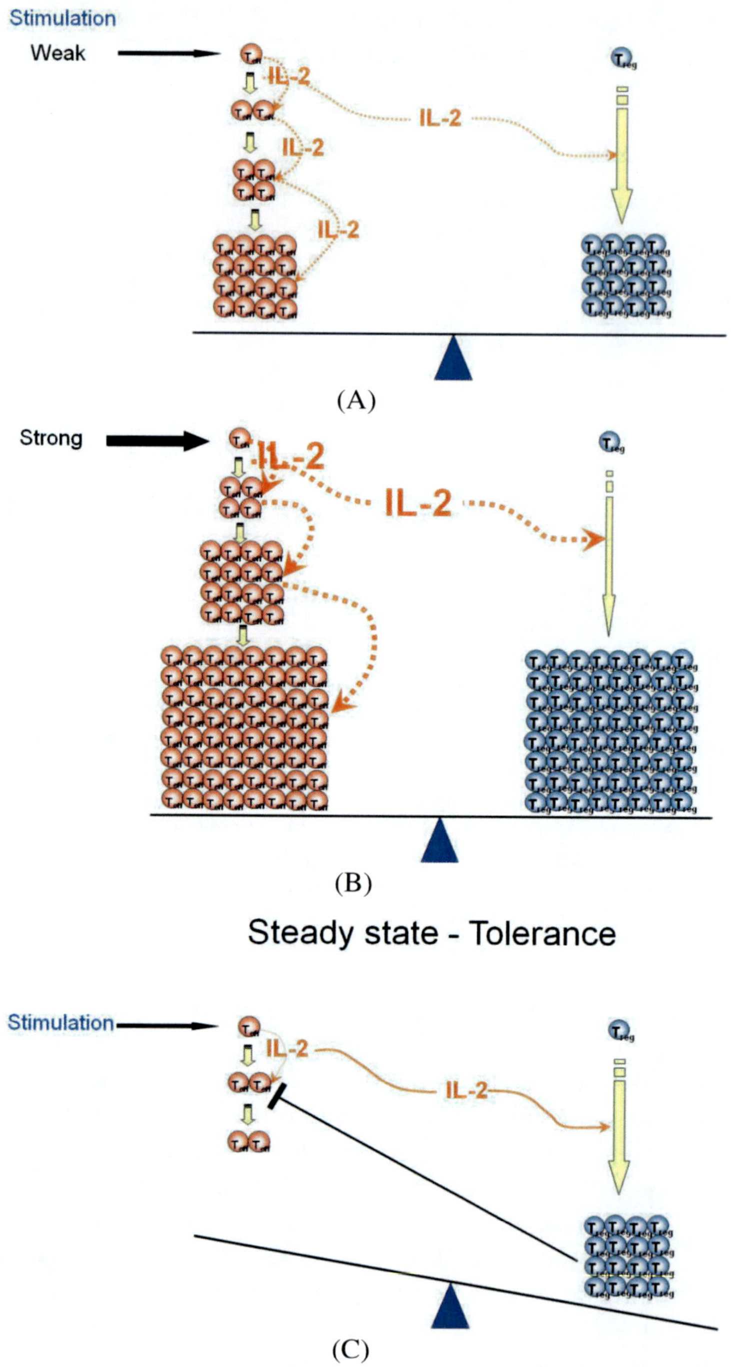

Figure 2. Feedback regulation of immune response by regulatory T cells (Tregs). Tregs proliferate only when necessary and only to the necessary extent. If the stimulating signal is weak, a small amount of IL-2 is produced from the activated or effector T cells (Teffs), and a small number of Tregs that can suppress the activated T cells, will proliferate (A). If the stimulating signal is strong, an abundant amount of IL-2 is produced and many Teffs will proliferate, but many Tregs, which can counterbalance the expanding Teffs, will also proliferate (B). In steady state, Tregs always win Teffs in the competition for IL-2, as they constitutively express a higher level of IL-2Rα. So, Treg proliferation always predominates over and blocks the proliferation of Teffs (C).

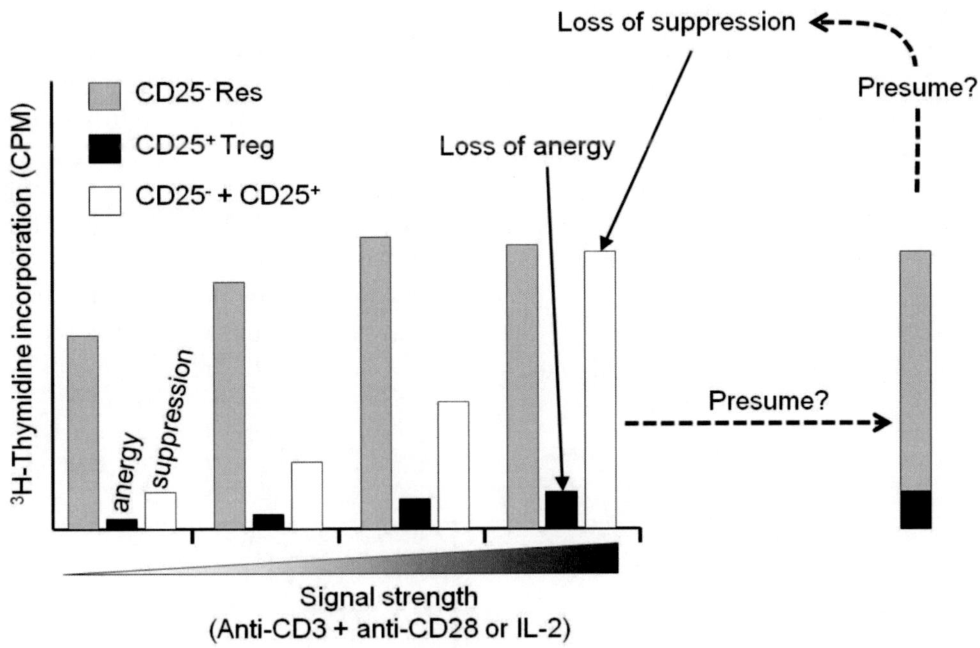

Figure 3. The relationship between anergy and suppressive activity of regulatory T cells (Tregs), presumed from the result of thymidine incorporation assay. As the strength of stimulating signal is increased by adding costimulation or exogenous IL-2, Tregs begin to proliferate (loss of anergy) while vigorous proliferation is observed in the coculture with Tregs and responder cells (Res). They presumed the cells that proliferated in the coculture should be responder cells, as Tregs do not proliferate well, although it is impossible to discriminate which cells proliferate in the coculture by thymidine incorporation assay. Accordingly, they interpreted that Tregs lose suppressive activity when they proliferate (lose anergy). Modified from [28].

2. Tregs Proliferate well *in vitro* as well as *in vivo*

For a long time, many investigators believed that Tregs do not proliferate well *in vitro* (anergy) [4, 22]. The characteristic of anergy was believed to be closely linked with the suppressive activity of Tregs. This bizarre characteristic of Tregs has been documented by S. Sakaguchi group [22] and reproduced by many investigators [18, 23, 24], and has become a solid dogma of Tregs [4]. Meanwhile, several groups observed that $CD4^{+}CD25^{+}$ Tregs proliferate as well as the $CD4^{+}CD25^{-}$ responder cells *in vivo* and preserve suppressive activities [25-27]. Although the observations *in vivo* were so clear, the dogma was too secure to break, and they reached a compromise that Tregs behave differently *in vivo* and *in vitro*; Tregs are proliferative *in vivo* but hypoproliferative *in vitro* [4]. At this point, we raised a doubt. How can a certain cell behave absolutely oppositely *in vitro* and *in vivo*? In our opinion, it would be more reasonable to interpret that the *in vitro* assay may not represent the *in vivo* situation appropriately. At that time, the method used for the *in vitro* suppressive test was thymidine incorporation assay (Figure 3) [22, 28]. As the strength of the stimulating signal is increased by adding costimulation or exogenous IL-2, $CD4^{+}CD25^{+}$ Tregs begin to proliferate (loss of anergy) while vigorous proliferation is observed in the coculture with Tregs and the $CD4^{+}CD25^{-}$ responder cells. They presumed the cells that proliferated in the coculture should be the responder cells, as Tregs do not proliferate well, although it is impossible to discriminate which cells proliferate in the coculture by thymidine incorporation assay. Accordingly, they interpreted that Tregs lose suppressive activities when they proliferate (lose anergy). This is the critical mistake because selective labeling of the responder cells with CFSE showed that the actual predominant cells in the coculture were Tregs instead of the responder cells (Figure 4) [21]. The mistake was due to inappropriate comparison of proliferative responsiveness and suppressive activities of Tregs in different settings; in the absence and presence of the responder cells, as at that time they were unaware of the dependence of Tregs on the responder cells for their proliferation (Figure 1). Furthermore, the proliferating Tregs in the coculture preserved suppressive activities, demonstrating that anergy is not related to the suppressive activities of Tregs (Figure 4).

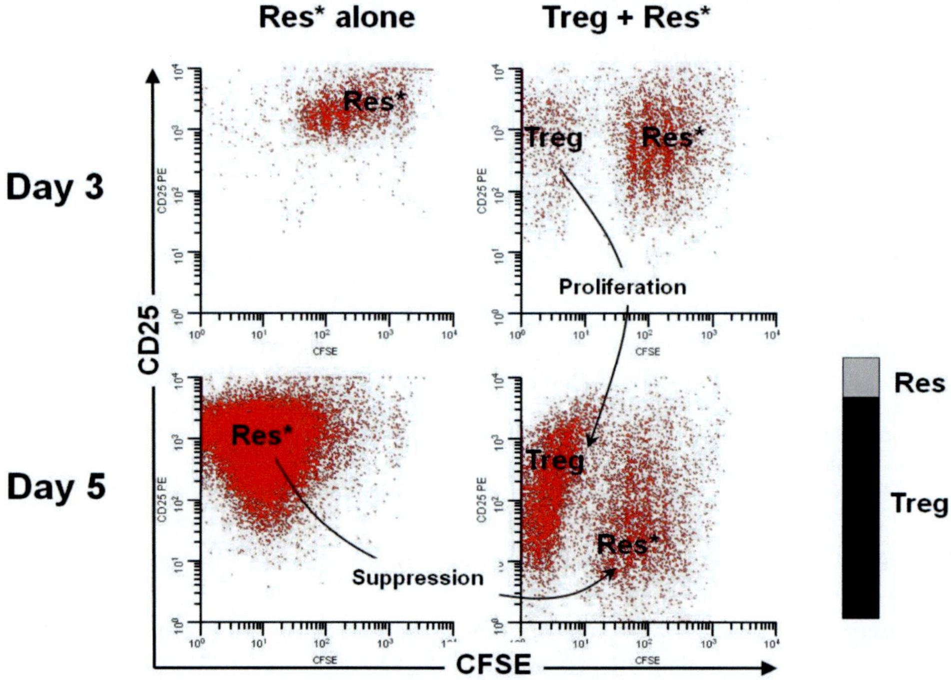

Figure 4. Proliferating regulatory T cells (Tregs) preserve suppressive activities. $CD4^+CD25^+$ Tregs, $CD4^+CD25^-$ responder cells (Res) and $CD11c^+$ dendritic cells (DCs) were isolated and cultured in the same way as in Figure 1. In this experiment, the responder cells were selectively labeled with CFSE and were cultured in the presence or absence of equal numbers of Tregs, with the stimulation of anti-CD3e (100 ng/mL) plus anti-CD28 (100 ng/mL). On the 3rd and 5th day after stimulation, the cells were harvested and stained for anti-CD4-PerCP and anti-CD25-PE. Flow cytometric analysis showed that Tregs proliferated from day 3 to day 5, while the proliferation of the responder cells in the coculture with Tregs was suppressed, compared with that in the absence of Tregs. Therefore, it can be argued that Tregs in the coculture suppressed the proliferation of the responder cells while they proliferated [21].

3. Distinct Modes of Suppression According to the Strength of Signal - A Quantitative Analysis

When we analyzed the suppressive activities of Tregs in response to varying concentrations of anti-CD3 and anti-CD28, distinct patterns of suppression were observed (Figure 5) [21]. Stimulation with anti-CD3 alone, or with very low concentrations of anti-CD3 plus anti-CD28, induced almost perfect suppression of the responder cells in the coculture with Tregs. By contrast, stimulation with medium or high concentrations of anti-CD3 plus and anti-CD28 induced suppression of the responder cells but in a different pattern; decrease in the IL-2Rα (CD25) expression level on the responder cells on day 3 and decrease in the late expansion until day 5 (Figure 5). For accurate analysis, we employed two quantitative parameters. Precursor frequency (Pf) is the fraction of the cells at the initial time point that have gone into cell cycle [29]. It can be easily calculated from the CFSE signals by using software (proliferation wizard of Modifit^{LT}). CD25 mean fluorescence intensity (MFI) of the responder cells on day 3 correlates very well with the cell number on day 5 (Figure 6) [21]. That is, CD25 MFI, acquired on day 3, may predict how many cells will proliferate until day 5, representing late proliferative responsiveness. In contrast, Pf, intrinsically, represents early proliferative responsiveness. The two parameters can be acquired by simple flow cytometric analysis on day 3.

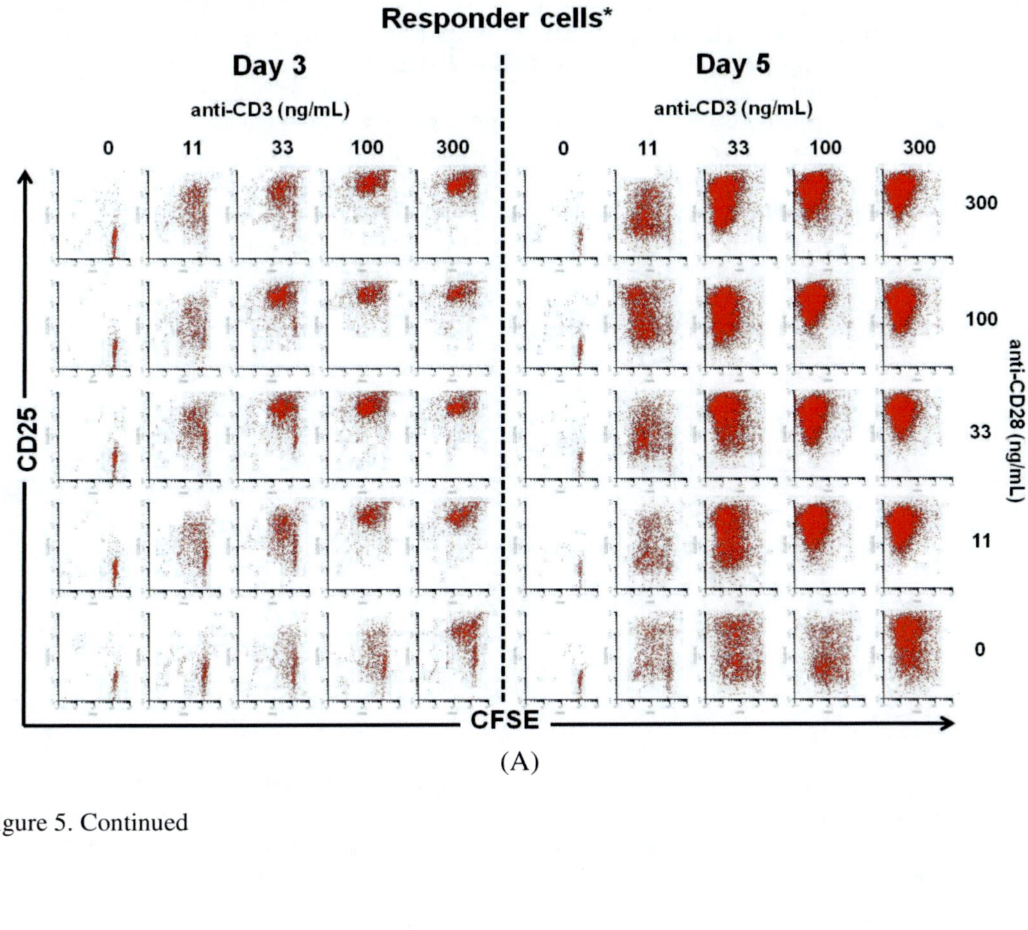

(A)

Figure 5. Continued

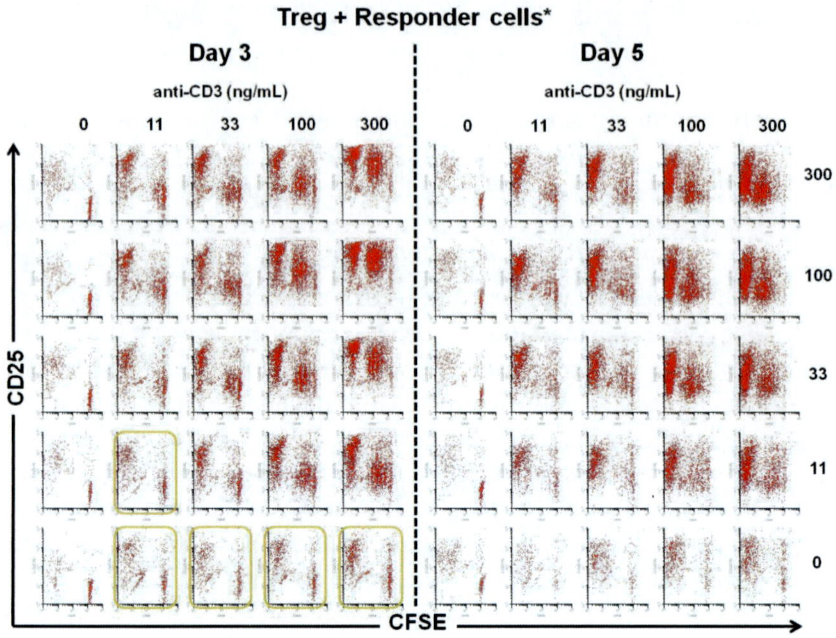

(B)

Figure 5. Distinct modes of suppression according to the strength of signal. CD4$^+$CD25$^+$ regulatory T cells (Tregs), CD4$^+$CD25$^-$ responder cells and CD11c$^+$ dendritic cells were isolated and cultured in the same way as Figure 1. Responder cells were stained with CFSE and were cultured in the absence (A) or presence (B) of equal number of Tregs, with varying concentrations of anti-CD3e and anti-CD28. The cells were harvested and analyzed on day 3 and day 5, in the same way as in Figure 4. Stimulation with anti-CD3e alone or very low concentrations of anti-CD3e plus anti-CD28 induced almost perfect suppression of the responder cells (yellow rectangles). By contrast, stimulation with medium or high concentrations of anti-CD3e plus and anti-CD28 induced suppression of the responder cells but in a different pattern; decrease in the IL-2Rα (CD25) **expression level** on the responder cells on day 3 and decrease in the late expansion until day 5.

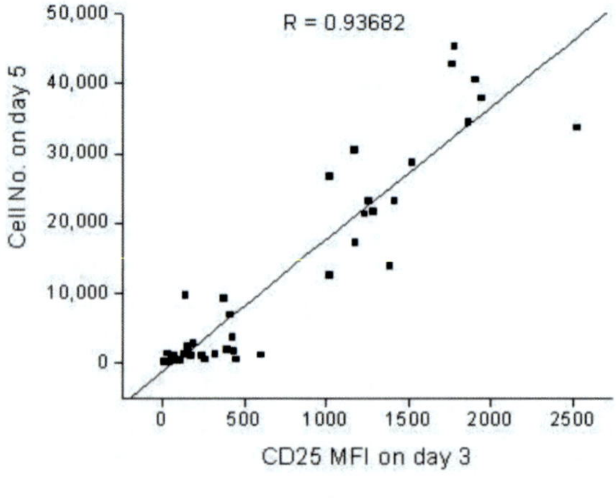

Figure 6. CD25 mean fluorescence intensity (MFI) of the responder cells on day 3 correlates well with the cell number on day 5. So, CD25 MFI, acquired on day 3, may predict how many cells will proliferate until day 5, and can be used as a quantitative parameter representing late proliferative responsiveness [21].

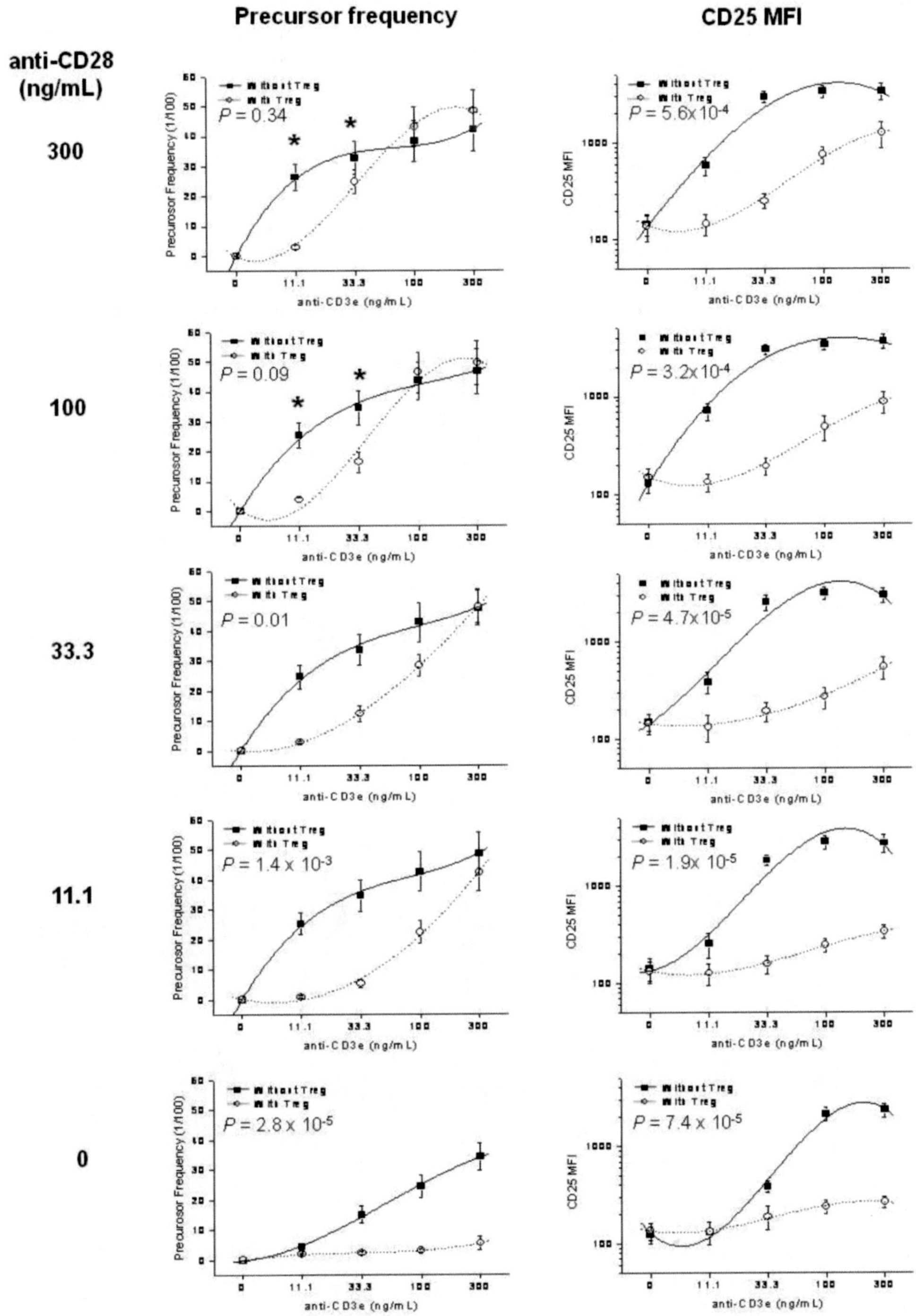

Figure 7. Quantitative analysis of the proliferative responsiveness of the responder cells in the absence or presence of regulatory T cells (Tregs) shows distinct modes of suppression according to the strength of the stimulating signals. Quantitative analysis of the experiments in Figure 5, left panel (day 3), employing the two parameters, precursor frequency (Pf) and CD25 mean fluorescence intensity (MFI). Stimulation with anti-CD3e alone induced almost perfect suppression of the responder cells in the coculture with Tregs in terms of both Pf and CD25 MFI (lowermost). Significant decrease in Pf in the

coculture with Tregs suggests that proliferation of the responder cells was suppressed from the beginning. When the cells were stimulated with anti-CD3e and low concentrations of anti-CD28 (11.1 or 33.3 ng/mL), Pf of the responder cells was also significantly decreased in the coculture with Tregs, suggesting initial proliferation was inhibited. By contrast, when stimulated with anti-CD3e and high concentrations of anti-CD28 (100 or 300 ng/mL), Pf of the responder cells was not significantly decreased, but CD25 MFI was significantly decreased in the coculture with Tregs, suggesting that late expansion was suppressed but initial proliferation was not. Although the overall values of Pf were not significantly different in the stimulation with anti-CD3e plus high concentrations of anti-CD28, Pf at two separate points of low concentrations of anti-CD3e (11.1 or 33.3 ng/mL) was significantly different (*), suggesting that initial proliferation of the responder cells was also inhibited at these points. Taken together, it seems the strength of the stimulating signals that determines the mode of suppression; in response to weak stimulating signals, Tregs suppress the initial proliferation of the responder cells from the beginning. By contrast, in response to strong signals, Tregs cannot suppress the initial proliferation, but suppress the late expansion of the responder cells. Pf values were calculated from the histograms of CFSE signals of the responder cells by the use of proliferation wizard module in Modfit LT™ software (Verity, Topsham, ME). Data represent mean ± SD of three separate experiments. *P*-values were calculated by one way ANOVA, and *P* values less than .05 were considered to be statistically significant.

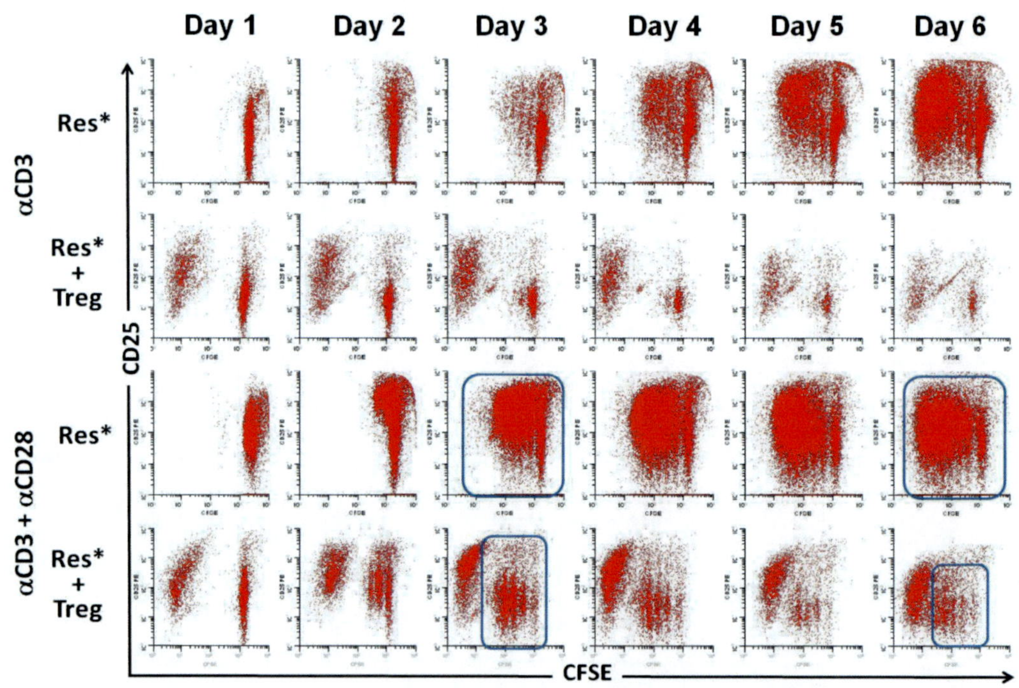

Figure 8. Proliferation and suppression patterns of the responder cells (Res*) in the absence and presence of equal number of Tregs in response to anti-CD3e (300 ng/mL) alone or together with anti-CD28 (300 ng/mL) show two distinct modes of suppression. In response to anti-CD3e alone, proliferation of the responder cells was suppressed from the beginning in the coculture with Tregs, whereas in response to anti-CD3e plus anti-CD28, initially the responder cells proliferated (from day 1 to day 3), but late expansion from day 4 was suppressed. The blue rectangles indicate the similar time points in Figure 9.

Quantitative analysis using the two parameters also showed distinct modes of suppression according to the strength of the stimulating signals (Figure 7). Stimulation with anti-CD3 alone induced almost perfect suppression of the responder cells in the coculture with Tregs in terms of both Pf and CD25 MFI. Significant decrease in Pf in the coculture with Tregs suggests that proliferation of the responder cells was suppressed from the beginning. When the cells were stimulated with anti-CD3 and low concentrations of anti-CD28 (11.1 or 33.3 ng/mL), Pf of the responder cells was also significantly decreased in the coculture with Tregs, suggesting initial proliferation was inhibited ($P < .05$). By contrast, when stimulated with anti-CD3 and high concentrations of anti-CD28 (100 or 300 ng/mL), Pf of the responder cells was not significantly decreased ($P > .05$), but CD25 MFI was significantly decreased in the coculture with Tregs. This result suggests that late expansion was suppressed but the initial proliferation was not. Although the overall values of Pf were not significantly different in the stimulation with anti-CD3 plus high concentrations of anti-CD28, Pf at two separate points of low concentrations of anti-CD3 (11.1 or 33.3 ng/mL) was significantly different, suggesting that initial proliferation of the responder cells was also inhibited at these points (*). Taken together, it seems the strength of the stimulating signals that determines the mode of suppression; in response to weak stimulating signals, Tregs suppress the initial proliferation of the responder cells from the beginning. By contrast, in response to strong signals, Tregs cannot suppress the initial proliferation, but suppress the late expansion of the responder cells.

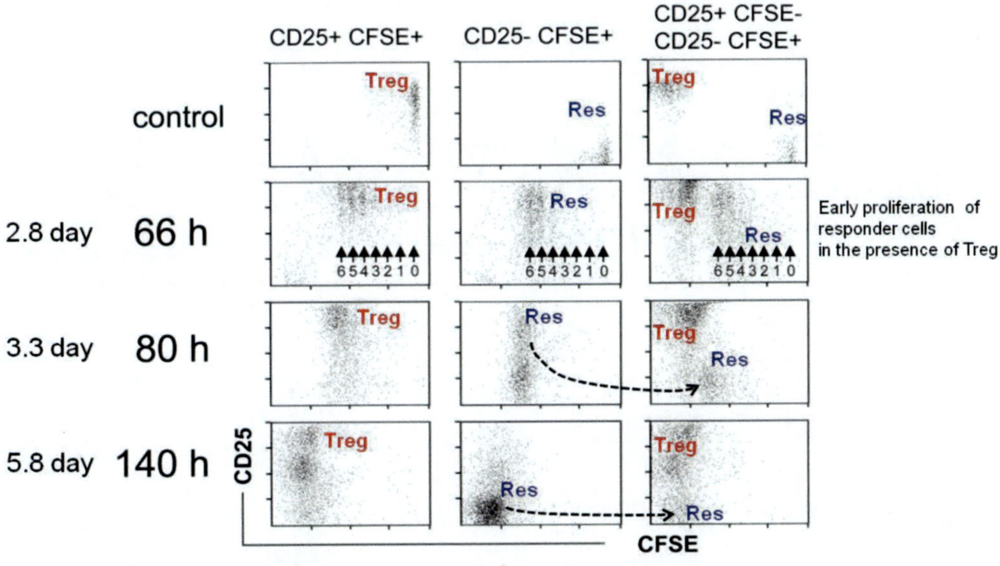

Figure 9. *In vivo* dynamics of Tregs are almost the same as observed *in vitro*, suggesting the *in vitro* assay represents the *in vivo* response appropriately. Reproduced and modified from [27]. Proliferation of adoptively transferred CD4$^+$CD25$^+$ T cells (Tregs) (left) or CD4$^+$CD25$^-$ T cells (Res) (middle) in the draining lymph nodes after immunization. Tregs and Res cells isolated from HA-specific TCR transgenic mice were selectively labeled with CFSE and adoptively transferred into syngeneic mice. Two days later, recipients were immunized with 100 ug of HA-peptide in incomplete Freund's adjuvant. Mice were killed at the indicated time points after immunization, and draining lymph node cells were harvested and stained for CD4, CD25, and the donor cell marker (Thy1.2). The dot plots show the expression of CD25 versus CFSE fluorescence intensity on gated donor-derived cells. Numbered arrows within the dot plots (66 h) indicate the number of divisions of CFSE-labeled cells. (Right) Representative analysis of a cotransfer of CFSE-labeled Res cells and equal number of unlabeled Tregs. Compared with the *in vivo* dynamics of Res and Tregs adoptively transferred alone, the proliferation pattern of the Res cells co-transferred with Tregs was different. In the early stage until 66 h or 80 h, Res cells proliferated but the expression of CD25 was downregulated, and at the late stage

until 140 h, expansion of Res cells was decreased. Copyright (2003) National Academy of Sciences, U.S.A.

Actual analysis of the proliferation of the responder cells alone or in the coculture with Tregs in response to anti-CD3 alone (300 ng/mL) or together with anti-CD28 (300 ng/mL) in time showed that the two distinct modes of suppression are exactly the same as expected by the quantitative analysis, demonstrating the validity of the quantitative analysis (Figure 8). In particular, the suppression pattern in response to anti-CD3 plus anti-CD28 is very similar to what had been observed *in vivo* (Figure 9) [27], supporting that the *in vitro* assay system closely reproduces *in vivo* responses. Based on our observations *in vitro* and the quantitative analysis, in association with others' observations *in vivo*, we formulated a feedback model of immune regulation by Treg, showing two distinct modes of suppression according to the strength of stimulating signal (Figure 10). Immune homeostasis through negative feedback regulation by Treg has also been proposed by others [30]. However, their model does not seem suitable to communicate their ideas as they did not overcome the dogma of anergy.

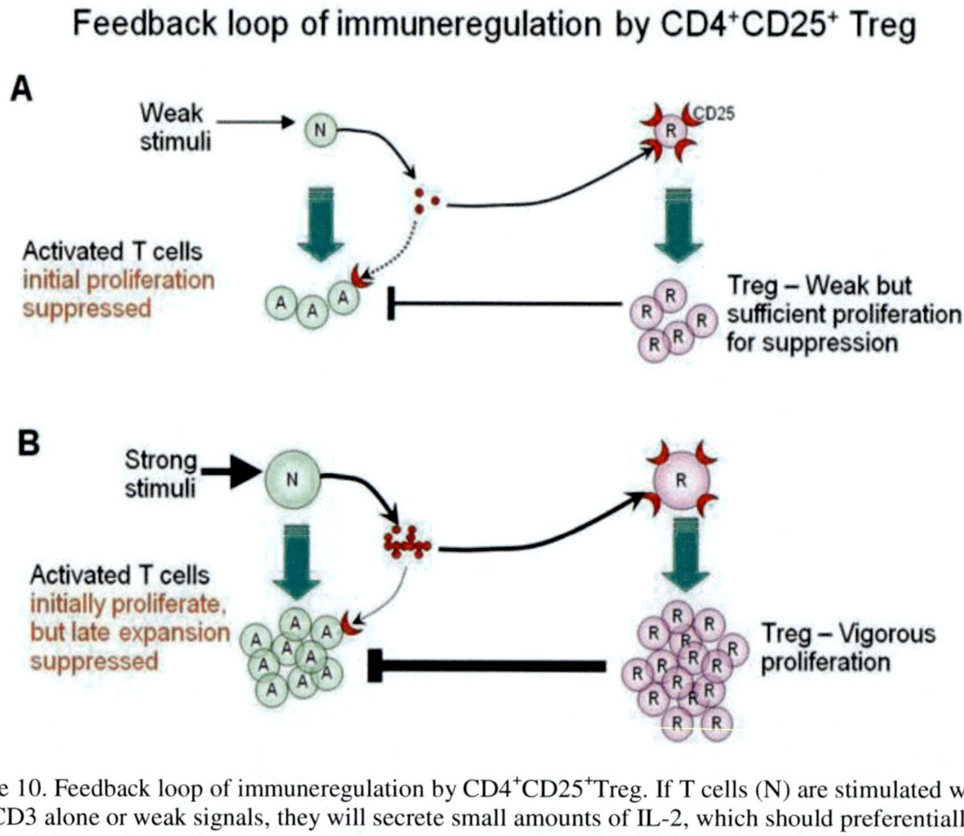

Figure 10. Feedback loop of immuneregulation by $CD4^{+}CD25^{+}$Treg. If T cells (N) are stimulated with anti-CD3 alone or weak signals, they will secrete small amounts of IL-2, which should preferentially bind to IL-2R on Tregs (A). Tregs will proliferate and suppress the proliferation of activated T cells (A) from the beginning. If T cells are stimulated with anti-CD3 plus anti-CD28 or strong signals, they will secrete large amounts of IL-2, which would bind to IL-2R on activated T cells themselves as well as on Tregs (B). Therefore, T cells initially proliferate while Tregs proliferate more vigorously as Tregs win in the competition for IL-2. The vigorously proliferating Tregs will suppress late expansion of the activated T cells by lowering CD25 expression [21].

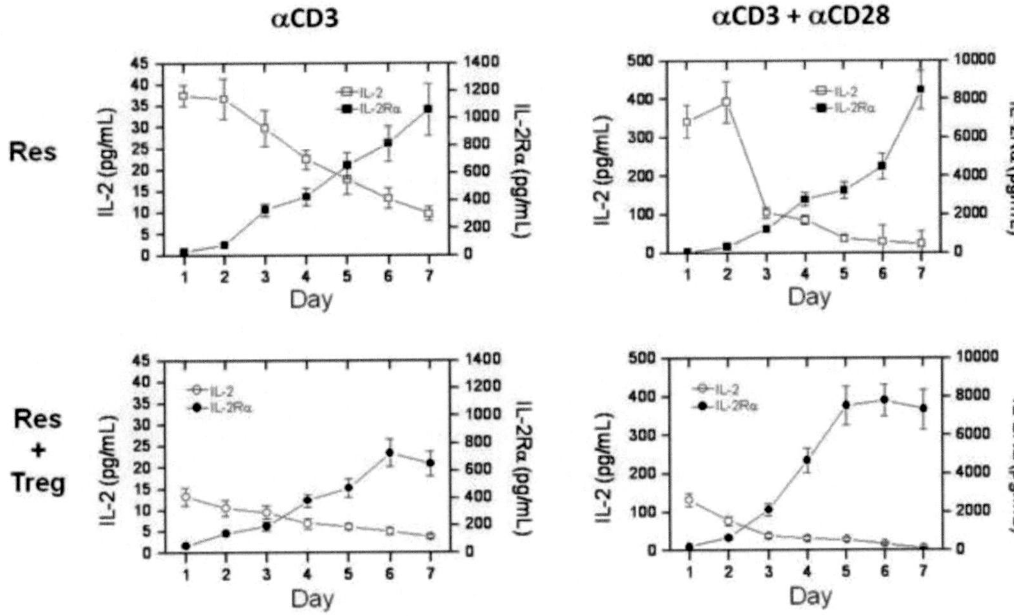

Figure 11. IL-2 and soluble IL-2Rα concentrations in the supernatants of the CD4$^+$CD25$^-$ responder cells (Res) alone or in the coculture with CD4$^+$CD25$^+$ regulatory T cells (Tregs) in response to anti-CD3e (300 ng/mL) alone or together with anti-CD28 (300 ng/mL). The cells were isolated and cultured in the same way as in Figure 1, and the supernatants were harvested daily, and stored at -70°C for later assay. IL-2 and IL-2Rα concentrations were measured by ELISA using OptEIA™ kits from BD Biosciences (San Diego, CA). Data represent mean ± SD of triplicates of three separate experiments showing the same pattern.

4. Tregs Lower CD25 Expression on the Responder Cells

It has already been documented that inhibition of the increase in the expression of IL-2Rα (CD25) on the responder cells is one of the suppressive mechanisms of Tregs [15]. However, it has not yet been clarified if the expression of IL-2Rα is decreased or IL-2Rα is shed from the membranes of the responder cells. To address this issue, we measured the concentrations of IL-2 and soluble IL-2Rα in the culture supernatants of the responder cells in the presence or absence of Tregs in response to anti-CD3 with or without anti-CD28 (Figure 11). At a glance, IL-2 concentrations are higher in the early stage than in the late stage, and those in the coculture with Tregs are lower than those in the responder cells alone, particularly at the early stage. Soluble IL-2Rα concentrations are higher in the late stage, reaching a very high level at the late stage, as reported previously [31]. In particular, in response to anti-CD3 plus anti-CD28, both the levels of IL-2Rα in the supernatants of responder cells alone and in the coculture with Tregs reached a very high level up to 8,000 pg/mL, but in a different pattern; IL-2Rα level in the supernatants of responder cells alone rises steeply in the late stage from day 6, whereas that in the coculture with Tregs rises steeply from the earlier stage and reaches a plateau at day 5. This result suggests that Tregs enhance secretion or shedding of IL-2Rα from the cells in the coculture at an early stage. Meanwhile, it has been already reported that IL-2 levels at the early stage are more important to determine the proliferative responsiveness [32]. Therefore, we precisely analyzed the levels at the early stage, in association with the IL-2Rα levels on the cell surface (Figure 12). Above all, the concentrations of IL-2Rα in the supernatants of Tregs alone are very low, and the levels of IL-2Rα on Treg surface are not decreased in the coculture with the responder cells. By contrast, IL-2Rα levels on the surface of the responder cells in the coculture with Tregs are lower than those of the responder cells alone, and IL-2Rα levels in the supernatants of the coculture are higher than those in the responder cells alone. Therefore, the increase in the level of IL-2Rα in the supernatants may be related to the decrease in the level of IL-2Rα on the surface of the responder cells in the coculture with Tregs. Western blot analysis showed smaller molecular weight, 36 kDa, of IL-2Rα in the supernatants than that, 55 kDa, on the cell membranes (data not shown). Taken together, these results raise the possibility that Tregs induce earlier shedding of IL-2Rα from the responder cells (than the usual spontaneous shedding at 7 – 8 days), probably by enzymatic cleavage. However, the exact mechanism of selective cleavage of IL-2Rα from the responder cells in the coculture with Tregs remains to be elucidated.

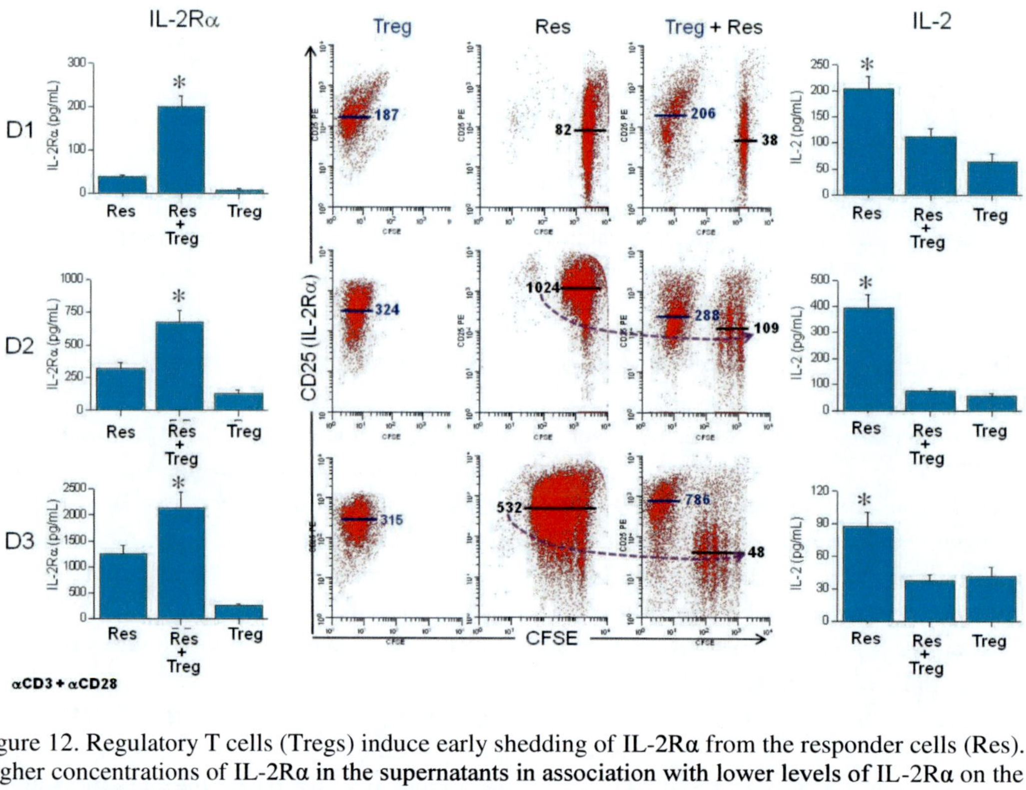

Figure 12. Regulatory T cells (Tregs) induce early shedding of IL-2Rα from the responder cells (Res). Higher concentrations of IL-2Rα in the supernatants in association with lower levels of IL-2Rα on the responder cell surface in the coculture suggest shedding of IL-2Rα from the responder cells. The shed IL-2Rα may contribute to the lower concentration of IL-2 in the supernatants of the coculture. Data represent mean ± SD of triplicates of three separate experiments showing the same pattern. *P*-values were calculated by independent Student's t test, and *P* values less than.05 were considered to be statistically significant (*).

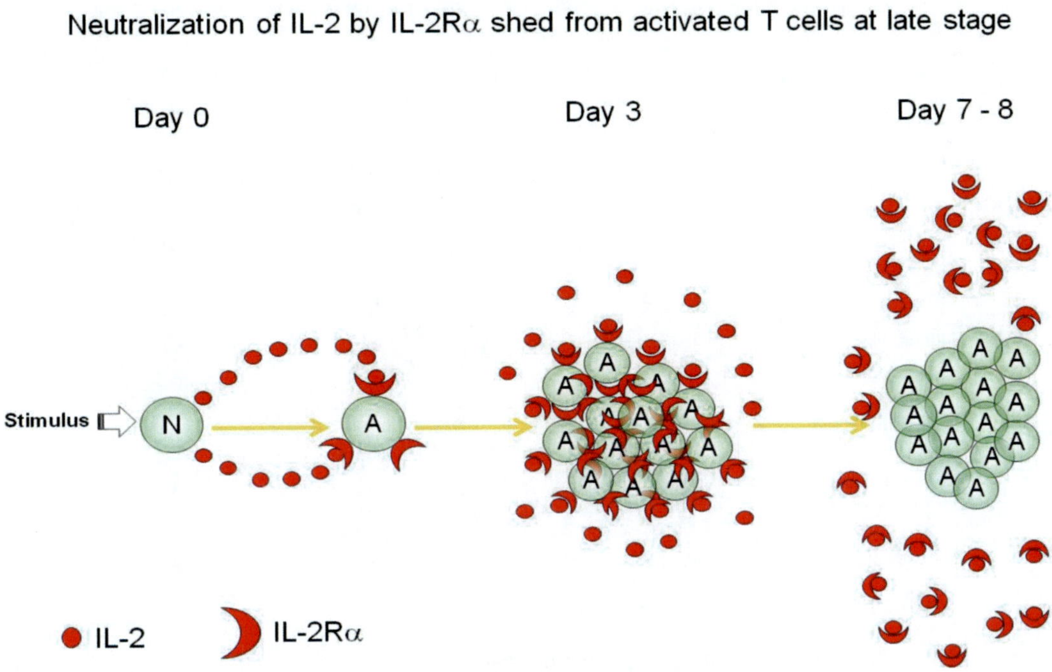

(A)

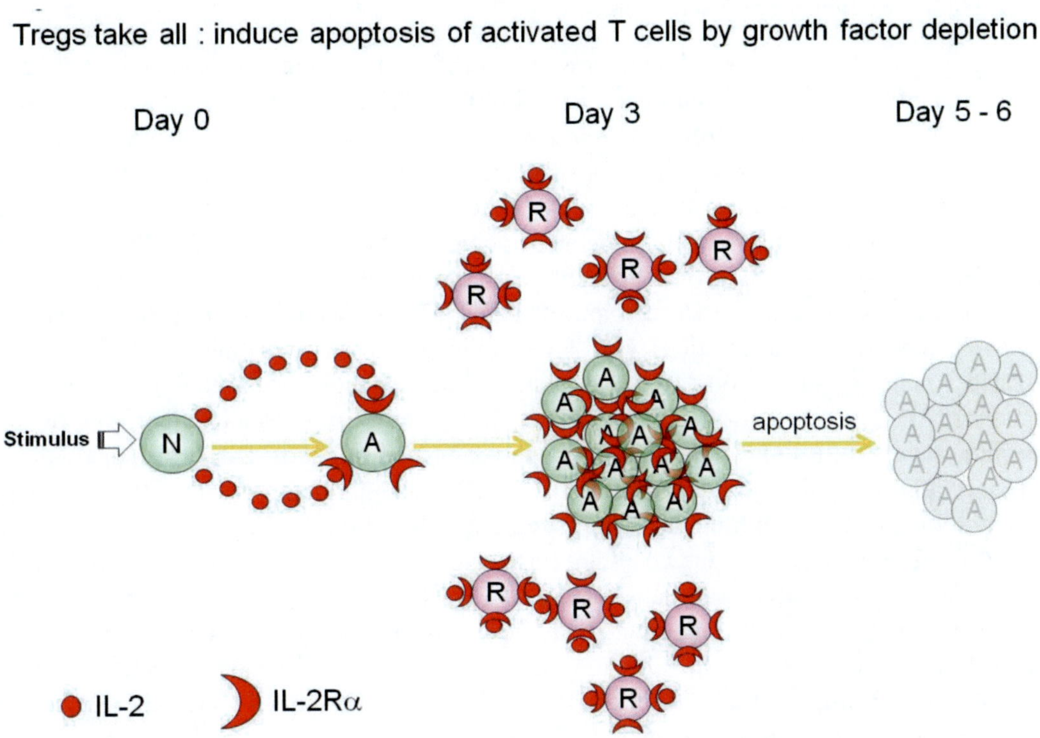

(B)

Figure 13. Continued

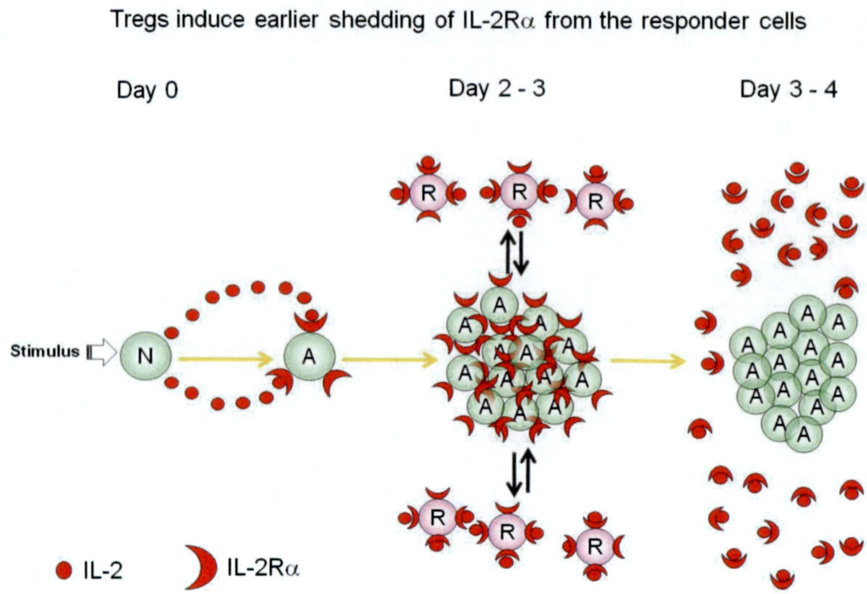

(C)

Figure 13. Evolution of the concepts on the immune-regulatory mechanism involving IL-2 and IL-2Rα. (A) Neutralization of IL-2 by IL-2Rα shed from activated T cells at late stage. (B) Growth factor (IL-2) depletion by regulatory T cells (Tregs) induces apoptosis of the activated T cells. (C) Tregs induce early shedding of IL-2Rα from the responder cells.

IL-2Rα shedding is one of the classical negative feedback mechanisms of T cell activation, that has been reported as early as in 1980's [33]. At the late stage of activation (ca. 7 – 8 days), T cells shed a profuse amount of IL-2Rα that can neutralize the residual IL-2 (Figure 13A). Recently, another regulatory mechanism involving IL-2 and IL-2Rα has been reported. As Tregs constitutively express a higher level of IL-2Rα, Tregs take all IL-2 secreted from the activated T cells, inducing apoptosis of the activated T cells by growth factor depletion (Figure 13B) [34]. In addition to the mechanisms already reported so far, based on our experimental evidence (Figure 11 and 12), we would like to propose another regulatory mechanism. Interaction with Tregs induces earlier shedding of IL-2Rα from activated T cells, as early as on day 2 or 3, than the usual spontaneous shedding on day 7 – 8 (Figure 13C). The shed IL-2Rα may bind IL-2 in the supernatants, and the higher concentration of IL-2Rα may contribute to the lower concentration of IL-2 in the supernatants of the coculture. Although the affinity of IL-2Rα single chain is very low, at high concentrations, IL-2Rα can bind and neutralize IL-2 [33]. Of course, the lower concentration of IL-2 in the supernatants of the coculture may be due to the decrease in the production of IL-2 because Tregs inhibit the production of IL-2 in the responder cells (Figure 14) [35]. Nonetheless, what is actually important is the local concentration of shed IL-2Rα and IL-2 near the activated cells *in situ*. At present, it is very difficult to estimate the local concentrations *in situ*, and the possibility of neutralizing IL-2 by IL-2Rα shed earlier through interaction with Tregs still remains to be excluded. Irrespective of the possibility of neutralization, downregulation of IL-2Rα on the target of regulation (responder cells), while competing together for IL-2, is an exquisite regulatory mechanism of Tregs that determines the later proliferative responsiveness of the target cells.

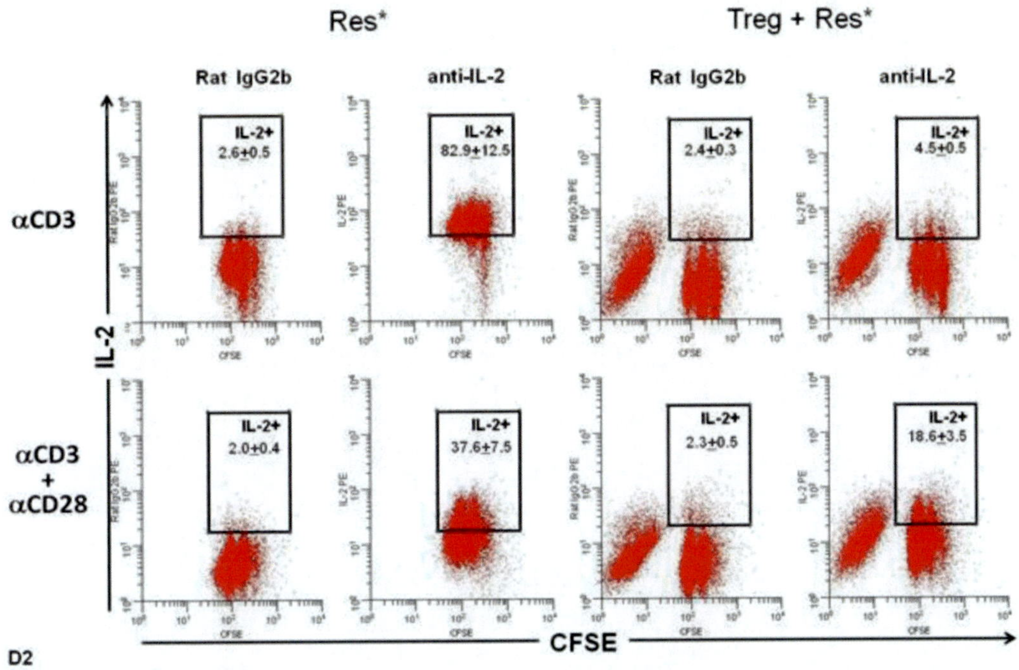

Figure 14. Regulatory T cells (Tregs) inhibit IL-2 production in the responder cells (Res). CFSE-labeled $CD4^{+}CD25^{-}$ responder cells and $CD4^{+}CD25^{+}$ Tregs were cultured in the same way as in Figure 1. On the 2nd day after stimulation with anti-CD3e (1 ug/mL) alone or together with anti-CD28 (1 ug/mL), the cells were re-stimulated with immobilized anti-CD3e (50 ng/well) and soluble anti-CD28 (1 ug/mL) in the presence of monensin (GolgiStop, BD Biosciences) for 4 hours. After staining for surface CD4, the cells were permeabilized and stained with anti-mouse IL-2-PE or isotype control (BD Biosciences).

5. Contra-Suppression

As we suggested earlier, there should be some mechanisms that block the suppressive function of Tregs for immune responses to occur when necessary [36]. Pasare C *et al* (2003) have reported that ligation of TLR4 or TLR9 on DCs induces secretion of cytokine(s) that overcome Treg-mediated suppression [37]. They found that IL-6 is critical for overcoming Treg-mediated suppression as neutralization of IL-6 abrogated the ability to block suppression. Contrary of their anticipation, they did not observe the direct blocking effect by adding recombinant (r) IL-6. On the other hand, we observed that Treg-mediated suppression was blocked by adding rIL-6 (10 ng/mL) or rIL-21 (20 ng/mL) (Figure 15). The discrepancy may be due to the difference between the conventional thymidine incorporation assay and our CFSE assay using two quantitative parameters.

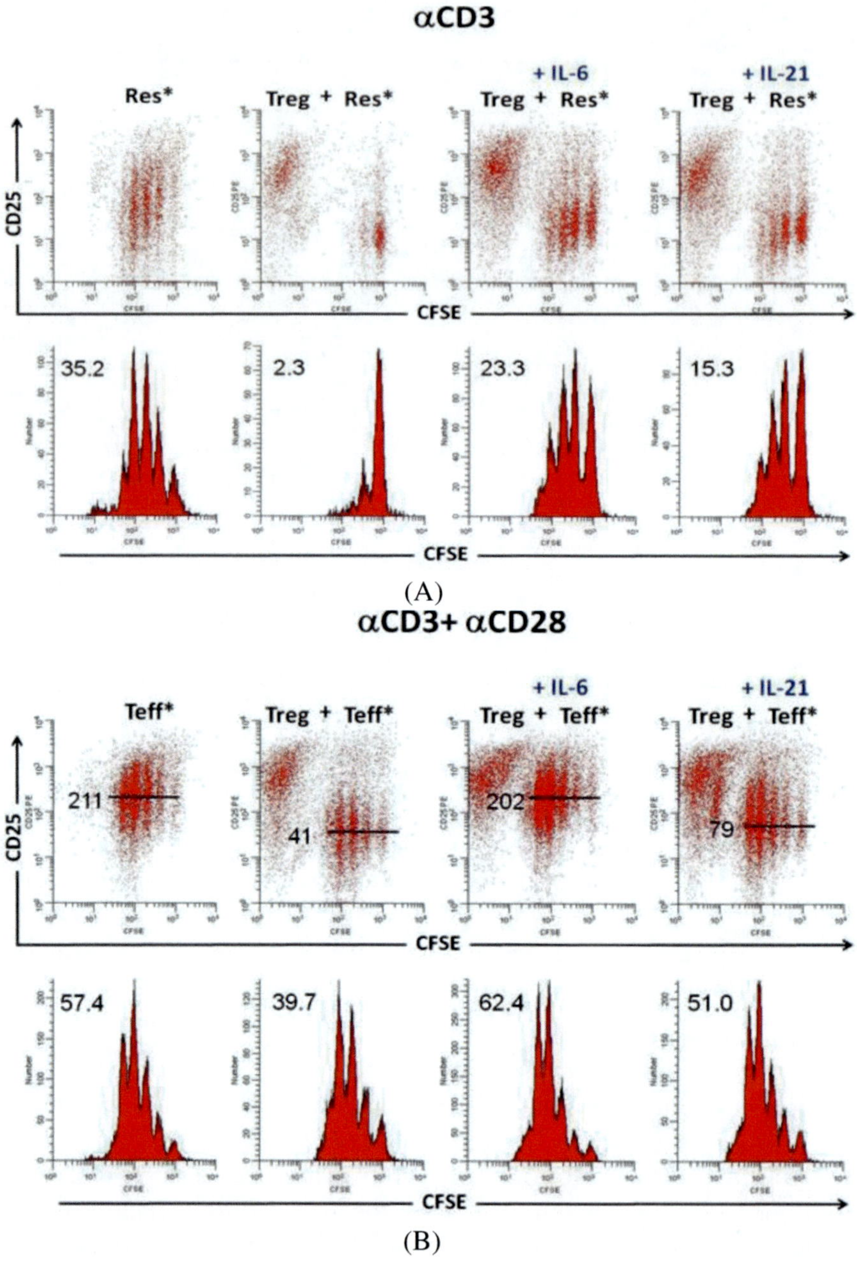

Figure 15. Contrasuppression by IL-6 or IL-21. $CD4^+CD25^-$ responder cells (Res) and $CD4^+CD25^+$ regulatory T cells (Tregs) were isolated and cultured in the same way as in Figure 1. The cells were stimulated with anti-CD3e (100 ng/mL) (A). Addition of recombinant (r) mouse (m) IL-6 (10 ng/mL, Chemicon) or rmIL-21 (20 ng/mL, eBiosciences) restored precursor frequency of the responder cells decreased in the coculture with Tregs, suggesting that initial proliferative responsiveness of the responder cells was restored. The values represent precursor frequency (Pf) of the responder cells. (B) When the cells were stimulated with anti-CD3e (100 ng/mL) plus anti-CD28 (100 ng/mL), addition of IL-6 restored the CD25 mean fluorescence intensity (MFI) of the responder cells decreased in the coculture with Tregs. The values in the upper and lower panels represent CD25 MFI and Pf of the responder cells, respectively.

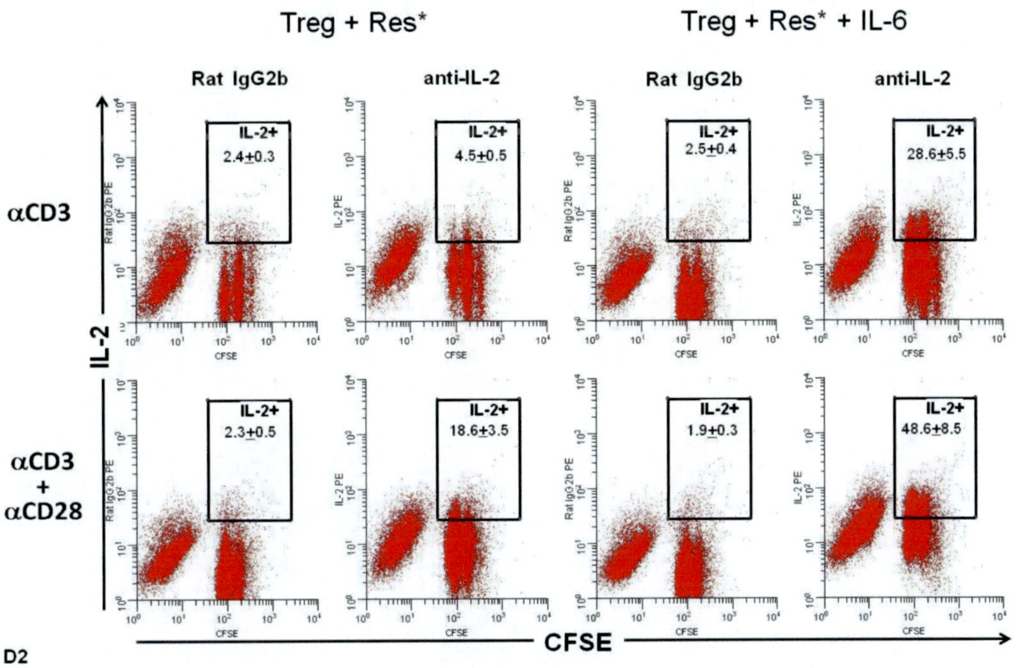

Figure 16. IL-2 production in the responder cells (Res) in the coculture with regulatory T cells (Tregs) was increased by adding IL-6. The methods are the same as in Figure 14, except for adding recombinant mouse IL-6 (10 ng/mL).

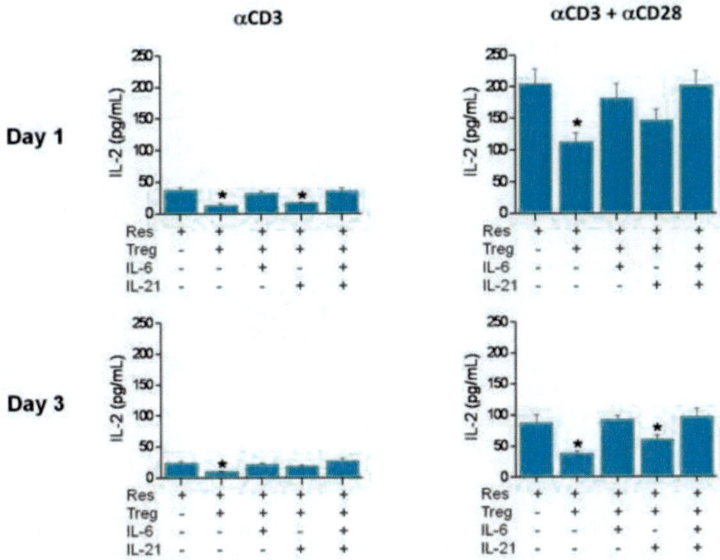

Figure 17. IL-2 concentration in the supernatants of the responder cells (Res) in the coculture with regulatory T cells (Tregs) was increased by adding IL-6 or IL-21. The methods are the same as in Figure 13, except for adding recombinant (r) mouse (m) IL-6 (10 ng/mL) or rmIL-21 (20 ng/mL). Data represent mean ± SD of triplicates of three separate experiments showing the same pattern. *P*-values were calculated by independent Student's t test, and *P* values less than .05 were considered to be statistically significant (*).

At present, we do not know the exact mechanism of the way IL-6 blocks the suppressive activity of Tregs. Nonetheless, concerning the IL-2 and IL-2Rα circuit, IL-6 seems to restore IL-2 production in the responder cells in the coculture with Tregs (Figure 16). Actually, the concentration of IL-2 in the culture supernatants of the coculture where rIL-6 was added, was comparable to that in the responder cells alone (Figure 17). In addition, the expression level of IL-2Rα on the responder cells in the coculture with Tregs was also restored by adding rIL-6, as already shown in Figure 15B.

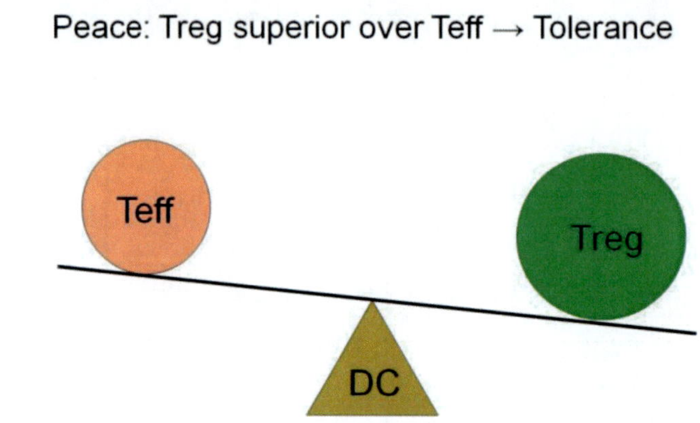

(A)

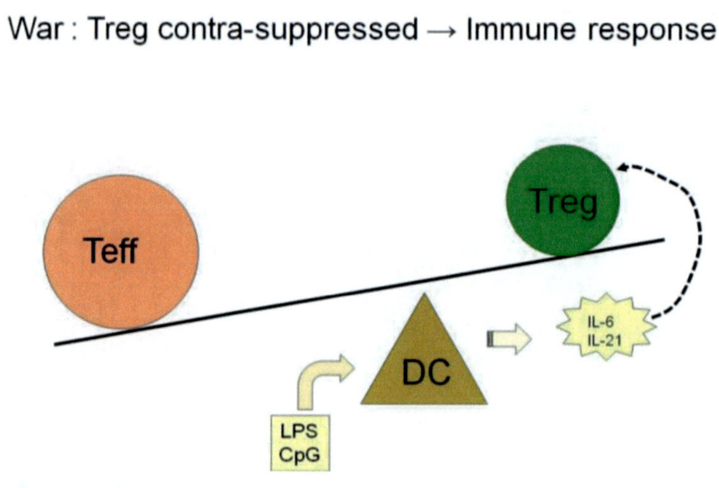

(B)

Figure 18. A conceptual model for immune balance *in vivo*. (A) In the steady state without any challenge, the suppressive activities of the regulatory T cells (Tregs) are superior to the proliferative activities of effector T cells (Teffs), as Tregs, constitutively expressing a higher level of IL-2Rα, always win Teffs in the competition for IL-2, hence the state of tolerance. (B) By contrast, in the state of infection or tissue damage when pathogen-associated molecules or danger signals are provided, innate immune cells, such as dendritic cells (DCs), secrete certain molecules (i.e. IL-6) that contra-suppress Treg activity, and Teffs escape from the suppressive control of Tregs and proliferate. Hence, immune response is elicited.

Conclusion

In summary, in the steady state, the suppressive activities of Tregs are superior to the proliferative activities of effector T cells (Teffs), as Tregs, constitutively expressing a higher level of IL-2Rα, always win over Teffs in the competition for IL-2, hence the state of tolerance (Figure 18A). By contrast, in the state of infection or tissue damage when pathogen-associated molecules or danger signals are provided, innate immune cells, such as dendritic cells, secrete certain molecules that contra-suppress Treg activity, and Teffs escape from the suppressive control of Tregs and proliferate (Figure 18B). Hence, immune response is elicited. In conclusion, regulation of IL-2 and IL-2Rα is an important mechanism of immune homeostasis maintained by Tregs.

Acknowledgment

This work was supported by the Korea Research Foundation (KRF) grant funded by the Korea government (MEST) (No. 2009-0073139) and by RP-Grant 2010 of Ewha Womans University to JYS and World Class University program through the National Research Foundation of Korea funded by the Ministry of Education, Science and Technology (R31-2008-000-10105-0) to MHJ.

References

[1] Piccirillo, C. A. (2008). Regulatory T cells in health and disease. *Cytokine, 43*, 395-401.

[2] Sakaguchi, S., Sakaguchi, N., Asano, M., Itoh, M. & Toda, M. (1995). Immunologic self-tolerance maintained by activated T cells expressing IL-2 receptor alpha-chains (CD25). Breakdown of a single mechanism of self-tolerance causes various autoimmune diseases. *J Immunol, 155*, 1151-1164.

[3] Sakaguchi, S., Toda, M., Asano, M., Itoh, M., Morse, S. S. & Sakaguchi, N. (1996). T cell-mediated maintenance of natural self-tolerance: its breakdown as a possible cause of various autoimmune diseases. *J Autoimmun, 9*, 211-220.

[4] Sakaguchi, S. (2004). Naturally arising CD4+ regulatory t cells for immunologic self-tolerance and negative control of immune responses. *Annu Rev Immunol, 22*, 531-562.

[5] Piccirillo, C. A. & Shevach, E. M. (2004). Naturally-occurring CD4+CD25+ immunoregulatory T cells: central players in the arena of peripheral tolerance. *Semin Immunol, 16*, 81-88.

[6] Piccirillo, C. A. & Shevach, E. M. (2001). Cutting edge: control of CD8+ T cell activation by CD4+CD25+ immunoregulatory cells. *J Immunol, 167*, 1137-1140.

[7] Lim, H. W., Hillsamer, P., Banham, A. H. & Kim, C. H. (2005). Cutting edge: direct suppression of B cells by CD4+ CD25+ regulatory T cells. *J Immunol, 175*, 4180-4183.

[8] Ralainirina, N., Poli, A., Michel, T., Poos, L., Andres, E., Hentges, F. & Zimmer, J. (2007). Control of NK cell functions by CD4+CD25+ regulatory T cells. *J Leukoc Biol 81*, 144-153.

[9] Wing, K., Onishi, Y., Prieto-Martin, P., Yamaguchi, T., Miyara, M., Fehervari, Z., Nomura, T. & Sakaguchi, S. (2008). CTLA-4 control over Foxp3+ regulatory T cell function. *Science, 322*, 271-275.

[10] Oliveira, V., Agua-Doce, A., Duarte, J., Soares, M. P. & Graca, L. (2006). Regulatory T cell maintenance of dominant tolerance: induction of tissue self-defense? *Transpl Immunol*, *17*, 7-10.

[11] Sakaguchi, S., Wing, K., Onishi, Y., Prieto-Martin, P. & Yamaguchi, T. (2009). Regulatory T cells: how do they suppress immune responses? *Int Immunol, 21*, 1105-1111.

[12] Scheffold, A., Huhn, J. & Hofer, T. (2005). Regulation of CD4+CD25+ regulatory T cell activity: it takes (IL-)two to tango. *Eur J Immunol, 35*, 1336-1341.

[13] Malek, T. R., Yu, A., Vincek, V., Scibelli, P. & Kong, L. (2002). CD4 regulatory T cells prevent lethal autoimmunity in IL-2Rbeta-deficient mice. Implications for the nonredundant function of IL-2. *Immunity*, *17*, 167-178.

[14] Malek, T. R. & Bayer, A. L. (2004). Tolerance, not immunity, crucially depends on IL-2. *Nat Rev Immunol, 4*, 665-674.

[15] Annunziato, F., Cosmi, L., Liotta, F., Lazzeri, E., Manetti, R., Vanini, V., Romagnani, P., Maggi, E. & Romagnani, S. (2002). Phenotype, localization, and mechanism of suppression of CD4(+)CD25(+) human thymocytes. *J Exp Med*, *196*, 379-387.

[16] Turka, L. A. & Walsh, P. T. (2008). IL-2 signaling and CD4+ CD25+ Foxp3+ regulatory T cells. *Front Biosci*, *13*, 1440-1446.

[17] Levings, M. K., Sangregorio, R. & Roncarolo, M. G. (2001). Human cd25(+)cd4(+) t regulatory cells suppress naive and memory T cell proliferation and can be expanded in vitro without loss of function. *J Exp Med*, *193*, 1295-1302.

[18] Thornton, A. M. & Shevach, E. M. (2000). Suppressor effector function of CD4+CD25+ immunoregulatory T cells is antigen nonspecific. *J Immunol*, *164*, 183-190.

[19] de la Rosa, M., Rutz, S., Dorninger, H. & Scheffold, A. (2004). Interleukin-2 is essential for CD4+CD25+ regulatory T cell function. *Eur J Immunol, 34*, 2480-2488.

[20] Ono, M., Yaguchi, H., Ohkura, N., Kitabayashi, I., Nagamura, Y., Nomura, T., Miyachi, Y., Tsukada, T. & Sakaguchi, S. (2007). Foxp3 controls regulatory T-cell function by interacting with AML1/Runx1. *Nature, 446*, 685-689.

[21] Jung, Y. J. & Seoh, J. Y. (2009). Feedback loop of immune regulation by CD4+CD25+ Treg. *Immunobiology, 214*, 291-302.

[22] Takahashi, T., Kuniyasu, Y., Toda, M., Sakaguchi, N., Itoh, M., Iwata, M., Shimizu, J. & Sakaguchi, S. (1998). Immunologic self-tolerance maintained by CD25+CD4+ naturally anergic and suppressive T cells: induction of autoimmune disease by breaking their anergic/suppressive state. *Int Immunol*, *10*, 1969-1980.

[23] Baecher-Allan, C., Brown, J. A., Freeman, G. J., Hafler, D. A. (2001). CD4+CD25high regulatory cells in human peripheral blood. *J Immunol*, *167*, 1245-1253.

[24] Taams, L. S., van Rensen, A. J., Poelen, M. C., van Els, C. A., Besseling, A. C., Wagenaar, J. P., van Eden, W. & Wauben, M. H. (1998). Anergic T cells actively suppress T cell responses via the antigen-presenting cell. *Eur J Immunol, 28*, 2902-2912.

[25] Yamazaki, S., Iyoda, T., Tarbell, K., Olson, K., Velinzon, K., Inaba, K. & Steinman, R. M. (2003). Direct expansion of functional CD25+ CD4+ regulatory T cells by antigen-processing dendritic cells. *J Exp Med*, *198*, 235-247.

[26] Walker, L. S., Chodos, A., Eggena, M., Dooms, H. & Abbas, A. K. (2003). Antigen-dependent proliferation of CD4+ CD25+ regulatory T cells in vivo. *J Exp Med*, *198*, 249-258.

[27] Klein, L., Khazaie, K. & von Boehmer, H. (2003). In vivo dynamics of antigen-specific regulatory T cells not predicted from behavior in vitro. *Proc Natl Acad Sci,* U S A, *100*, 8886-8891.

[28] Oberg, H. H., Wesch, D., Lenke, J. & Kabelitz, D. (2006). An optimized method for the functional analysis of human regulatory T cells. *Scand J Immunol, 64*, 353-360.

[29] Suchin, E. J., Langmuir, P. B., Palmer, E., Sayegh, M. H., Wells, A. D., Turka, L. A. (2001). Quantifying the frequency of alloreactive T cells in vivo: new answers to an old question. *J Immunol*, *166*, 973-981.

[30] Yarkoni, S., Kaminitz, A., Sagiv, Y., Yaniv, I. & Askenasy, N. (2008). Involvement of IL-2 in homeostasis of regulatory T cells: the IL-2 cycle. *Bioessays 30*, 875-888.

[31] Rubin, L. A., Kurman, C. C., Fritz, M. E., Biddison, W. E., Boutin, B., Yarchoan, R. & Nelson, D. L. (1985). Soluble interleukin 2 receptors are released from activated human lymphoid cells in vitro. *J Immunol*, *135*, 3172-3177.

[32] Sojka, D. K., Hughson, A., Sukiennicki, T. L. & Fowell, D. J. (2005). Early kinetic window of target T cell susceptibility to CD25+ regulatory T cell activity. *J Immunol*,, *175*, 7274-7280.

[33] Rubin, L. A., Jay, G. & Nelson, D. L. (1986). The released interleukin 2 receptor binds interleukin 2 efficiently. *J Immunol*, *137*, 3841-3844.

[34] Pandiyan, P., Zheng, L., Ishihara, S., Reed, J. & Lenardo, M. J. (2007). CD4+CD25+Foxp3+ regulatory T cells induce cytokine deprivation-mediated apoptosis of effector CD4+ T cells. *Nat Immunol, 8*, 1353-1362.

[35] Thornton, A. M. & Shevach, E. M. (1998). CD4+CD25+ immunoregulatory T cells suppress polyclonal T cell activation in vitro by inhibiting interleukin 2 production. *J Exp Med*, *188*, 287-296.

[36] Lehner, T. (2008). Special regulatory T cell review: The resurgence of the concept of contrasuppression in immunoregulation. *Immunology*, *123*, 40-44.

[37] Pasare, C. & Medzhitov, R. (2003). Toll pathway-dependent blockade of CD4+CD25+ T cell-mediated suppression by dendritic cells. *Science, 299*, 1033-1036.

In: Regulatory T Cells
Editor: Ren S. Hayashi, pp. 225-237

ISBN: 978-1-61761-554-2

Chapter 12

Regulating Foxp3 by Cis Elements

Peter C. J. Janson and Ola Winqvist
Department of Medicine, Clinical Allergy Research Unit,
Karolinska Institutet, Stockholm, Sweden

Foxp3 expression is required for the development and function of T regulatory cells. The last years intensive research focused on elucidating the transcriptional mechanisms behind Foxp3 expression have revealed a gene strictly regulated by epigenetic mechanisms, where multiple enhancer elements coordinates the expression of Foxp3. In this short communication article we will summarize the recent advances in studies of Foxp3 expression control.

Introduction

Development of alternative CD4+ T cells provides flexibility to the immune system. This is crucial for the initiation of appropriate effector mechanisms to protect against various pathogens such as bacteria, viruses, tumors and parasites. Research on the epigenetic regulation of T cell subsets during the past years have increased our understanding of how the alternative effector populations arise and how their identity is maintained during clonal expansion [1].

The regulatory T cell subset is essential for the maintenance of peripheral tolerance. Naturally occurring regulatory T cells (referred to in this article as nTregs) develop from CD4-single positive thymocytes, and migrate to the periphery where they prevent autoimmunity by suppressing the activity of self-reactive T cells that have escaped elimination during thymic maturation [2, 3]. A regulatory phenotype can also be induced from naïve T cells, either when antigens are encountered during sub-immunogenic conditions, or in the presence of TGF-β, or Retinoic Acid. These are the so called induced regulatory T cells (iTregs), and the function of these cells is believed to be especially important for acquired tolerance to food antigens [4].

The gene expression profile of regulatory T cells, and their effector functions are dependent on the transcription factor Foxp3 [5]. Retroviral transduction or ectopic expression of Foxp3 in conventional murine T cells converts them to a regulatory phenotype [6, 7], and mutations in the human *FOXP3* gene is responsible for the disease Immune dysregulation, Polyendocrinopathy, Enteropathy, X-linked syndrome (IPEX) [8]. The *Foxp3* mutant *Scurfy* mouse model displays a similar pathology as IPEX patients with dysregulated CD4+ T cell infiltration and activation [9].

Given the importance for Foxp3 in maintaining peripheral tolerance, it becomes important to understand the molecular mechanisms behind Foxp3 expression. This is important since increased knowledge on how Foxp3 expression is regulated, what factors that turns on transcription and what conditions results in stable expression opens up for Treg based immune therapies. An intensive research during the last years has focused on clarifying the role of conserved regulatory elements surrounding the *Foxp3* gene. The *FOXP3* promoter region was first described by Mantel et al [10, 11]. Since then, additional enhancer elements surrounding the *Foxp3* locus have been identified including the *CNS1*/TGF-β response element [12], the *CNS2/TSDR* (Treg Specific Demethylated Region) [13] and the *CNS3* [14]. Analysis of epigenetic status, transcription factor binding, and reporter construct activity of these segments initially provided insight into the accessibility and enhancer activity of these elements in different cell types. Lately, specific deletion of these elements in mouse knock-out models has revealed the individual role of these elements for the induction, and maintenance of nTregs and iTregs.

nTregs

The regulatory fate of a T cell is determined by TCR specificity for self-antigens during thymic development. There are evidence that the both antigen presentation from cortical thymic epithelias cells (cTECs) as well as AIRE expressing medullary thymic epithelial cells (mTECs) can support the development of Tregs [3, 15]. During thymic development, CD4+ T cells equipped with a T cell receptor with high affinity for self-peptide-MHC complexes undergo a selection process that is distinct from positive selection as well as negative selection to become CD4+CD25^{hi} T regulatory cells [16]. The importance of certain TCR specificities for the initiation of the T regulatory developmental program is supported by observations that T cells of the effector arm and naturally occurring T regulatory cells have non-overlapping TCR repertoires [17-19]. In the process of thymic nTreg development, CD28 co-stimulation is also required. The cytosolic tail of CD28 requires an intact Lck-binding motif for efficient Treg generation, suggesting that Lck-binding enhance the intensity and duration from TCR-specific interactions, thereby contributing to Foxp3 induction [20, 21].

The requirements of TCR and CD28 stimulation to initiate the Treg program during thymic maturation has been known for quite some time, but the molecular mechanisms and the transcription factors involved in Foxp3 induction downstream of TCR activation have been less clear. Several downstream signaling components are required for nTreg induction including TAK1, Bcl10, CARMA1, PKCθ, and IKK-β [22-26]. It was recently shown that enhancement of NF-κB signaling can rescue the defective nTreg development in TAK1 and CARMA1 deficient mice, demonstrating that NF-κB signaling is a crucial component of the TCR downstream signaling events responsible for Foxp3 induction. In OT-II TCR transgenic animals the absence of TCR diversity and high-affinity ligands for the T-cells during thymic maturation results in defective nTreg generation. Enhanced NF-κB signaling in OT-II transgenic thymocytes however results in Foxp3 expression suggesting that NF-κB participates in the signaling events that determines Treg vs conventional T cell development in [27]. The individual role of NF-κB family members C-rel, p65, p50 and RelB was recently investigated by Ruan et al. Whereas RelB or p50 deficiency only had minor effects on Treg development, mice deficient of C-rel had severely impaired Treg development. Ruan et al also showed that C-rel and p65 together have strong promoter activating function and that C-rel directly binds to the Foxp3 promoter upon TCR engagement together with NF-AT, p65 and CREB [28].

Several reports have addressed the importance of cytokine signaling for the development of T regulatory cells. nTregs constitutively express high levels of the IL-2 receptor alpha chain but IL-2Rα or IL-2 cytokine is not required for nTreg development or function. Mice lacking the IL-2 receptor beta chain (IL-2Rβ) or the common common γ-chain (γ_c) on the other hand, have fewer nTregs [29, 30]. IL-2Rβ is also a component of the IL-15 receptor and deletion of IL-2 in combination with IL-15 results in fewer Tregs [30]. The common signaling molecule downstream of both IL-2Rβ as well as γ_c is Stat5, and Stat5 deficient mice have impaired Treg development [31, 32]. The Stat5 protein binds to several regions in the *Foxp3* locus, including the promoter region and the *CNS2* region [30, 31, 33]. Interestingly Stat5 expression can replace the need for co-stimulatory signals and override TCR instructive signals, thereby initiating the Treg developmental program in T cells destined to become naïve T cells [34]. Thus in the current model of nTreg development, multiple delicate signals contribute to the induction of a stable nTreg phenotype. In a two-step process TCR signals first induce cytokine responsiveness by increasing chromatin accessibility at regulatory elements. In this process, the NF-κB transcription factor C-Rel is likely to play a crucial role by binding to both the promoter region as well as *CNS3*. In the second step of this model, the Treg developmental program is completed by Stat5-dependent cytokine signaling [34, 35].

iTregs

As mention in the introduction of this paper, regulatory T cells can also be induced in the periphery from the same peripheral naïve T cell pool as the effector T helper cells, either by the presence of TGF-β or during exposure to the vitamin A metabolite Retinoic Acid during antigen recognition [4, 36, 37]. TGF-β activation results in Smad3 activation and retinoic acid increases the expression and phosphorylation of Smad3, thereby enhancing the Treg inducing effects of TGF-β. Smad3 in turn increases the expression of Foxp3 by binding to several elements in the Foxp3 locus including the promoter region as well as the *CNS1* enhancer [12, 28].

The TGF-β induced iTregs differ from the nTregs in terms of Foxp3 transcription stability. Whereas nTregs constitutively express Foxp3 transcript at a high level, the expression pattern of Foxp3 and the suppressive ability of the TGF-β induced iTregs is transient [38-40]. Besides though the exposure to TGF-β or retinoic acid, iTregs can also be generated when naïve T-cells encounter antigens during sub-immunogenic conditions. By targeting medium doses of peptide agonist ligads to DC's in mice with TCRs that specifically recognizes the injected peptide Kretschmer et al demonstrated that a stable Treg phenotype can be induced from peripheral naïve T cells *in vivo* [41]. Lately, this discrepancy between nTregs and iTregs induced in different settings has been linked to different patterns of epigenetic modifications at the *CNS* elements surrounding *Foxp3* [10, 13, 42-45]. It is now believed that the stable expression of *Foxp3* requires permissive modifications of certain elements in the *Foxp3* locus, which in turn allows for binding of Foxp3 protein to the *Foxp3* gene. This feedback loop sustains the Foxp3 expression in nTregs and facilitates the heritable maintenance of the nTreg phenotype [14].

Promoter

The human *FOXP3* promoter contains binding sites for transcription factors downstream of the TCR such as nuclear factor of activated T-cells (NFAT) and activator protein 1 (AP-1] [11]. Disruption of these elements reduces the *trans* activation of the promoter after TCR stimulation, demonstrating the importance of TCR engagement for the induction of FOXP3. As mentioned earlier, C-Rel also binds to the promoter region during TCR engagement, The promoter region also binds Stat5 in murine Tregs, which is in agreement with the requirement for Stat5 signaling for both development of Tregs in the thymus as well as the generation of iTregs *in vitro* [30, 31].

When we investigated the methylation status of the human *FOXP3* promoter we found that the promoter is completely demethylated in freshly isolated $CD25^{hi}$ T regulatory cells, whereas stimulated CD4+ T cells with a transient FOXP3 expression profile have a partially methylated promoter regions that remains methylated during transient Foxp3 expression. On the other hand, histone H4 acetylation analysis of the human *FOXP3* locus showed that the promoter region is accessible to a certain level already in naïve CD4+ T cells, and that this accessibility is increased upon activation CD4+ cells [11]. Thus the promoter region seems to be regulated by epigenetic mechanisms on multiple levels, where permissive flexible epigenetic marks (i. e. histone acetylation) are present both in committed nTregs, as well as in conventional T cells with transient FOXP3 expression and to a certain extent also in naïve T cells with the capacity to express FOXP3 upon stimulation [11]. CpG demethylation, on the other hand, is only observed in *ex vivo* isolated Tregs, observations which probably reflect the nature of DNA methylation as less flexible and a more permanent epigenetic modification as compared to histone modifications [10]. Interestingly, in the recent report by Zheng et al, permissive histone moficications (H3K4me1] was observed at the promoter region in single positive thymocytes, suggesting a combined activity of the promoter region and the *CNS3* region during thymic maturation of Tregs [14]. Since binding sites for C-rel have been identified both in the *CNS3* as well as the promoter region it is possible that these two elements together are responsible for the TCR mediated initiation of Treg development during thymic maturation.

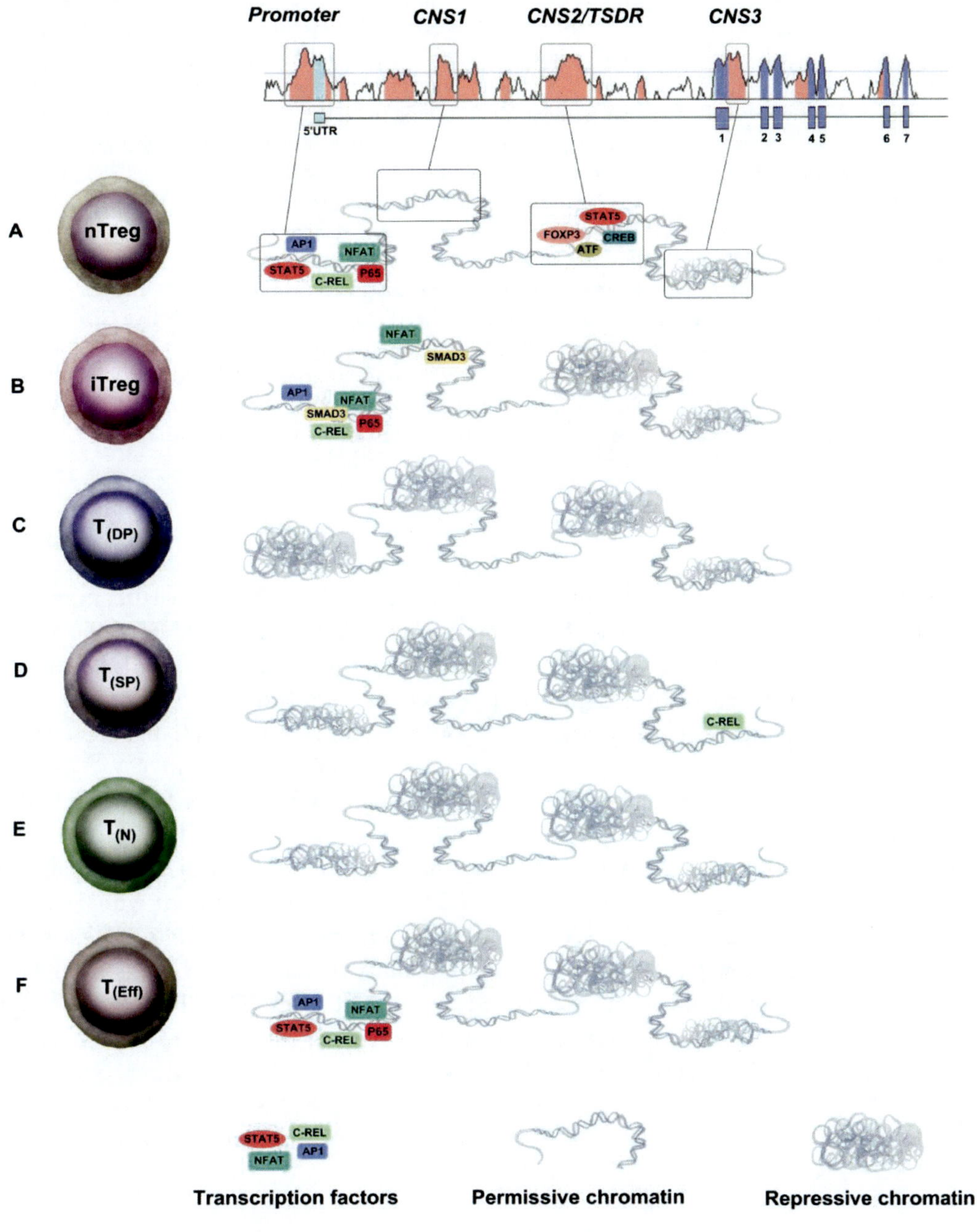

Figure 1. Chromatin structure and transcription factor binding to conserved elements of the *Foxp3* gene in various cell types. Top illustration is an mVista alignment showing level of conservation between human and murine genome at the *Foxp3* locus. Conserved Non-coding Sequences (*CNS*) are in red and *CNS* regions corresponding to exons are in light blue. The promoter region and the three enhancer elements (*CNS1*, *CNS2* and *CNS3]* identified in the *Foxp3* locus are marked by boxes below the schematic drawing of the Foxp3 transcript. The illustrations below the Foxp3 transcript show chromatin accessibility of the *Foxp3* locus in various cell types; nTregs - naturally occurring thymus derived T regulatory cells, iTregs - TGF-β stimulated naïve T-cells, $T_{(DP)}$ - CD4+CD8+ double positive thymocytes, $T_{(SP)}$ - CD4+ single positive thymocytes, $T_{(N)}$ - naïve CD4+ T-cells, $T_{(Eff)}$ - CD4+ effector T-cells. Permissive chromatin is defined as demethylated CpG DNA and permissive histone modifications whereas repressive chromatin is defined as methylated CpG DNA

and histone modifications associated with heterochromatin. The increased accessibility of DNA allows for transcription factor binding to the *Foxp3* locus and stabilization of Foxp3 transcription; NFAT – nuclear factor of activated T-cells, AP1 – activator protein 1, C-rel – C-rel proto-oncogene protein, p65 – transcription factor p65, Stat5 – signal transducer and activator of transcription 5, Smad3 – Sma- and Mad-related protein 3, CREB – cAMP reponse element-binding protein, ATF – activating transcription factor. FOXP3 – forkhead box P3 protein. (A) In nTregs, an open configuration of the *CNS2* region supports stable expression of Foxp3 through the binding of Foxp3 and the induction of a self-sustainable feedback-loop. (B) In response to TGF-β, *CNS1* becomes accessible for transcription factor binding and Smad3 binds to the *CNS1* region as well as the promoter. (C) Double-positive thymocytes display repressive chromatin marks at the entire *Foxp3* locus. (D) The *CNS3* is crucial for initiation of Foxp3 transcription in single positive thymocytes. The NF-κB member C-rel binds *CNS3*, and increased accessibility of the Foxp3 promoter suggests a combined activity of these two elements during early nTreg development. (E) Naïve CD4+ T-cells do not express Foxp3 and the *CNS* elements are occupied by repressive chromatin marks. (F) Effector CD4+ cells transiently express FOXP3. The expression is driven by the transcription factor binding to the promoter region. Repressive chromatin marks at the *CNS2* region and hypermethylation of CpG dinucleotides in the *CNS2* region as well as the promoter region results in an unstable Foxp3 transcription profile.

CNS1

The TGF-β sensor, or *CNS1*, is located approximately 2 kb downstream of the transcription start site. The ability of TGF-β to induce transcription of Foxp3 has been known for many years [37], and evidence that Smad3 binds to *CNS1* upon TGF-β receptor engagement provided mechanistic insight to the effects of TGF-β on Foxp3 induction [12]. Whereas Smad3 binding is required for chromatin remodeling of *CNS1*, maintenance of activity in this region is dependent on the binding of NFAT. Histones in *CNS1* become acetylated in response to TGF-β and TCR stimulation, indicating higher accessibility of DNA during these conditions.

Deletion of *CNS1* does not affect the thymic maturation of Tregs or the maintenance of Foxp3 expression in Tregs, however the ability to differentiate Tregs *in vitro* with TGF-β is severely impaired. In addition the *in vivo* conversion from naïve T cells into Tregs is also defective in *CNS1* deficient T cells [14]. The reduced ability to convert into Tregs from naïve T cells is especially prominent in gut associated lymphoid tissues (GALT) and mesenteric lymph nodes (MLN), which is in coherence with the previous established requirements of TGF-β for MLN derived DC generation of iTregs [4].

CNS2/TSDR

CNS2 or the Treg-cell specific demethylated region (*TSDR*) is a conserved element located within intron 1, approximately 4. 6 kb downstream of the *Foxp3* transcription start site. The *CNS2* is completely demethylated in murine nTregs, and hypermethylated in conventional effector T cells. TGF-β induced iTregs with an unstable Foxp3 expression profile display partial demethylation of the *CNS2* but complete demethylation is only observed in nTregs, observations which suggested that the *CNS2* is involved in stabilization of the Foxp3 expression in thymic derived T regulatory cells [13]. The *CNS2* displays complete demethylation also in freshly isolated human $CD4^{+}CD25^{hi}$ cells, whereas Foxp3 expression from TCR stimulated naïve CD4+ T cells or TGF-β treated CD4+ T cells is associated with a hypermethylated *CNS2*. These results suggest that the *CNS2* has a similar function for regulating *FOXP3* expression in human Tregs [43]. The *in vitro* enhancer function of the *CNS2* is increased by co-transfection of Stat3 or Stat5 expressing plasmids, and reporter activity is function of the *CNS2* is reduced when binding sited for Stat proteins are mutated in reporter constructs [33], indicating that Stat proteins are crucial for the transcriptional activity of the *CNS2*. The binding of Stat5 to the *CNS2* has also been demonstrated by ChIP-experiments [31]. The *CNS2* enhancer activity is impaired by hypermethylation [42, 45], observations which are most likely explained by the methylation dependent binding of cyclic AMP response element (CREB) [42].

The *in vivo* effects of the *CNS2* on Treg stability was demonstrated in the recent paper by Zheng et al. Mice lacking the *CNS2* show reduced frequency of Foxp3+ Tregs. This reduction is only observed in 6-months and older mice, but not younger. In addition, transferred Foxp3$^+$ cells lacking the *CNS2* fail to maintain Foxp3 expression and Foxp3 expression is progressively lost during *in vitro* culturing of *CNS2* deficient Tregs. Furthermore, the *CNS2* seems to be dispensable iTreg induction as *CNS2* deficient mice. These observations supports the suggested role of the *CNS2* as an element responsible for the maintenance of Foxp3 expression in cells already committed to the Treg lineage [14].

Even though Stat5 and CREB have shown to bind to the *CNS2*, and have the ability to induce expression in reporter assays, the activity of these transcription factors are not necessary for the heritability of Tregs. Stat5 deficient Tregs are capable of maintaining high Foxp3 expression and retrovirally transduced dominant negative CREB doesn't affect Foxp3 expression in Tregs. Instead the protein responsible for maintaining high Foxp3 expression in committed T regulatory cells seems to be Foxp3 itself [46]. Foxp3-null T cells display a similar reduced frequency of Foxp3 activity as *CNS2* deficient T cells, and Foxp3 bind to the *CNS2* in Tregs but not to the TGF-β sensor/*CNS1*. Interestingly, the *CNS2* is demethylated in Tregs with a null-mutation of Foxp3, suggesting that a functional protein is not required for the induction of demethylation of the *CNS2*. In addition, Foxp3 does not occupy the *CNS2* in iTregs where the *CNS2* is hypermethylated. Thus Foxp3 seems to be responsible for the maintenance of Foxp3 expression in Tregs through binding to its *CNS2*, but not until after demethylation of the *CNS2* [14].

CNS3

CNS3 is located downstream of the first exon of *Foxp3*. Its role in the regulation of *Foxp3* was recently revealed by Zheng et al. When patterns of histone modifications of the *Foxp3* locus was analyzed during thymic maturation of T-cells it was shown that the *CNS3* region was marked with H3K4me1 patterns already at the double positive (DP) state, observations which pointed towards an involvement of *CNS3* during thymic maturation of Tregs. Indeed, deletion of *CNS3* results in decreased frequencies of Foxp3+ CD4 single-positive thymocytes. *CNS3* contains a binding motif for c-Rel, and c-Rel deficient thymocytes display a similar phenotype as the *CNS3* deficient cells. Based upon these observations, the *CNS3* is believed to act as a pioneering element which binds c-Rel in response to TCR and CD28 triggering during thymic maturation of T regulatory cells [14].

Foxp3 Methylation and Treg Stability

It has now become clear that nTreg differs from iTregs in terms of Foxp3 messenger RNA stability. The mechanisms behind induction of a stable nTregs most likely involves [1] Precise and delicate signals through the T cell receptor, the CD28 co-receptor and Stat5 in single–positive thymocytes. [2] Increased accessibility of *CNS* regions in the *Foxp3* locus which in turn allows for transcription factor biding and increased Foxp3 expression. [3] Maintenance of high expression through the formation of a self-sustainable feedback loop where Foxp3 protein binds to the *CNS3* region in the Foxp3 gene and enhances its own expression. A prerequisite for Foxp3 to sustain its own expression in this manner is that the *CNS3* region remains demethylated during cell cycle progression. We and others have previously observed that a subset of cells in a culture of CD4+CD25^{hi} T-cells lose Foxp3 expression in long term *in vitro* culture [10, 13]. Recently, it was demonstrated by Hoffman et al that a Treg population expressing low level of the IL-7R (CD4+CD127^{low}CD25^{hi}) becomes partially remethylated at the *CNS3* region during long term culture. A subpopulation of CD45RA^{-} CD25^{hi} Tregs, which undergoes *de novo* methylation is the most likely reason for these observations. Whereas CD45RA^{+}CD25^{hi} T cells retain high Foxp3 expression levels, CD45RA^{-}CD25^{hi} Tregs gradually lose Foxp3 expression and start expressing pro-inflammatory cytokines in long-term cultures [47]. These observations clearly demonstrate that a subtype of Tregs lose their Treg phenotype *in vitro* despite complete demethylation of the Foxp3 enhancer element *CNS2*.

Even though Tregs seems able to lose their phenotype in *in vitro* cultures, it has been a matter of debate whether this also happens *in vivo*. Upon transfer of Foxp3^{+} cells into lymphopenic hosts, about 50% of the transferred cells lose Foxp3 expression, suggesting that exogenous factors are required for lineage integrity of Tregs [48]. In an elegant system where former Foxp3 expressing cells can be traced due to a Foxp3-cre mediated excision of a stop codon upstream of the yellow fluorescent protein (YFP) gene in the Rosa26 locus, Zhou et al investigated the lineage stability of Foxp3 expressing cells in healthy and inflammatory conditions [49]. It was demonstrated that about 20% of the cells that turn on the Foxp3-diven cre-recombinase actually lose their Foxp3 expression in the periphery (exFoxp3 cells). Interestingly, 13% of these exFoxp3 cells display a partial methylated *CNS2* which is not seen in neither effector T cells nor nTregs. The question remain whether these former Foxp3 expressing cells lost Foxp3 expression or if they never completely entered the nTreg program because they did not fulfill the signaling criteria's to become fully developed Tregs. It is possible that these cells never expressed sufficient Foxp3 to initiate the self-sustaining Treg program, and therefore lost their Foxp3 expression. TCR repertoire analysis of the exFoxp3 cells demonstrated that these cells share TCR specificity with both nTregs as well as effector T-cells, suggesting that at least a portion of the exFoxp3 cells are derived from nTregs. Zhou et al observed that transfer of exFoxp3 cells into Rag2$^{-/-}$ mice resulted in development of diabetes, suggesting that Foxp3 instability can lead to the generation of pathogenic T cells that promote autoimmunity.

Future Perspective

Being able to manipulate T regulatory cells for the use of adoptive transfer is considered the holy grail of T-cell immune based therapies for the treatment of autoimmunity. A prerequisite for the Treg based immune-therapies is the establishment of a stable Treg phenotype in *in vitro* cultures, and also ensuring continued stability after adoptive transfer. Lineage instability, and production of pro-inflammatory cytokines from T-cells with a Treg TCR specificity may have hazardous effects, as has been demonstrated in murine models [49]. Recent years studies of the epigenetic regulation of Foxp3 have revealed mechanisms which support stable expression of Foxp3. These mechanisms include the induction of chromatin accessibility and establishment of self-sustainable feedback loops [14]. Still, $Foxp3^{+}$ cells fulfilling the epigenetic criteria for stable expression are not entirely unimpressionable to surrounding signals as certain conditions and cytokine surroundings have shown to delineate the Treg phenotype. One such scenario is when Tregs are exposed to IL-6 or deprived of TGF-β [50], which leads to Stat3 upregulation and Foxp3 inhibition [51]. One plausible explanation for plasticity of nTregs during certain conditions is the existence of bivalent histone modifications at transcription factor loci [52]. When Wei et al investigated the histone pattern at transcription factor and cytokine loci in nTregs, they found that the Rorc locus, encoding the key lineage transcription factor of Th17 cells, was simultaneously occupied by H3K4me3 and H3K27me3 marks. Such bivalent histone modifications are commonly associated with ′poised` genes, indicating that the Th17 pathway is not permanently silenced in T regulatory cells. The inconsistency in reports regarding nTreg stability and lineage integrity calls for further studies on epigenetic regulation of Foxp3.

Acknowledgments

This work was supported by the Swedish cancer society, the Cancer and Allergy society, the Hedlunds foundation, the Lundberg foundation, Gustav V. foundation, the Selander foundation and the Söderbergs foundation.

References

[1] Wilson, C. B., Rowell, E. & Sekimata, M. (2009). Epigenetic control of T-helper-cell differentiation. *Nat Rev Immunol, 9*, 91-105.

[2] Fontenot, J. D., Dooley, J. L., Farr, A. G. & Rudensky, A. Y. (2005). Developmental regulation of Foxp3 expression during ontogeny. *J Exp Med, 202,* 901-906.

[3] Aschenbrenner, K., D'Cruz, L. M., Vollmann, E. H., Hinterberger, M., Emmerich, J., Swee, L. K., Rolink, A. & Klein, L. (2007). Selection of Foxp3+ regulatory T cells specific for self antigen expressed and presented by Aire+ medullary thymic epithelial cells. *Nat Immunol 8,* 351-358.

[4] Coombes, J. L., Siddiqui, K. R., Arancibia-Carcamo, C. V., Hall, J., Sun, C. M., Belkaid, Y. & Powrie, F. (2007). A functionally specialized population of mucosal

CD103+ DCs induces Foxp3+ regulatory T cells via a TGF-beta and retinoic acid-dependent mechanism. *J Exp Med 204,* 1757-1764.

[5] Gavin, M. A., Rasmussen, J. P., Fontenot, J. D., Vasta, V., Manganiello, V. C., Beavo, J. A. & Rudensky, A. Y. (2007). Foxp3-dependent programme of regulatory T-cell differentiation. *Nature, 445,* 771-775.

[6] Fontenot, J. D., Gavin, M. A. & Rudensky, A. Y. (2003). Foxp3 programs the development and function of CD4+CD25+ regulatory T cells. *Nat Immunol, 4,* 330-336.

[7] Hori, S., Nomura, T. & Sakaguchi, S. (2003). Control of regulatory T cell development by the transcription factor Foxp3. *Science, 299,* 1057-1061.

[8] Gambineri, E., Torgerson, T. R. & Ochs, H. D. (2003). Immune dysregulation, polyendocrinopathy, enteropathy, and X-linked inheritance (IPEX), a syndrome of systemic autoimmunity caused by mutations of FOXP3, a critical regulator of T-cell homeostasis. *Curr Opin Rheumatol, 15,* 430-435.

[9] Brunkow, M. E., Jeffery, E. W., Hjerrild, K. A., Paeper, B., Clark, L. B., Yasayko, S. A., Wilkinson, J. E., Galas, D., Ziegler, S. F. & Ramsdell, F. (2001). Disruption of a new forkhead/winged-helix protein, scurfin, results in the fatal lymphoproliferative disorder of the scurfy mouse. *Nat Genet, 27,* 68-73.

[10] Janson, P. C., Winerdal, M. E., Marits, P., Thorn, M., Ohlsson, R. & Winqvist, O. (2008). FOXP3 promoter demethylation reveals the committed Treg population in humans. *PLoS ONE, 3,* e1612.

[11] Mantel, P. Y., Ouaked, N., Ruckert, B., Karagiannidis, C., Welz, R., Blaser, K. & Schmidt-Weber, C. B. (2006). Molecular mechanisms underlying FOXP3 induction in human T cells. *J Immunol, 176,* 3593-3602.

[12] Tone, Y., Furuuchi, K., Kojima, Y., Tykocinski, M. L., Greene, M. I. & Tone, M. (2008). Smad3 and NFAT cooperate to induce Foxp3 expression through its enhancer. *Nat Immunol, 9,* 194-202.

[13] Floess, S., Freyer, J., Siewert, C., Baron, U., Olek, S., Polansky, J., Schlawe, K., Chang, H. D., Bopp, T., Schmitt, E., et al. (2007). Epigenetic control of the foxp3 locus in regulatory T cells. *PLoS Biol, 5,* e38.

[14] Zheng, Y., Josefowicz, S., Chaudhry, A., Peng, X. P., Forbush, K. & Rudensky, A. Y. Role of conserved non-coding DNA elements in the Foxp3 gene in regulatory T-cell fate. *Nature, 463,* 808-812.

[15] Liston, A., Nutsch, K. M., Farr, A. G., Lund, J. M., Rasmussen, J. P., Koni, P. A. & Rudensky, A. Y. (2008). Differentiation of regulatory Foxp3+ T cells in the thymic cortex. *Proc Natl Acad Sci,* U S A, *105,* 11903-11908.

[16] Jordan, M. S., Boesteanu, A., Reed, A. J., Petrone, A. L., Holenbeck, A. E., Lerman, M. A., Naji, A. & Caton, A. J. (2001). Thymic selection of CD4+CD25+ regulatory T cells induced by an agonist self-peptide. *Nat Immunol, 2,* 301-306.

[17] Hsieh, C. S., Zheng, Y., Liang, Y., Fontenot, J. D. & Rudensky, A. Y. (2006). An intersection between the self-reactive regulatory and nonregulatory T cell receptor repertoires. *Nat Immunol, 7,* 401-410.

[18] Pacholczyk, R., Ignatowicz, H., Kraj, P. & Ignatowicz, L. (2006). Origin and T cell receptor diversity of Foxp3+CD4+CD25+ T cells. *Immunity, 25,* 249-259.

[19] Hsieh, C. S., Liang, Y., Tyznik, A. J., Self, S. G., Liggitt, D. & Rudensky, A. Y. (2004). Recognition of the peripheral self by naturally arising CD25+ CD4+ T cell receptors. *Immunity, 21,* 267-277.

[20] Tai, X., Cowan, M., Feigenbaum, L. & Singer, A. (2005). CD28 costimulation of developing thymocytes induces Foxp3 expression and regulatory T cell differentiation independently of interleukin 2. *Nat Immunol, 6,* 152-162.

[21] Salomon, B., Lenschow, D. J., Rhee, L., Ashourian, N., Singh, B., Sharpe, A. & Bluestone, J. A. (2000). B7/CD28 costimulation is essential for the homeostasis of the CD4+CD25+ immunoregulatory T cells that control autoimmune diabetes. *Immunity, 12,* 431-440.

[22] Medoff, B. D., Sandall, B. P., Landry, A., Nagahama, K., Mizoguchi, A., Luster, A. D. & Xavier, R. J. (2009). Differential requirement for CARMA1 in agonist-selected T-cell development. *Eur J Immunol, 39,* 78-84.

[23] Gupta, S., Manicassamy, S., Vasu, C., Kumar, A., Shang, W. & Sun, Z. (2008). Differential requirement of PKC-theta in the development and function of natural regulatory T cells. *Mol Immunol, 46,* 213-224.

[24] Schmidt-Supprian, M., Courtois, G., Tian, J., Coyle, A. J., Israel, A., Rajewsky, K. & Pasparakis, M. (2003). Mature T cells depend on signaling through the IKK complex. *Immunity, 19,* 377-389.

[25] Wan, Y. Y., Chi, H., Xie, M., Schneider, M. D. & Flavell, R. A. (2006). The kinase TAK1 integrates antigen and cytokine receptor signaling for T cell development, survival and function. *Nat Immunol, 7,* 851-858.

[26] Schmidt-Supprian, M., Tian, J., Grant, E. P., Pasparakis, M., Maehr, R., Ovaa, H., Ploegh, H. L., Coyle, A. J. & Rajewsky, K. (2004). Differential dependence of CD4+CD25+ regulatory and natural killer-like T cells on signals leading to NF-kappaB activation. *Proc Natl Acad Sci,* U S A, *101,* 4566-4571.

[27] Long, M., Park, S. G., Strickland, I., Hayden, M. S. & Ghosh, S. (2009). Nuclear factor-kappaB modulates regulatory T cell development by directly regulating expression of Foxp3 transcription factor. *Immunity, 31,* 921-931.

[28] Ruan, Q., Kameswaran, V., Tone, Y., Li, L., Liou, H. C., Greene, M. I., Tone, M. & Chen, Y. H. (2009). Development of Foxp3(+) regulatory t cells is driven by the c-Rel enhanceosome. *Immunity, 31,* 932-940.

[29] Fontenot, J. D., Rasmussen, J. P., Gavin, M. A. & Rudensky, A. Y. (2005). A function for interleukin 2 in Foxp3-expressing regulatory T cells. *Nat Immunol, 6,* 1142-1151.

[30] Burchill, M. A., Yang, J., Vogtenhuber, C., Blazar, B. R. & Farrar, M. A. (2007). IL-2 receptor beta-dependent STAT5 activation is required for the development of Foxp3+ regulatory T cells. *J Immunol, 178,* 280-290.

[31] Yao, Z., Kanno, Y., Kerenyi, M., Stephens, G., Durant, L., Watford, W. T., Laurence, A., Robinson, G. W., Shevach, E. M., Moriggl, R., et al. (2007). Nonredundant roles for Stat5a/b in directly regulating Foxp3. *Blood, 109,* 4368-4375.

[32] Passerini, L., Allan, S. E., Battaglia, M., Di Nunzio, S., Alstad, A. N., Levings, M. K., Roncarolo, M. G. & Bacchetta, R. (2008). STAT5-signaling cytokines regulate the expression of FOXP3 in CD4+CD25+ regulatory T cells and CD4+CD25- effector T cells. *Int Immunol, 20,* 421-431.

[33] Zorn, E., Nelson, E. A., Mohseni, M., Porcheray, F., Kim, H., Litsa, D., Bellucci, R., Raderschall, E., Canning, C., Soiffer, R. J., et al. (2006). IL-2 regulates FOXP3

expression in human CD4+CD25+ regulatory T cells through a STAT-dependent mechanism and induces the expansion of these cells in vivo. *Blood, 108,* 1571-1579.

[34] Burchill, M. A., Yang, J., Vang, K. B., Moon, J. J., Chu, H. H., Lio, C. W., Vegoe, A. L., Hsieh, C. S., Jenkins, M. K. & Farrar, M. A. (2008). Linked T cell receptor and cytokine signaling govern the development of the regulatory T cell repertoire. *Immunity, 28,* 112-121.

[35] Lio, C. W. & Hsieh, C. S. (2008). A two-step process for thymic regulatory T cell development. *Immunity, 28,* 100-111.

[36] Xiao, S., Jin, H., Korn, T., Liu, S. M., Oukka, M., Lim, B. & Kuchroo, V. K. (2008). Retinoic acid increases Foxp3+ regulatory T cells and inhibits development of Th17 cells by enhancing TGF-beta-driven Smad3 signaling and inhibiting IL-6 and IL-23 receptor expression. *J Immunol, 181,* 2277-2284.

[37] Chen, W., Jin, W., Hardegen, N., Lei, K. J., Li, L., Marinos, N., McGrady, G. & Wahl, S. M. (2003). Conversion of peripheral CD4+CD25- naive T cells to CD4+CD25+ regulatory T cells by TGF-beta induction of transcription factor Foxp3. *J Exp Med, 198,* 1875-1886.

[38] Pillai, V., Ortega, S. B., Wang, C. K. & Karandikar, N. J. (2007). Transient regulatory T-cells, a state attained by all activated human T-cells. *Clin Immunol, 123,* 18-29.

[39] Tran, D. Q., Ramsey, H. & Shevach, E. M. (2007). Induction of FOXP3 expression in naive human CD4+FOXP3 T cells by T-cell receptor stimulation is transforming growth factor-beta dependent but does not confer a regulatory phenotype. *Blood, 110,* 2983-2990.

[40] Walker, M. R., Kasprowicz, D. J., Gersuk, V. H., Benard, A., Van Landeghen, M., Buckner, J. H. & Ziegler, S. F. (2003). Induction of FoxP3 and acquisition of T regulatory activity by stimulated human CD4+CD25- T cells. *J Clin Invest, 112,* 1437-1443.

[41] Kretschmer, K., Apostolou, I., Hawiger, D., Khazaie, K., Nussenzweig, M. C. & von Boehmer, H. (2005). Inducing and expanding regulatory T cell populations by foreign antigen. *Nat Immunol, 6,* 1219-1227.

[42] Kim, H. P. & Leonard, W. J. (2007). CREB/ATF-dependent T cell receptor-induced FoxP3 gene expression, a role for DNA methylation. *J Exp Med, 204,* 1543-1551.

[43] Baron, U., Floess, S., Wieczorek, G., Baumann, K., Grutzkau, A., Dong, J., Thiel, A., Boeld, T. J., Hoffmann, P., Edinger, M., et al. (2007). DNA demethylation in the human FOXP3 locus discriminates regulatory T cells from activated FOXP3(+) conventional T cells. *Eur J Immunol, 37,* 2378-2389.

[44] Nagar, M., Vernitsky, H., Cohen, Y., Dominissini, D., Berkun, Y., Rechavi, G., Amariglio, N. & Goldstein, I. (2008). Epigenetic inheritance of DNA methylation limits activation-induced expression of FOXP3 in conventional human CD25-CD4+ T cells. *Int Immunol, 20,* 1041-1055.

[45] Polansky, J. K., Kretschmer, K., Freyer, J., Floess, S., Garbe, A., Baron, U., Olek, S., Hamann, A., von Boehmer, H. & Huehn, J. (2008). DNA methylation controls Foxp3 gene expression. *Eur J Immunol, 38,* 1654-1663.

[46] Williams, L. M. & Rudensky, A. Y. (2007). Maintenance of the Foxp3-dependent developmental program in mature regulatory T cells requires continued expression of Foxp3. *Nat Immunol, 8,* 277-284.

[47] Hoffmann, P., Boeld, T. J., Eder, R., Huehn, J., Floess, S., Wieczorek, G., Olek, S., Dietmaier, W. &reesen, R. & Edinger, M. (2009). Loss of FOXP3 expression in natural human CD4+CD25+ regulatory T cells upon repetitive in vitro stimulation. *Eur J Immunol, 39,* 1088-1097.

[48] Duarte, J. H., Zelenay, S., Bergman, M. L., Martins, A. C. & Demengeot, J. (2009). Natural Treg cells spontaneously differentiate into pathogenic helper cells in lymphopenic conditions. *Eur J Immunol, 39,* 948-955.

[49] Zhou, X., Bailey-Bucktrout, S. L., Jeker, L. T., Penaranda, C., Martinez-Llordella, M., Ashby, M., Nakayama, M., Rosenthal, W. & Bluestone, J. A. (2009). Instability of the transcription factor Foxp3 leads to the generation of pathogenic memory T cells in vivo. *Nat Immunol, 10,* 1000-1007.

[50] Xu, L., Kitani, A., Fuss, I. & Strober, W. (2007). Cutting edge, regulatory T cells induce CD4+CD25-Foxp3- T cells or are self-induced to become Th17 cells in the absence of exogenous TGF-beta. *J Immunol, 178,* 6725-6729.

[51] Yang, X. O., Nurieva, R., Martinez, G. J., Kang, H. S., Chung, Y., Pappu, B. P., Shah, B., Chang, S. H., Schluns, K. S., Watowich, S. S., et al. (2008). Molecular antagonism and plasticity of regulatory and inflammatory T cell programs. *Immunity, 29,* 44-56.

[52] Wei, G., Wei, L., Zhu, J., Zang, C., Hu-Li, J., Yao, Z., Cui, K., Kanno, Y., Roh, T. Y., Watford, W. T., et al. (2009). Global mapping of H3K4me3 and H3K27me3 reveals specificity and plasticity in lineage fate determination of differentiating CD4+ T cells. *Immunity, 30*, 155-167.

In: Regulatory T Cells
Editor: Ren S. Hayashi, pp. 239-246

ISBN: 978-1-61761-554-2

Chapter 13

Genetic Engineering of Antigen-Specific Regulatory T Cells

***Eran Elinav*[1] *and Zelig Eshhar*[2]**
[1]Section on Immunobiology, Yale University School of Medicine
and [2]Department of Immunology, The Weizmann Institute, Rehovot, Israel

Abstract

Regulatory T cells (Tregs) potently suppress a variety of innate and adoptive immune responses mediated by T effector cells (Teff). However, because of their low number (comprising only minor subset (4-10%) of the total CD4 T cell population) and their multiple specificities, large numbers of expanded antigen-specific Tregs are required in order to suppress an ongoing inflammatory response. This scarcity of Tregs greatly limits their clinical use against autoimmune inflammatory conditions. To overcome the scarcity of antigen-specific Tregs, several attempts have been made to redirect the specificity of Tregs to a pre-defined target antigen.

In this chapter, we review three approaches to endow Tregs with the required novel specificity; all rely on the ability to genetically engineer the recognition of Treg by their transduction with any receptor-gene of interest. The first approach utilizes a chimeric receptor whose specificity is comprised of a class-II-peptide complex; this receptor redirects the genetically modified Tregs against pathogenic antigen-specific effector T cells, resulting in suppression of an experimental model of multiple sclerosis, experimental allergic encephalomyelitis (EAE). The second approach makes use of TCR α and β chain gene transduction to generate MHC-1:peptide specific Tregs that were shown to suppress arthritis and rejection of a non-HLA-matched transplant. The third approach is based on an antibody-based chimeric receptor, which we have used to generate murine Tregs that suppress TNBS colitis in a murine model. In this review, we describe the activity of each of these modified Tregs, compare their reactivity to unmodified nTreg, and discuss their potential as well as obstacles towards future clinical application.

Introduction

Thymic central tolerance to self is mediated by the negative selection of autoreactive T cells. Yet, a minority of self-reactive thymocytes overcome this selective barrier and escape to the periphery, where they can potentially induce autoimmunity. Such autoreactive T cells are usually suppressed in the periphery by several subsets of regulatory T cells (Tregs), including thymic derived naturally-occurring $CD4^{+}CD25^{+}Foxp3^{+}$ cells, IL-10 producing Tr1 cells, TGF-β producing TH3 cells, NKT cells, γδ T cells, and $CD8^{+}CD25^{-}$ T cells[1-4]. In addition to central induction of Tregs, peripheral activation of mature T cells under tolerogenic conditions (e.g. in the presence of TGFβ) or sub-optimal antigen presentation with concurrent costimulation, results in development of inducible Tregs (iTreg) [5].

Classical CD4+CD25+ Tregs are a unique subpopulation of anergic peripheral regulatory T cells that comprise 5-10% of peripheral $CD4^{+}$ cells in mice and 3-6% of the $CD4^{+}$ population in humans[6]. They are characterized by constitutive expression of the transcription factor Foxp3 and of the inhibitory costimulatory molecule, CTLA-4. $CD4^{+}CD25^{+}$ cells are potent inducers of self tolerance in animals and humans [7, 8]. In-vitro, when co-cultured with effector $CD4^{+}CD25^{-}$ lymphocytes, they promote significant suppression of lymphocyte proliferation and IL2 secretion, a process mediated by cell-cell contact and TGF-β secreted by the Treg cell[9-11]. In vivo, adoptive transfer of Treg has been shown to suppress the development of autoimmunity including auto-inflammatory colitis, experimental allergic encephalomyelitis, autoimmune thyroiditis, gastritis, insulitis, and arthritis[1]. In addition, Tregs prevent or ameliorate acute rejection of transplanted solid organs and mediate suppression of anti-tumor immunity by promoting immune tolerance [11-13].

The repertoire and mode of action of Treg has been extensively investigated. Most studies suggest that Treg activation is antigen-specific and that their interaction with autoantigens is required for their maintenance in-vivo. On the other hand, the suppressive effect of Tregs is not antigen-specific and is mediated by suppressive cytokines such as TGF-β and IL-10, which are released following antigen recognition by the Treg's TCR [11, 14, 15].

The clinical use of Tregs in auto-immune disorders or graft rejection has been limited by the extreme scarcity of Treg specific for a given antigen, and the low numbers of these cells in peripheral blood. Consequently, polyclonal Tregs were used in experimental models, but were found to be less suppressive as compared to antigen-specific Tregs. While polyclonal Tregs can prevent the onset of diabetes in NOD mice, reversion of disease was only demonstrated when using Tregs specific for pancreatic antigens [16-18]. Even in a preventative setting, the antigen-specific Tregs were 100-fold more efficacious in controlling diabetes when compared to polyclonal Tregs. In similar studies, antigen-specific Tregs isolated from TCR-transgenic mice were more effective than polyclonal Tregs in controlling murine models of autoimmune gastritis and multiple sclerosis [19, 20]. A different approach to obtain antigen specific Treg involves massive expansion of antigen-specific Tregs using MHC II:peptide/anti-CD28 beads, as was reported by Masteller et al. [21]. However, antigen-specific Tregs expanded in this way were not as efficient as Tregs expanded from TCR-transgenic mice in controlling autoimmune diabetes.

These studies demonstrated the advantages of antigen-specific Tregs for clinical trials. Nevertheless, they also highlighted a major concern that was not evident in polyclonal populations of non antigen-specific Tregs. This concern is the possible conversion of ex-vivo manipulated Tregs into effector subpopulations, such as TH17 cells, under the effect of inflammatory cytokines such as IL6 and IL23 [22]. Such polarization can occur following the accumulation of the adoptively transferred Tregs at the site of inflammation, and may result in deleterious clinical effects.

Tregs Redirected with Antigen-Specific Receptors

The most effective approach to obtain therapeutic quantities of antigen-specific Treg is by the modification of autologous Treg with target-specific receptors. The technology involves ex-vivo transduction of isolated Tregs with retrovectors expressing the desired recombinant genes. Tregs are usually isolated from peripheral blood lymphocytes (PBL) or from the spleens of mice. The most common isolation protocol uses magnetic beads or columns selecting for either $CD4^+CD25^+$ or $CD4^+CD25^-$ T cells. The $CD4^+CD25^+$ nTreg subpopulation is then activated and transduced with retrovectors harboring antigen-specific receptor constructs. When transfection is followed by in vitro expansion, this antigen-specific Treg population provides a large and potent cell population for the suppression of detrimental immune or inflammatory responses. Such an approach bears obvious advantages: relatively small numbers of antigen-specific Tregs can achieve a potent effect, and the target antigen is predetermined. Drawbacks of this approach include the use of retroviral or lentiviral gene delivery systems in a cell population that is in general anergic and does not readily undergo activation . While the genetic manipulation of Tregs is currently in its early stages of development, three types of receptors that differ in their antigen recognition moiety have been tested in experimental models and have demonstrated proof of concept in different models of auto-immune disorders.

I. Treg Redirected with MHCII-Peptide Chimeric Receptor

The first approach, developed by the group of Geiger [23] is exemplified in Tregs derived from transgenic mice (Tg) [24] harboring T cells expressing a myelin basic protein (MBP)-specific chimeric receptor made of an extracellular MBP peptide linked to MHCII sequences and to the trans-membrane and cytoplasmic domains of CD3ζ. In this approach, the extracellular moiety of the chimeric receptor of the $CD4^+CD25^+$ Tregs engages the TCR of potentially pathologic antigen-specific autoreactive $CD4^+CD25^-$ cells, activating the redirected Tegs to produce TGFβ and IL-10, which suppress the Teff cells [23]. As shown in both in vitro and in vivo studies [23-25], relatively small numbers of redirected Tregs were able to potently suppress EAE, proving that the chimeric receptor may serve as surrogate for the endogenous TCR. Interestingly, redirected Tregs ameliorated EAE even when administered at a late stage of the disease, after the pathologic T-cell response featured epitope spreading that included additional specificities not directly targeted by the Tregs [25]. This indicated the possibility of 'bystander' T-cell inhibition by redirected Tregs that can induce suppression even when directed against an antigen different from the one that caused the inflammatory process. The major limitation of this approach as a potential clinical application is that it necessitates the identification and construction of patient-tailored chimeric receptor with a recognition unit comprising an MHC-antigen complex.

II. Treg Redirected with Ectopic TCR

The second approach makes use of TCR α and β chain transduction to generate antigen-specific Tregs. This approach relies on the expression of a given pair of TCR chains specific for a predetermined peptide derived from the target antigen associated with the individual's HLA haplotype. Similar redirection of effector T cells expressing ectopic TCR genes has been developed to retarget T effector cells to cancer antigens and is being applied in clinical trials to treat melanoma patients [23]. Two studies by the groups of Lechler and Stauss applied this approach to redirect nTreg in an attempt to suppress experimental arthritis in a murine system[24, 25]. Tsang et al studied the ability of Treg redirected to allo-MHC specificity to prolong the survival of an allograft [24]. The study consisted first of selection of CD4+CD25+ Treg by their stimulation with foreign antigen-specific (MHC class II) dendritic cells, followed by retroviral transduction of these allo-specific Tregs with α and β TCR chains that recognize endogenous a foreign (K^d) MHC-I molecule. Using this approach, the authors created a Treg population that is capable of targeting both the direct and indirect Teff allo-responses (responsible for acute and chronic graft rejection, respectively). These dual allo-specific Tregs retained their suppressive capabilities, migrated to draining lymph node sites of antigen availability, and improved allogeneic heart transplant survival in vivo. The combination of redirected Treg with immuno-suppressive therapy induced even greater long-term graft survival. While in fully-mismatched donor organs, the clinical response to Tregs with direct allospecificity (class II-specific) was comparable to that of Tregs with dual specificity, the authors reported a marked histological improvement in graft appearance in dual-Treg treated mice. This elegant study provided proof of concept that dual antigen-specificity can be engineered to redirect the specificity of Treg, and that such an approach is effective in the transplantation setting.

In another study, from the group of Stauss, Wright et al [25] described the redirection of the specificity of Tregs using retroviral transduction of OVA–specific TCR and demonstrated that TCR-gene modified Tregs maintained their phenotype, and suppressed established arthritic joint inflammation in an antigen-dependent manner. As suppression of arthritis occurred even when the regulatory T cells and the pathogenic T cells recognized two distinct antigens (OVA and BSA, respectively), the authors suggested that similar 'bystander' approaches can be employed to treat autoimmune conditions in which the specificity of disease-causing autoreactive T cells is unknown. In contrast to the transplantation setting described above, treatment of arthritis did not require additional lymphoablative pre-conditioning. Interestingly, co-transduction of conventional CD4 lymphocytes with a construct encoding both the OTII-TCR and Foxp3 resulted in partial response, with less pronounced histopathological evidence of disease amelioration, and no effect on the accumulation of TH17 cells within the diseased knee. This finding is still controversial, and several groups have demonstrated that transduction of $CD4^+$ Teff cells with Foxp3 reverted these cells to an iTreg phenotype (which are somewhat less potent suppressors compared to nTreg (e.g. [28]). Altogether, the major drawback of future clinical use of the TCR based redirection approach is that, as a treatment of either in transplantation or for autoimmunity, it does not provide means of breaching the HLA barrier in the human setting.

III. Redirection of Treg in an MHC Independent Manner

The two previous approaches demonstrated that Tregs can be genetically modified to express receptors that recognize or are recognized in an MHC restricted manner. This notion narrows the spectrum of application of the approach only to known, HLA/peptide complexes.

In the third approach, the authors of this review have used an antibody-based chimeric receptor (CAR) to generate antigen-specific Tregs that alleviate autoimmunity, and can be used for any individual regardless of HLA haplotype [26]. In a pivotal study, we demonstrated that Tregs derived from transgenic mice expressing an engineered transgene made of anti-TNP scFv-CD28-FcRγ tripartite chimeric receptor (TPCR), specific to TNP, the colitis-associated antigen, induced a significant improvement in experimental 2,4,6-trinitrobenzene sulphonic acid (TNBS) induced colitis. Such TPCR promoted antigen-specific, MHC-non-restricted, and costimulation-independent Treg accumulation and activation at the inflammatory site. When adoptively transferred into mice in which TNBS colitis was induced, these Tregs could indeed alleviate the degree of disease severity, and rescue a significant proportion of mice from colitis-induced death. Importantly, nontransfected Tregs from wild type mice (WT) have no effect. Moreover, the results of this study support a "bystander" effect by demonstrating that when a different eliciting antigen (oxazolone) is used, Tregs specific to TNP can still ameliorate acute oxazolone colitis, providing that the Treg-specific antigen is also present at the inflamed site.

In a follow up study, an efficient protocol was developed for retroviral transduction of nTreg from normal WT mice with a vector encoding the same TNP specific TPCR followed by enrichment and expansion of murine nTregs, to enable antigen-specific alleviation of acute experimental colitis [27]. The TNP-transduced nTregs maintained high Foxp3 expression and demonstrated antigen-specific suppressive activity in vitro that was TGFβ, and IL10-independent.

Altogether, these studies demonstrated that (i) Tregs expressing an anti-TNP TPCR are specifically and functionally activated through their engineered receptor; (ii) this activation leads to a potent suppression of T effector cell proliferation in a manner independent of the B7 costimulation ligand; (iii) naïve, TNP-specific Tregs can protect mice from TNBS-induced colitis; (iv) when activated by their pre-determined antigen (e.g. TNP), redirected Tregs can, through a 'bystander effect', cure colitis induced with a different antigen, (e.g.oxazolone); (v) adoptive transfer of naïve TNP-specific Tregs leads to their accumulation in colonic mucosa and suppression of colitis. Thus, we suggested that these studies lay the foundations for a novel approach to the treatment not only of colitis, but of other types of autoimmune inflammatory diseases.

Conclusion

The ability to efficiently transduce purified nTregs with retroviral vectors has both scientific and therapeutic implications. As a research tool, the retroviral gene transfer protocol of engineered receptors provides an efficient means to obtain large amounts of a pure population of $CD4^{+}CD25+$ murine Tregs, providing useful models for the study of nTreg generation, activation and homeostatic proliferation. From a therapeutic point of view, the chimeric receptor approach is advantageous over the other approaches described in this review, by allowing for redirection of nTreg with specificity against any predetermined antigen, in an MHC independent manner. In fact, it is preferable that the target antigen is not the disease causing pathogen, since in many diseases in which the causative antigen is unknown. These, along with the ensuing ex-vivo expansion and purification protocols, pave the road for future therapeutic application of Treg therapy in IBD, as well as for other organ-specific and non organ -specific autoimmune disorders.

Nevertheless, the adoptive transfer of antigen-specific Tregs for the suppression of immune inflammatory responses must still be viewed with caution. The major risk is the polarization of Treg to antigen-specific TH17 effector cells by certain inflammatory cytokines such as IL6, IL23, and IL21 [22]. The results of such a process are currently unpredictable. Generation of 'effector' TH17 cells may potentially exacerbate the inflammation rather than suppressing it. In contrast, generation of regulatory TH17 cells, which were recently suggested to be protective in models of auto-inflammatory colitis [28], may potentate suppression of auto-inflammation. Means to prevent this potentially harmful conversion to TH17 cells are needed. One possible approach may include stable ectopic expression of FoxP3, down regulation of a common receptor chain shared by the undesired cytokines, or knockdown of a down-stream component essential for effector polarization (such as the transcription factor, RORgammaT) that will maintain the suppression and tolerance.

Acknowledgment

We are grateful to Dr Shelly Schwarzbaum for editing the manuscript and to MD Moross Center at the Weizmann Institute for a research grant.

References

[1] Sakaguchi, S; Sakaguchi, N; Asano, M; Itoh, M; Toda, M. Immunologic self-tolerance maintained by activated T cells expressing IL-2 receptor alpha-chains (CD25). Breakdown of a single mechanism of self-tolerance causes various autoimmune diseases. *J Immunol*, 11995, 155, 1151-64.

[2] Pearson, CI; McDevitt, HO. Redirecting Th1 and Th2 responses in autoimmune disease. *Curr Top Microbiol Immunol*, 11999, 238, 79-122.

[3] Sharif, S; Arreaza, GA; Zucker, P; Mi, QS; Delovitch, TL. Regulation of autoimmune disease by natural killer T cells. *J Mol Med l*, 2002, 80, 290-300.

[4] Hayday, A; Tigelaar, R. Immunoregulation in the tissues by gammadelta T cells. *Nat Rev Immunol*, 2003, 3, 233-42.

[5] Bluestone, JA; Abbas, AK. Natural versus adaptive regulatory T cells. *Nat Rev Immunol l*, 2003, 3, 253-7.

[6] Jonuleit, H; Schmitt, E; Stassen, M; Tuettenberg, A; Knop, J; Enk, AH. Identification and functional characterization of human CD4(+)CD25(+) T cells with regulatory properties isolated from peripheral blood. *J Exp Med l*, 2001, 193, 1285-94.

[7] Chen, W; Jin, W; Wahl, SM. Engagement of cytotoxic T lymphocyte-associated antigen 4 (CTLA-4) induces transforming growth factor beta (TGF-beta) production by murine CD4(+) T cells. *J Exp Med l*, 1998, 188, 1849-57.

[8] Read, S; Malmstrom, V; Powrie, F. Cytotoxic T lymphocyte-associated antigen 4 plays an essential role in the function of CD25(+)CD4(+) regulatory cells that control intestinal inflammation. *J Exp Med l*, 2000, 192, 295-302.

[9] Thornton, AM; Shevach, EM. CD4+CD25+ immunoregulatory T cells suppress polyclonal T cell activation in vitro by inhibiting interleukin 2 production. *J Exp Med l*, 1998, 188, 287-96.

[10] Thornton, AM; Shevach, EM. Suppressor effector function of CD4+CD25+ immunoregulatory T cells is antigen nonspecific. *J Immunol l*, 2000, 164, 183-90.

[11] Shevach, EM. CD4+ CD25+ suppressor T cells: more questions than answers. *Nat Rev Immunol l*, 2002, 2, 389-400.

[12] Sakaguchi, S; Sakaguchi, N; Shimizu, J; Yamazaki, S; Sakihama, T; Itoh, M; Kuniyasu, Y; Nomura, T; Toda, M; Takahashi, T. Immunologic tolerance maintained by CD25+ CD4+ regulatory T cells: their common role in controlling autoimmunity; tumor immunity; and transplantation tolerance. *Immunol Rev l*, 2001, 182, 18-32.

[13] Maloy, KJ; Powrie, F. Regulatory T cells in the control of immune pathology. *Nat Immunol l*, 2001, 2, 816-22.

[14] Hsieh, CS; Zheng, Y; Liang, Y; Fontenot, JD; Rudensky, AY. An intersection between the self-reactive regulatory and nonregulatory T cell receptor repertoires. *Nat Immunol l*, 2006, 7, 401-10.

[15] Asano, M; Toda, M; Sakaguchi, N; Sakaguchi, S. Autoimmune disease as a consequence of developmental abnormality of a T cell subpopulation. *J Exp Med l*, 1996, 184, 387-96.

[16] Gregori, S; Giarratana, N; Smiroldo, S; Adorini, L. Dynamics of pathogenic and suppressor T cells in autoimmune diabetes development. *J Immunol l*, 2003, 171, 4040-7.

[17] Tang, Q; Henriksen, KJ; Bi, M; Finger, EB; Szot, G; Ye, J; Masteller, EL; McDevitt, H; Bonyhadi, M; Bluestone, JA. In vitro-expanded antigen-specific regulatory T cells suppress autoimmune diabetes. *J Exp Med l*, 2004, 199, 1455-65.

[18] Tarbell, KV; Petit, L; Zuo, X; Toy, P; Luo, X; Mqadmi, A; Yang, H; Suthanthiran, M; Mojsov, S; Steinman, RM. Dendritic cell-expanded; islet-specific CD4+ CD25+ CD62L+ regulatory T cells restore normoglycemia in diabetic NOD mice. *J Exp Med l*, 2007, 204, 191-201.

[19] DiPaolo, RJ; Brinster, C; Davidson, TS; Andersson, J; Glass, D; Shevach, EM. Autoantigen-specific TGFbeta-induced Foxp3+ regulatory T cells prevent autoimmunity by inhibiting dendritic cells from activating autoreactive T cells. *J Immunol l*, 2007, 179, 4685-93.

[20] Stephens, LA; Malpass, KH; Anderton, SM. Curing CNS autoimmune disease with myelin-reactive Foxp3+ Treg. *Eur J Immunol l*, 2009, 39, 1108-17.

[21] Masteller, EL; Warner, MR; Tang, Q; Tarbell, KV; McDevitt, H; Bluestone, JA. Expansion of functional endogenous antigen-specific CD4+CD25+ regulatory T cells from nonobese diabetic mice. *J Immunol l*, 2005, 175, 3053-9.

[22] Littman, DR; Rudensky, AY. Th17 and regulatory T cells in mediating and restraining inflammation. *Cell*, 140, 845-58.

[23] Johnson, LA; Morgan, RA; Dudley, ME; Cassard, L; Yang, JC; Hughes, MS; Kammula, US; Royal, RE; Sherry, RM; Wunderlich, JR; Lee, CC; Restifo, NP; Schwarz, SL; Cogdill, AP; Bishop, RJ; Kim, H; Brewer, CC; Rudy, SF; VanWaes, C; Davis, JL; Mathur, A; Ripley, RT; Nathan, DA; Laurencot, CM; Rosenberg, SA. Gene therapy with human and mouse T-cell receptors mediates cancer regression and targets normal tissues expressing cognate antigen. *Blood* l, 2009, 114, 535-46.

[24] Tsang, JY; Tanriver, Y; Jiang, S; Xue, SA; Ratnasothy, K; Chen, D; Stauss, HJ; Bucy, RP; Lombardi, G; Lechler, R. Conferring indirect allospecificity on CD4+CD25+ Tregs by TCR gene transfer favors transplantation tolerance in mice. *J Clin Invest l*, 2008, 118, 3619-28.

[25] Wright, GP; Notley, CA; Xue, SA; Bendle, GM; Holler, A; Schumacher, TN; Ehrenstein, MR; Stauss, HJ. Adoptive therapy with redirected primary regulatory T cells results in antigen-specific suppression of arthritis. *Proc Natl Acad Sci*, U S A l 2009, 106, 19078-83.

[26] Elinav, E; Waks, T; Eshhar, Z. Redirection of regulatory T cells with predetermined specificity for the treatment of experimental colitis in mice. *Gastroenterology l*, 2008, 134, 2014-24.

[27] Elinav, E; Adam, N; Waks, T; Eshhar, Z. Amelioration of colitis by genetically engineered murine regulatory T cells redirected by antigen-specific chimeric receptor. *Gastroenterology l*, 2009, 136, 1721-31.

[28] O'Connor, W; Jr; Kamanaka, M; Booth, CJ; Town, T; Nakae, S; Iwakura, Y; Kolls, JK; Flavell, RA. A protective function for interleukin 17A in T cell-mediated intestinal inflammation. *Nat Immunol l*, 2009, 10, 603-9.

In: Regulatory T Cells
Editor: Ren S. Hayashi, pp. 247-250

ISBN: 978-1-61761-554-2

Chapter 14

Regulatory T Cells in Allergic Disorders and Immunomodulation

Kei-ichi Yamanaka *
Department of Dermatology, Mie University,
Graduate School of Medicine, Edobashi, Tsu, Mie, Japan

The immune system protects the host from infection and exogenous stimuli using innate and acquired immune systems and is critical to maintaining health. These stimuli lead to type 1 helper T cell (Th1) and/or Th2 type cytokine secretion according to the character of the exogenous or endogenous stimuli. In addition, FoxP3$^+$(forkhead box P3$^+$) CD4$^+$CD25$^+$regulatory T cells: natural regulatory T cell (nTreg), IL-10 producing regulatory T cells (Tr1), and Th17 cells play a pivotal role in this immune balance triggering immune suppression and stimulation. Altogether they form a four-way balance.

In allergic diseases including atopic dermatitis (AD) and nasal allergy, it is understood that the cytokine profile comprises Th2 dominant, and this milieu gives rise to phenotypical abnormality and pathologic conditions including cellular infiltration and overproduction of IgE.

In order to ameliorate this immune imbalance, the most appropriate solution is the induction of Th1 cells. Several approaches have been tried recently. Various types of bacteria have been used, including *propionibacterium acnes (p. acnes),* a facultative anaerobic gram-positive rod. This is thought to be a potent Th1-type cytokine inducer; which converts Th2 cytokine milieu to Th1 depending on antigen presenting cells and with the involvement of IL-12 and IFN-γ. In the study using intrinsic AD mouse model [1], *p. acnes* treatment improves the manifestations dramatically showing limited skin eruption, and lymphocytes skin infiltration including mast cells [2]. The number of IFN-γ^+ T cells is increased in the spleen cells and in the skin lesions in *p.acnes*-treated AD mice [2]. It is noted that both nTreg and Tr1 cells are also induced by *p.acnes*-treatment. The mycobacterium 30-kD, a major secretory

* Corresponding author: TEL: +81-59-231-5025, FAX: +81-59-231-5206, E-mail yamake@clin.medic.mie-u.ac.jp,

protein, antigen 85B (Ag85B), is a well-known potent Th1 inducer [3]. Furthermore the plasmid DNA encoding Ag85B gene inhibited an antigen dependent hypersensitivity dermatitis through the induction of nTreg [4]. The precise mechanism why these Th1 inducers also attribute to the generation of nTregs is not yet clearly understood. However, a bacterial TLR2 agonist is known to act as nTreg inducer in vivo; therefore, infectious agents associated with the Th1 inducer may increase nTreg via TLR2 signaling. In human study, the number of nTreg is inversely correlated with the severity of atopic dermatitis [5, 6] and other kinds of allergic disorders [7, 8]. Contrarily, the number of local nTreg is increased in the nasal mucosa after successful allergen-specific immunotherapy (SIT) [9, 10], therefore nTreg induction may be a potential therapeutic approach for allergic disorders.

On the other hand, IL-10 is a regulatory cytokine secreted during inflammatory responses where they inhibit T cell proliferation. IL-10 plays a pivotal role in the acquisition and maintenance of specific T cell tolerance [11]. IL-10 producing Tr1 is one type of acquired regulatory T cell, and upon activation, it plays a critical role in the suppression of allergic diseases perhaps together with TGF-β. Like nTreg, there is an inverse correlation between the IL-10 level and the incidence of allergic disease [12-14]. However, contrary to nTreg, Tr1 inhibit the cellular proliferation not via cell-to-cell contact manner, but by a cytokine-mediated mechanism; therefore this suppression is remote controllable. Specific peptide-based SIT is currently being used in the treatment of several allergic diseases all over the world, and Tr1 induction is thought to be critical in allergen desensitization. SIT often elicits allergen specific IgG responses (mostly IgG1 and IgG4), and prevents the seasonal rise in IgE antibodies associated with natural exposure to allergens [15, 16]. Tr1 is increased in patients after successful immunotherapy [10, 17-20], and T cell receptor repertoire diversity in increased Tr1 population is fully preserved [20], suggesting the up-regulated Tr1 population has the potential to attend to multiple T cell epitopes from various allergens. This result may explain why some patients undergoing immunotherapy show reduced sensitivity to other allergens in addition to specific peptide used in SIT [20]. To avoid an adverse effect during the process of SIT, allergen-derived T cell epitope peptides may be better compared to native allergens, as they do not cross-link with IgE [14]. Of course the modulation of immunological imbalance by applying recombinant IL-10 protein holds great promise, however the effect of recombinant protein is temporary and is also difficult to deliver to the inflamed areas. Contrarily, excessive administration of IL-10 may lead to immunosuppression [21], therefore natural induction by SIT may be the safest and most efficient method, at present, for the appropriate amount of IL-10.

References

[1] Yamanaka, K., Tanaka, M., Tsutsui, H., Kupper, T. S., Asahi, K., Okamura, H., Nakanishi, K., Suzuki, M., Kayagaki, N., Black, R. A., Miller, D. K., Nakashima, K., Shimizu, M. & Mizutani, H. (2000). Skin-specific caspase-1-transgenic mice show cutaneous apoptosis and pre-endotoxin shock condition with a high serum level of IL-18. *J Immunol.*, *165*, 997-1003.

[2] Kitagawa, H., Yamanaka, K., Kakeda, M., Inada, H., Imai, Y., Gabazza, E., Kurokawa, I. & Mizutani, H. (2010). Propionibacterium acnes vaccination induces regulatory T

cells and Th1 immune responses and improves mouse atopic dermatitis *Experimental Dermatology*. in press.

[3] Takatsu, K. & Kariyone, A. (2003). The immunogenic peptide for Th1 development. *Int Immunopharmacol., 3*, 783-800.

[4] Mori, H., Yamanaka, K., Matsuo, K., Kurokawa, I., Yasutomi, Y. & Mizutani, H. (2009). Administration of Ag85B showed therapeutic effects to Th2-type cytokine-mediated acute phase atopic dermatitis by inducing regulatory T cells. *Arch Dermatol Res., 301*, 151-157.

[5] Orihara, K., Narita, M., Tobe, T., Akasawa, A., Ohya, Y., Matsumoto, K. & Saito, H. (2007). Circulating Foxp3+CD4+ cell numbers in atopic patients and healthy control subjects. *J Allergy Clin Immunol., 120*, 960-962.

[6] Reefer, A. J., Satinover, S. M., Solga, M. D., Lannigan, J. A., Nguyen, J. T., Wilson, B. B. & Woodfolk, J. A. (2008). Analysis of CD25hiCD4+ "regulatory" T-cell subtypes in atopic dermatitis reveals a novel T(H)2-like population. *J Allergy Clin Immunol., 121*, 415-422 e413.

[7] Van Bruaene, N., Perez-Novo, C. A., Basinski, T. M., Van Zele, T., Holtappels, G., De Ruyck, N., Schmidt-Weber, C., Akdis, C., Van Cauwenberge, P., Bachert, C. & Gevaert, P. (2008). T-cell regulation in chronic paranasal sinus disease. *J Allergy Clin Immunol., 121*, 1435-1441, 1441 e1431-1433.

[8] Jartti, T., Burmeister, K. A., Seroogy, C. M., Jennens-Clough, M. L., Tisler, C. J., Salazar, L. P., Dasilva, D. F., Evans, M. D., Vrtis, R. F., Wallace, P. K., Ruiz-Perez, B., Gangnon, R. E., Lemanske, R. F. Jr., & Gern, J. E. (2007). Association between CD4(+)CD25(high) T cells and atopy in children. *J Allergy Clin Immunol., 120*, 177-183.

[9] Radulovic, S., Jacobson, M. R., Durham, S. R. & Nouri-Aria, K. T. (2008). Grass pollen immunotherapy induces Foxp3-expressing CD4+ CD25+ cells in the nasal mucosa. *J Allergy Clin Immunol., 121*, 1467-1472, 1472 e1461.

[10] Bohle, B., Kinaciyan, T., Gerstmayr, M., Radakovics, A., Jahn-Schmid, B. & Ebner, C. (2007). Sublingual immunotherapy induces IL-10-producing T regulatory cells, allergen-specific T-cell tolerance, and immune deviation. *J allergy Clin Immunol.*, 120, 707-713.

[11] Akdis, C. A., Blesken, T., Akdis, M., Wuthrich, B. & Blaser, K. (1998). Role of interleukin 10 in specific immunotherapy. *J Clin Invest., 102*, 98-106.

[12] Akdis, M., Verhagen, J., Taylor, A., Karamloo, F., Karagiannidis, C., Crameri, R., Thunberg, S., Deniz, G., Valenta, R., Fiebig, H., Kegel, C., Disch, R., Schmidt-Weber, C. B., Blaser, K. & Akdis, C. A. (2004). Immune responses in healthy and allergic individuals are characterized by a fine balance between allergen-specific T regulatory 1 and T helper 2 cells. *J Exp Med., 199*, 1567-1575.

[13] Heaton, T., Rowe, J., Turner, S., Aalberse, R. C., de Klerk, N., Suriyaarachchi, D., Serralha, M., Holt, B. J., Hollams, E., Yerkovich, S., Holt, K., Sly, P. D., Goldblatt, J., Le Souef, P. & Holt, P. G. (2005). An immunoepidemiological approach to asthma, identification of in-vitro T-cell response patterns associated with different wheezing phenotypes in children. *Lancet., 365*, 142-149.

[14] Wu, K., Bi, Y., Sun, K. & Wang, C. (2007). IL-10-producing type 1 regulatory T cells, & allergy. *Cell Mol Immunol., 4*, 269-275.

[15] Hakansson, L., Heinrich, C., Rak, S. & Venge, P. (1998). Activation of B-lymphocytes during pollen season. Effect of immunotherapy. *Clin Exp Allergy.*, *28*, 791-798.

[16] Mothes, N., Heinzkill, M., Drachenberg, K. J., Sperr, W. R., Krauth, M. T., Majlesi, Y., Semper, H., Valent, P., Niederberger, V., Kraft, D. & Valenta, R. (2003). Allergen-specific immunotherapy with a monophosphoryl lipid A-adjuvanted vaccine, reduced seasonally boosted immunoglobulin E production and inhibition of basophil histamine release by therapy-induced blocking antibodies. *Clin Exp Allergy.*, *33*, 1198-1208.

[17] Frew, A. J. (2008). Sublingual immunotherapy. *N Engl J Med.*, *358*, 2259-2264.

[18] Akdis, M. & Akdis, C. A. (2007). Mechanisms of allergen-specific immunotherapy. *J Allergy Clin Immunol.*, *119*, 780-791.

[19] Taylor, A., Verhagen, J., Blaser, K., Akdis, M. & Akdis, C. A. (2006). Mechanisms of immune suppression by interleukin-10 and transforming growth factor-beta, the role of T regulatory cells. *Immunology.*, *117*, 433-442.

[20] Yamanaka, K. I., Yuta, A., Kakeda, M., Sasaki, R., Kitagawa, H., Gabazza, E. C., Okubo, K., Kurokawa, I. & Mizutani, H. (2009). Induction of IL-10-producing regulatory T cells with TCR diversity by epitope-specific immunotherapy in pollinosis. *J allergy Clin Immunol.*, *124*, 842-845.

[21] Mosser, D. M. & Zhang, X. (2008). Interleukin-10, new perspectives on an old cytokine. *Immunol Rev.*, *226*, 205-218.

In: Regulatory T Cells
Editor: Ren S. Hayashi, pp. 251-268

ISBN: 978-1-61761-554-2

Chapter 15

The Number of Regulatory T Cells: Pursuit of the Golden Mean

Alexander Pukhalsky[a, b]**, *Galina Shmarina***[a, b]
and Vladimir Alioshkin[b]
[a]Research Centre for Medical Genetics, Moscow, Russia
[b]G.N. Gabrichevsky Institute of Epidemiology and Microbiology, Moscow Russia

Abstract

Regulatory T cells (Tregs) are an important part of the complex adaptive system of the body that is able to respond to environmental challenges. Tregs maintain peripheral tolerance and play an important role in inflammatory reaction control. Several subsets of Tregs have been described. Naturally occurring $CD4^+CD25^+$ Tregs are recognized as a major subset of immune cells maintaining immune self-tolerance in the periphery. Inhibitory potential of these cells against self-reacting T cells is linked to the thymic origin of Tregs. Another subtype of Tregs that is inducible. Such Tregs are generated in the periphery and exert their suppressive potential substantially in the form of anti-inflammatory activity. The latter plays an important role in cooperation of three principal anti-inflammatory mechanisms, which have been developed in the course of evolution: macrophages possessing the suppressive activity (most ancient), Tregs, and stress hormones (phylogenetically youngest). Normally, all three mechanisms of inflammation control are well equalized. However the balance may be disturbed by age due to repeated episodes of stress and HPA axis activation. Consequently, stress hormone release coupled with antigen overload results in Treg accumulation. In time HPA axis activation replaces by its depletion that is manifested by both basal cortisol level decrease and the reduction of stress-induced cortisol response. So, cortisol being in low concentration in the blood is not able any more to control inflammation and Tregs become a principal mechanism of anti-inflammatory machinary. Superfluous Treg accumulation results in development functional somatic syndromes, such as chronic fatigue syndrome, and in some patients, by reason of anticancer immunity suppression, to tumor growth. On the other hand, in childhood the lack of adequate antigen loading may lead to delay of Treg maturation. Allergy and asthma manifestation may be a consequence of such Treg insufficiency. Thus, both excess and deficiency in Treg number may be at the bottom of morbid conditions. The advances in modern pharmacology allow devising the methods of Treg level control.

Ability to respond to environmental changes is a fundamental property of all living things from bacteria to human society. If life is an ability to adapt, a loss of such ability brings the threat of death. In the course of biological evolution, the ancient adaptive mechanisms do not disappear but become an integrated part of a complex adaptive system presented in mammalia as a triad including the brain, immune and endocrine systems. The body is secured against stressors, chemical agents, and infection until the work of all three systems is well equalized. Such balance may be disturbed by age due to repeated episodes of stress and HPA axis activation. Consequently, stress hormone release coupled with antigen overload results in regulatory T cell (Treg) accumulation. In time HPA axis activation replaces by its depletion and cortisol, being in low concentration in the blood, is not able any more to control inflammation. In this issue Tregs become a principal mechanism of anti-inflammatory machinery. In this Review, we discuss the quantitative and qualitative changes within Treg subset observed through human life and the capacity of medicamental control over Treg function.

Heterogeneity of Treg Subset and the Change of Treg Repertoire with the Course of Time

There has been a voluminous literature focusing on the role of Tregs in several settings including microbial and viral immunity, autoimmunity, cancer immunity, transplantation tolerance, and allergic response [1-6]. After the renewal interest in suppressive function of T cells connected with the work of Sakaguchi [7, 8], previous investigations of Tregs, which were initially described by Gershon et al. in the early 1970s and were called suppressor T cells [9, 10] deserve new reading. Now the terms “suppressor T cells” and “regulatory T cells” are used interchangeably, but the term “regulatory T cells” is preferable because it reflects the function of such cells more accurately. Tregs of a $CD4^+$ phenotype have been categorized into two main classes based on their ontogeny: naturally occurring Tregs (nTregs), which are developed in the thymus and present in mice and humans from an early postnatal period, and adoptive or inducible Tregs (iTregs) which are developed in the periphery from naïve T cells under the influence of various inductive signals, essentially TGF-β, NO or IL-10 [11] (Fig. 1).

Naturally occurring Treg cells are generated during development of T lymphocytes [12]. These Tregs, constituting 5–10% total lymphocytes, then enter in peripheral circulation and are widely distributed in peripheral compartment including lymph nodes and spleen [13]. Apparently, with the lapse of time such thymus-derived Tregs become a self-sustained population of cells, which undergo constant homeostatic division in the steady state [14, 15].

nTregs are characterized by their high expression of CD25 ($CD4^+CD25^{+HIGH}$) and co-expressed of the forkhead box P3 transcription factor ($FoxP3^+$). These $CD4^+CD25^+FoxP3^+$ nTregs have been shown to suppress proliferation of other T cells *in vitro* and to protect from autoimmune diseases in several animal models [3, 8, 16].

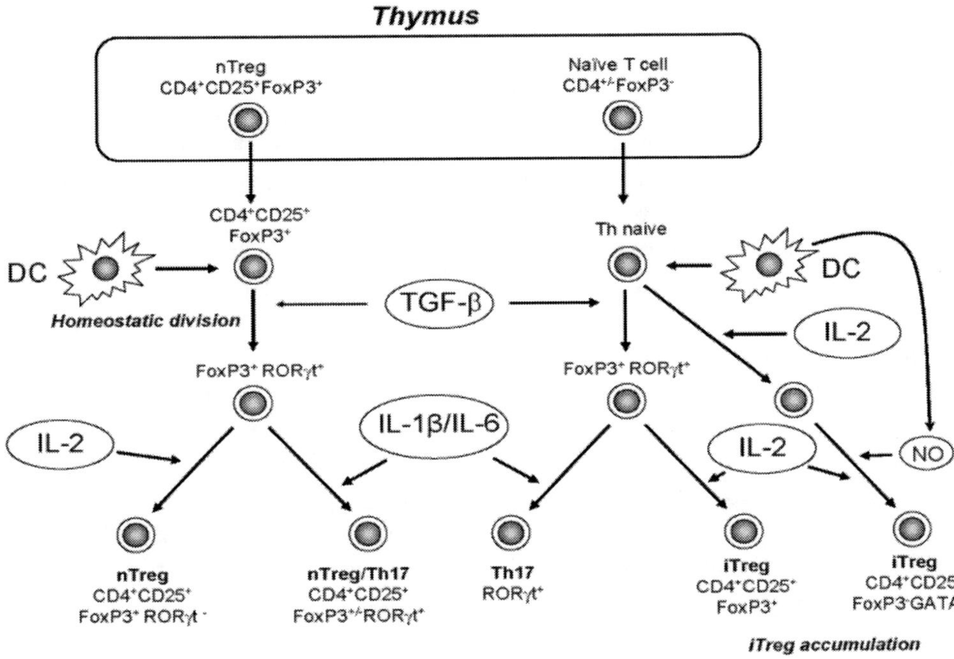

Figure 1. Generation of natural (nTregs) and inducible (iTregs) regulatory T cells. $CD4^+CD25^+FoxP3^+$nTregs differentiate in the thymus and migrate in peripheral tissues. Their survival in the periphery is maintained by constant homeostatic division in response to self-antigens. Such process requires antigen presentation by dendritic cells (DC) and co-stimulation through the CD28 and B7 pathways. $FoxP3^+$ Tregs and Th17 cells share overlapping pathways of differentiation and a subset of FoxP3 cells also express ROR-γt . Thus, the factors known to promote $FoxP3^+$ expression (including TGF-β) can also promote Th17 differentiation. Adoptive $FoxP3^+$ and $FoxP3^-$ Treg cells (iTregs) are developed in the periphery from naïve T cells under the influence of various inductive signals including DC-presented bacterial antigens. $FoxP3^+$ iTreg formation also requires TGF-β. Meanwhile, $FoxP3^-$ iTreg cells are developed under the influence of nitric oxide (NO) and IL-10. Both nTregs and iTregs are in need of IL-2 and may enter into competition for this cytokine. Thus, homeostatic division of nTregs in low IL-2 and high IL-6 environment results in Th17 cell dominating formation.

On the contrary, adoptive or inducible Tregs (iTregs) are developed in the periphery from naïve T cells under the influence of various inductive signals such as bacterial antigens, TGF-β, and IL-10 [11, 17, 18]. Most researchers subdivide iTregs into two putatively separate subsets of Th3 and Tr1 cells [18, 20]. However, we support the opinion of Corotto de Lafaille and Lafaille [11] that dividing of iTregs into FoxP3+ and $FoxP3^-$ cells is more justified. First of all such division is more justified in terms of change of Treg repertoire down the stream of life. The human is born with immature immunity. Indeed, nTregs are generated in the thymus through MHC class II-dependent T cell receptor interactions resulting in high avidity selection and are present at birth in quantity, whereas iTregs are formed outside the thymus at postnatal period under a variety of conditions. iTreg cell induction by foreign antigens has been described by several groups [11, 20-22]. The mucosal delivery of antigen, in particular the oral route, as a means of generation of $FoxP3^+$ iTreg cells plays a principal role in formation of tolerance to food-born antigens [20-22]. At the same time, insufficient stimulation of Th1 arm of immune system associated with a habitation in too sterile an environment (hygiene hypothesis) also delays iTreg generation, which in turn may lead to allergic diseases [23-25]. Thus, gradual adjustment of child to the environment is associated in part with iTreg accumulation. After a short space of time of the immaturity of immune system and maladaptation to psychological stress a "golden period" of adaptation comes and the body acquires an ability to respond to environmental challenges most efficiently. Such response is characterized by cooperation of three principal anti-inflammatory mechanisms, which were developed in the course of evolution: macrophages possessing the suppressive activity, both natural and inducible Tregs demonstrating overlapping activities, and high activity of hypothalamic-pituitary-adrenal (HPA) axis (Fig. 2). Normally, all three mechanisms of the control over inflammation are well equalized. However, the balance may be disturbed by age due to repeated episodes of stress and HPA axis hyper activation. Multiple episodes of stress and/or immune system activation occurring down the stream of life result in gradual transition of control over the inflammation from stress hormones to Tregs. Thus, in time HPA axis activation is replaced by its depletion that is manifested by both basal cortisol level decrease and the reduction of stress-induced cortisol response [26-27]. Since both lymphoid and non-lymphoid cytokine-producing cells such as T cells, dendritic cells (DC), macrophages (Mφ), endothelial and epithelial cells are no longer under adequate glucocorticoid control, the production of increased amounts of pro-inflammatory factors in response to environmental challenges takes place. Central and/or local imbalance of cytokines including the pain mediator IL-8 and pro-inflammatory cytokines TNF-α, IL-1β, and IL-6 is associated with the onset of such symptoms as chronic fatigue, fever, fibromialgia, joint pains, and skin disorders. Among the lymphocyte subsets included in mechanism of "fast response" in the first place it should be pointed out T memory cells and Tregs. In the zone of inflammation Tregs are activated by TLR, CD28, and MHC II. At the time of division Tregs lose their ability to suppress DC activity. Out of Treg control DC progenitors are differentiated in mature DCs and Mφ [15-28].

So, cortisol being in low concentration in the blood is not able any more to control inflammation, and Tregs become a principal anti-inflammatory instrument of the immunity. This mode of regulation of inflammatory reaction being more slow and rough does not allow to fine tune immune system and each new environmental challenge promotes further Treg accumulation that leads to stable immunosuppression. It is quite possible that this fact may hold the key why old age is the greatest cancer-risk factor in humans. Besides Treg accumulation [29], other age-dependent alterations of immune regulation have been described (Fig. 3). Such alterations include DC and Mφ depletion, as well as Th17 cell accumulation [30-33]. Looking at the transcriptional mechanisms of Th17 cells and Tregs in mice, Zhou and colleagues have showed that TNF-β alone is sufficient to induce RORγt and FoxP3 in naïve $CD4^+$ T cells, where FoxP3 inhibits RORγt directly through a physical interaction, repressing Th17 programme [34]. Upon addition of exogenous IL-6 or IL-21, the inhibitory effect of FoxP3 on IL-17 induction is circumvented and FoxP3 levels are reduced [34, 35]. The intimate relationship described between murine Treg and Th17 cells is also present in humans, as $FoxP3^+$ Tregs capable of IL-17 production have now been identified in humans [36, 37]. Pro-inflammatory cytokines, in particular IL-1β, also promote IL-17 production by human Tregs [36-40]. It is also known that exogenous factors such as retinoic acid [41-43] and aryl-hydrocarbon receptor agonist [44, 45] regulate the balance between the Treg and Th17 generation, established an especially close relationship between these subsets. It should be noted that Tregs show obligate requirement, both *in vivo* and *in vitro*, for IL-2 and the structurally related IL-2 family members, IL-15 and IL-7, for maintenance of FoxP3 expression and suppressive function [46-48]. Specifically, in mice, IL-2 inhibits Th17 differentiation from IL-6 and TGF-β stimulated naïve Th cells due to STAT-5 activation and direct induction of FoxP3 [50]. STAT-5 deficiency in humans results in loss of Tregs and immune dysregulation, while overexpression of STAT-5 in $CD4^+CD25^-$ cells leads to elevated levels of FoxP3. In this sense, individual characteristics of age-dependent immune abnormalities depend on genetic background including the capacity for IL-2 production. Results from several laboratories have demonstrated that the ability to produce IL-2 declines with age [30, 33]. Thus, under IL-2 deficiency Tregs transdifferentiate into Th17-like cells, which result in immune dysregulation [33]. The phenomenon of loss in Treg function and induction of Th17 cells has been demonstrated in the pathogenesis of human autoimmune diseases such as multiple sclerosis [52], rheumatoid arthritis [53], Crohn's disease [54, 55] and psoriasis [56].

Thus, homeostatic division of nTregs under the condition of IL-2 deficiency results in dominating Th17 cell accumulation (see Fig. 1) that in the sequel may be associated with tissue fibrosis and autoimmune diseases [14, 33, 47]. On the contrary, the subjects with genetically high IL-2 production form a risk group in respect of cancer and chronic infection development [56]. In addition, surplus accumulation of Tregs results in aberrant course of inflammation. Owing to competition for growth factors a part of accumulated Tregs cannot enter into proliferation and maintains the functional activity. Active Tregs prevent early DC progenitors from differentiating. Such immature DC characterized by low MHC II expression. Although these cells are not capable of phagocytosis, they produce high amounts of NO, TGF-β and IL-10 [57, 58]. The latter inhibits Th1 cells, including autoimmune lymphocytes in the brain. These cells are suggested to be necessary at least in the CNS for neural maintenance and repair, possibly by controlling the local microglial response [59]. Malfunction of Th1 cells and clogging of the brain with waste products of cells leads to the appearance of neurodegenerative foci [60], which seems to be an anatomic substrate for such cognitive and psychological symptoms as difficulty in speech, insomnia, changes in memory, difficulty in concentration, mood swings, depression, anxiety, and difficulty learning [61]. The events described above are not only typical of physiological aging. Early immunosenescence and aging of other adaptive mechanisms may be observed in chronic diseases accompanied with permanent activation of neuro-immuno-endocrine system. However, in many case cardinal symptoms of the disease mask the signs of the immune system aging.

Therefore, the number of Treg cells defines to a marked degree the ability to resist unfavourable effect of an environment. Tregs as the immune system matures become more numerous achieving optimum when balance between Treg number and HPA axis activity should protect the body against environmental challenges in most effective manner. Subsequently, in the process of aging or severe disease development the balance may be disturbed in the form of decrease of HPA axis activity and further Treg amplification. We consider below feasible therapeutic approaches aimed at the Treg number normalization.

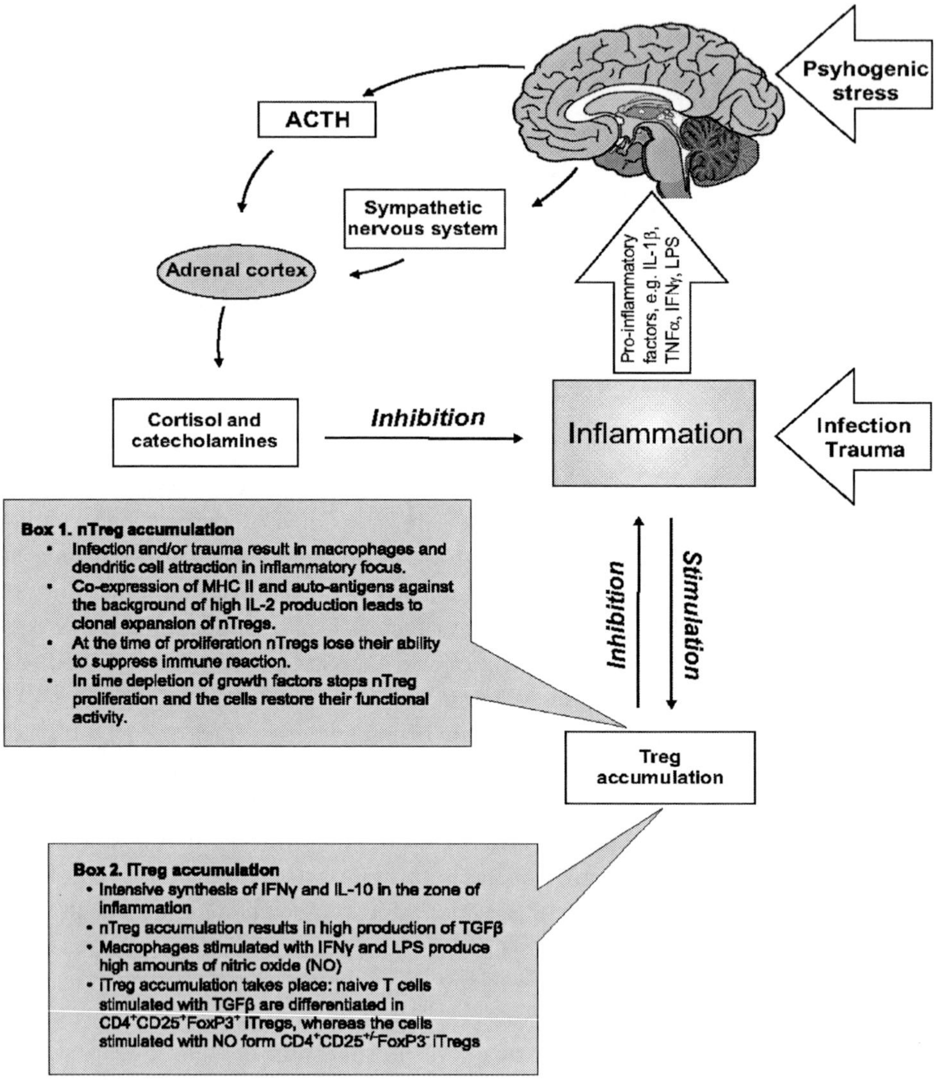

Figure 2. Induction of HPA axis activation and Treg accumulation during immune response. Activation of pituitary glands and sympathetic nervous system by stress and pro-inflammatory cytokines leads to systemic release of glucocorticoids and catecholamines, which in turn influence immune/inflammatory responses in a major fashion. Normally, anti-inflammatory mechanisms are capable to damp the consequences of antigen stimulation and the brain no longer receives the signals via pro-inflammatory cytokines. Multiple episodes of stress and/or immune system activation occurring down the stream of life lead to gradual depletion of HPA axis, which is manifested by cortisol level decrease. Under the circumstances Tregs begin to play a principal role in anti-inflammatory mechanisms and each new pro-inflammatory stimulus increases their number.

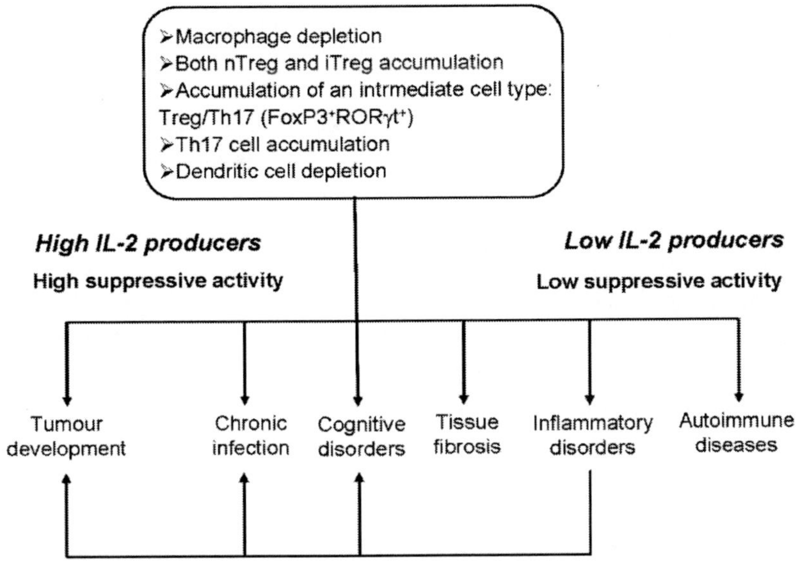

Figure 3. Age-related changes in immune regulation. A constellation of signs typical of elders are the same that may be observed in patients with chronic fatigue syndrome and in almost all cases results from HPA axis failure. Cognitive disorders, as well as other neurological symptoms including chronic fatigue, headaches, sleep disturbance, low mood, and general unwellness are manifested in most of patients. Type of disease developing with age depends in part on genetically conditioned IL-2 production. It may be expected relatively rapid increase of Treg levels in high IL-2 producers. Stable immunosuppression in these subjects threatens with tumor development and chronic infection. On the contrary, in low IL-2 producers Treg number is unlikely to reach a high level due to deficiency of the key factor for Treg development. Consequently a drifting towards Th17 cells will take place. Thus, there is strong possibility that tissue fibrosis, autoimmune diseases, and inflammatory disorders will be observed in such individuals. At the same time inflammatory disorders may become a secondary cause of the cancer and chronic infections.

Treg Deficiency

As stated above iTreg deficiency in childhood reflects an immaturity of the infant immunity. There is reason to believe that from early 19th century a period of such immune immaturity progressively increases [62]. That may be explained by loss of exposure to environmental saprophytes, disappearance of helminthes, less orofecally transmitted organisms, less *Toxoplasma*, restricted exposure to gut microbiota of other individuals, and disturbance of gut microbiota by antibiotics [62, 63]. In addition, recent changes in microbial flora of skin also contribute to delay of the immune system maturation. Before the invention of modern soaps and detergents, skin was colonized by ammonia-oxidizing bacteria. The latter are ubiquitous in soil, but they are exquisitely sensitive to alkyl-benzene sulphonate detergents. These organisms convert the high concentrations of urea and ammonia found in human sweat into nitrite and NO. The latter is essential for iTreg development and thereby plays an important role in peripheral tolerance establishment in the skin [64].

Despite the fact that the "hygiene hypothesis" wins more and more supporters and gradually turns into "hygiene theory", we clearly understand that it is impossible to set back the clock and to return to mud and epidemics. The efforts aimed at the making of artificial human environment are not vain and result in significant increase of age and quality of life. Methods intended for more rapid immunity maturation have to be a component of such system. Unfortunately, in this respect we are at the beginning of the way. Here we can only list several approaches each being in need of additional investigations:

1. Planned immunization with virus and microbial vaccines may accelerate immune maturation [65]. However it is necessary to specially investigate effect of vaccination on iTreg formation as well as to correct the immunization schedule.
2. It is necessary to return to investigations of bacterial immunomodulators that contain killed bacteria, their lysate or components of bacterial cells such as Luivac, Biostim, Ribomunyl, Imudon, Bronchovaxom [66]. Consequently the indications for these drug prescriptions have to become more rigorous and be based on laboratory examination results.
3. Further studies are necessary for substantiation of usage of dietary supplements containing retinoic acid and/or vitamin D. It has been shown that that increased vitamin D in maternal diet during pregnancy may reduce risk of wheeze and other symptoms of astma in children [67, 68].
4. In-depth study of probiotic influence on immune system maturation is also desirable [69].
5. It is deserved special consideration the recent studies of therapy with intestinal helminthes such as *Trichuris suis* affords protection from immunological diseases [70].

Superfluous Treg Accumulation

It is becoming increasingly clear that Tregs play an active and significant role in the progression of many diseases, mostly characteristic of senior age groups. Indeed it was recently shown that the peripheral blood level of $CD4^{+}CD25^{+}$ Treg cells is significantly higher in elderly than in young subjects [29]. A large body of data in the literature concerns the role of Tregs in suppressing tumor-specific immunity and progression of cancer (for review see T.Y. Ha [4]). Treg accumulation was also demonstrated in chronic bowel disease, asthma, COPD, rheumatoid arthritis and some other diseases including functional somatic syndromes with cognitive disorders [71-75]. Apart from Treg cell plasticity the list of such diseases would be considerably extended. However, facility of Treg transformation into Th17 cells may prevent to discover increased Treg level in many cases. So, Treg cell depletion might be assumed as a basis of treatment of a full range of diseases.

Different methods of Treg depletion have been proposed. In our opinion alkylating drugs are most promising among them. Alkylating drugs (ADs) belonging to nitrogen mustard family are commonly used as cytostatic and immunosuppressive agents. The effect of the drugs is mainly associated with cross-linking of DNA double strands [76] and, at higher concentrations, with induction of DNA strand breaks [77]. Although DNA is not a unique target for alkylation in the cell the others (RNA and some proteins) do not play any role in the cytostatic effect realization if the drug is used at a DNA-altering dose. However, when the dose is gradually decreased, the number of targets for alkylation will also be reduced. Studies in *in vitro* lymphocyte proliferation model show that the scenario varies with the ADs concentration decrease. If the concentration of ADs is high (100 μg/ml or more), cells die within few hours due to irreversible DNA damage [78]. If the concentrations of ADs vary in the range from 30 to 100 μg/ml numerous sites of DNA are also alkylated, but DNA damaged segments restored during DNA repair. Nevertheless, the affected cells are died due to apoptosis induction. Moderate concentrations of ADs (1-10 μg/ml) do not kill the cells but prevent IL-2 production making the cells resistant to proliferative stimuli [79, 80]. Ultra-low concentrations of ADs (0.3 μg/ml and lower) selectively inhibit Tregs due to disruption of the signal transduction by IL-2R. Among a large variety of T cell subsets only Tregs constitutively express high affinity receptor for IL-2, the cytokine, which is the factor of their growth and survival [81]. IL-2R is not a unique receptor, which may be blocked with ADs. In addition similar effect has been shown for at least two other surface receptors: TNFR and Fas [82, 83].

ADs in non-cytotoxic doses are already successfully used in treatment of patients with acute steroid refractory bowel disease [84] and severe steroid-dependent asthma [85]. The difference between two protocols is just this, that in the former cyclophosphamide (Cy) pulse-therapy is used while in the latter short course of inhalation with melphalan in ultra-low dose has been prescribed. The usual single dose of Cy is 10 mg/kg administered intravenously over 5 months in divided doses, the dosage interval is one month [84]. It should be noted that Cy in contrast to other ADs is biotransformed principally in the liver to active alkylating metabolites by a mixed function microsomal oxidase system. Another AD melphalan, which does not need metabolic activation, has been used in asthmatic patients (five daily inhalations of 0.1 mg melphalan) [85]. Moreover we have our own modest, but successful experience of chronic fatigue syndrome treatment with melphalan (data not published). Melphalan prescription to these patients has been based on the opinion that the pathogenesis of chronic fatigue syndrome is associated with Treg superfluous accumulation [86, 87]. The drug has been prescribed *per os* in a single dose of 10 mg. In whole each patient received five such doses with the dosage interval one month. Amelioration in the patients treated in accordance with such protocol has been observed during at least 18 months. The amelioration concerns essentially such symptoms as chronic fatigue, headaches, sleep disturbance, low mood, changes in memory, difficulty in concentration, mood swings, and depression.

The greatest number of both experimental and clinical investigations concern to depletion or reduction of Tregs for the purpose of restoration of antitumor immunity. As a rule Treg depletion precedes the therapeutic vaccination [4, 88-93]. Besides Cy administration the following methods of Treg depletion have been proposed: (a) treatment with CD25 depleting antibodies (anti-CD25 mAb) [88-91] and (b) using IL-2 diphtheria toxin conjugate DAB_{389}IL-2 (denileukin difitox, ONTAK) [91, 92]. However most of recent works come down in favour of Cy because the latter is not only capable of Treg elimination but also resets dendritic cell (DC) homeostasis (Fig. 4). Ability of Cy to rebound myelopoiesis and perturb DC homeostasis is a distinctive feature of this drug and possibly of other members of this drug family. The resulting DC turnover leads to the emergence of tumor-infiltrating DCs that secrete more IL-12 and loss IL-10, in other words the switching to pro-inflammatory phenotype takes place [93, 94].

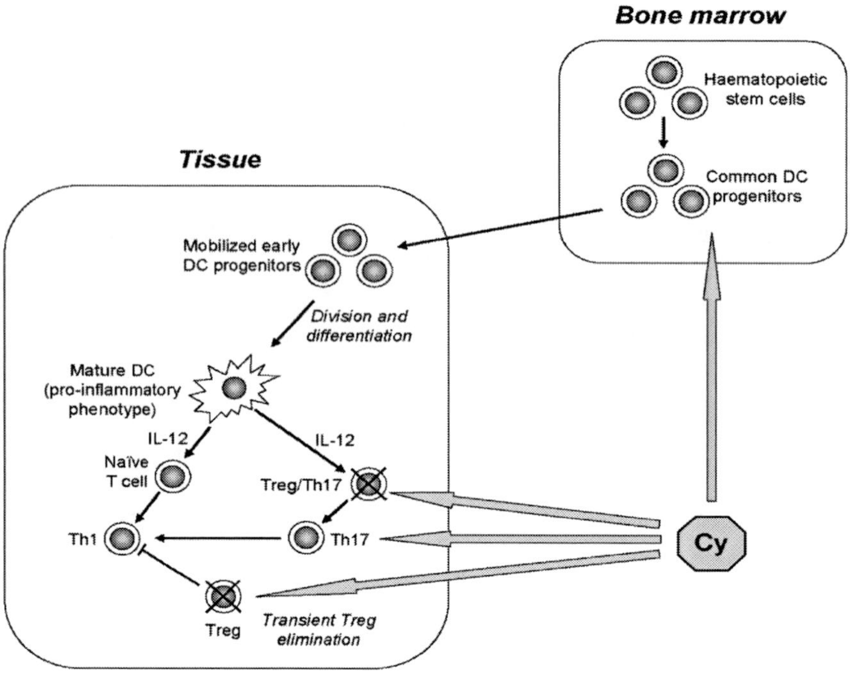

Figure 4. Cellular targets for alkylating drugs applied in non-cytotoxic doses illustrated by pulse-therapy with cyclophosphamide (Cy). Altered immune regulation in elderly age is characterized by Treg and Th17 cell accumulation (see Figs 1 and 2). Alkylating metabolites of CY find their cellular targets both in bone marrow and in periphery. In bone marrow Cy mobilizes DC progenitors, which are translocated into peripheral tissues and differentiated there in mature DCs, which demonstrate high pro-inflammatory activity including intensive production of IL-12. It is known that IL-12 plays an important role in promoting the differentiation of naïve T cells to Th1 cells. In addition, Cy influences directly upon immune cells in periphery. Thus, Cy selectively eliminates Tregs and promotes transition of Th17 cells to Th1. The latter in contrast to Th17 cells are susceptible to Treg-mediated suppressive signals.

So, ADs have a number of characteristics, which allow using them for the treatment of diseases based on Treg superfluous accumulation. First of all, ADs have a unique capacity for selective elimination of Tregs. Such elimination may be achieved with low doses of the drug, which do not affect other cell subsets. Optimal effect may be obtained when ADs are used in the form of pulse-therapy: repeated dose after one month recovery. In addition the most of alkylation drugs has elimination half-life of 3 to 12 hours, whereas such remedy as anti-CD25 mAb persists in the body for an indefinite time (in any event a few weeks or longer). Thereby the concept of pulse-therapy with ADs may be formulated as "hit and run away".

Conclusion

Regulatory T cells being an important part of complex adaptive system play a significant role in human pathology. Both surplus and deficiency in Treg number may be the reason for disturbance of peripheral tolerance. However, if in childhood the disturbance of peripheral tolerance is a consequence of delayed iTreg maturation, in old age surplus accumulation of

Treg results in their transformation into Th17 cells. Moreover, in the elderly a redundancy activity of Tregs entails stable immunosuppression including the breach of antitumor immunity. One more consequence of Treg hyperactivity is an inhibition of Th1 autoimmune lymphocytes in the brain. These cells are necessary for neural maintenance and repair, possibly by controlling the local microglial response. Since any Treg deviation may take its toll on the health, modern therapy has to have for a goal the restoration of optimal Treg number. Unfortunately, generally accepted approaches to acceleration of iTreg maturation have not been suggested until now. With respect to Treg surplus, different methods of such cell selective elimination have been proposed. Comparative analysis shows that pulse-therapy with low doses of alkylating drugs should be considered the method of choice.

References

[1] Belkaid, Y. (2007). Regulatory T cells and infection: a dangerous necessity. *Nat. Rev. Immuno.l, 7*, 875–88.

[2] Lund, J.M., Hsing, L., Pham, T.T., and Rudensky, A.Y. (2008). Coordination of early protective immunity to viral infection by regulatory T cells. *Science, 320,* 1220–1224.

[3] Wan, Y.Y, Flavell, R.A. (2008). TGF-β and Regulatory T Cell in Immunity and Autoimmunity. *J. Clin. Immunol., 28,* 647–659.

[4] Ha, T-Y. (2009). The Role of Regulatory T Cells in Cancer. *Immune Network, 9*, 209-235.

[5] Cobbold, S.P., Adams, E., Graca, L., Daley, S., Yates, S., Paterson, A., Robertson, N.J., Nolan, K.F., Fairchild, P.J., Waldmann, H. (2006). Immune privilege induced by regulatory T cells in transplantation tolerance. *Immunol. Rev., 213*, 239–255.

[6] Hawrylowicz, C.M,, O'Garra, A. (2005). Potential role of interleukin-10-secreting regulatory T cells in allergy and asthma. *Nat. Rev. Immuno.11l, 5*, 271–283.

[7] Sakaguchi, S., Sakaguchi, N., Asano, M., Itoh, M., Toda, M. (1995). Immunologic self-tolerance maintained by activated T cells expressing IL-2 receptor alpha-chains (CD25). Breakdown of a single mechanisms of self-tolerance causes various autoimmune diseases. *J. Immunol., 155*, 1151-1164.

[8] Sakaguchi, S. (2004). Naturally arising CD4+ regulatory T cells for immunologic self-tolerance and negative control of immune response. *Ann. Rev. Immunol., 22*, 531-562.

[9] Gershon, R.K., Kondo, K. (1971). Infectious immunological tolerance. *Immunology, 21*, 903-914.

[10] Gershon, R.K., Cohen, P., Hencin, R., Liebhaber, S.A. (1972) Suppressor T cells. *J. Immunol., 108*, 586-590.

[11] Curotto de Lafaille, M.A., Lafaille, J.J. (2009). Natural and Adaptive Foxp3+ Regulatory T Cells: More of the Same or a Division of Labor? *Immunity, 30*, 626-635.

[12] Klein, L., Hinterberger, M., Wirnsberger, G. & Kyewski, B. (2009). Antigen presentation in the thymus for positive selection and central tolerance induction. *Nat. Rev. Immunol. 9,* 833–844.

[13] Baecher-Allan, C., Brown, J.A., Freeman, G.J., Hafler, D.A. (2001). CD4+CD25+high regulatory cells in human peripheral blood. *J Immunol, 167*, 1245-1253.

[14] Setoguchi, R., Hori,S., Takahashi, T., Sakaguchi, S. (2005). Homeostatic maintenance of natural Foxp3+ CD25+ CD4+ regulatory T cells by interleukin (IL)-2 and induction of autoimmune disease by IL-2 neutralization. *J. Exp. Med. 201,* 723–735.

[15] Darrasse-Jeze, G., Deroubaix, S., Mouquet, H., Victora, G.D., Eisenreich, T., Yao, K-H., Masilamani, R.F., Dustin, M.L., Rudensky, A., Liu, K., and Nussenzweig, M.C. (2009). Feedback control of regulatory T cell homeostasis by dendritic cells in vivo. *J. Exp. Med., 206*, 1853-1862.

[16] Adeegbe, D., Levy, R.B., Malek, T.R. (2010). Allogeneic T regulatory cell-mediated transplantation tolerance in adoptive therapy depends on dominant peripheral suppression and central tolerance. *Blood, 115*, 1932-1940.

[17] Horwitz, D.A., Zheng, S.G., Gray, J.D. (2008). Natural and TGF-beta-induced Foxp3(+)CD4(+) CD25(+) regulatory T cells are not mirror images of each other. *Trends Immunol., 29*, 429-435.

[18] Nouri-Aria, K.T. (2009). Foxp3 expressing regulatory T-cells in allergic disease. *Adv. Exp. Med. Biol., 665*, 180-194.

[19] Saurer, L., Mueller, C. (2009). T cell-mediated immunoregulation in the gastrointestinal tract. *Allergy, 64*, 505-519.

[20] Faria, A.M., Weiner, H.L. (2006). Oral tolerance: therapeutic implications for autoimmune diseases. *Clin. Dev. Immunol., 13,* 143-57.

[21] Mizrahi, M., Ilan, Y. (2009). The gut mucosa as a site for induction of regulatory T-cells. *Curr. Pharm. Des., 15*, 1191-1202.

[22] Dimeloe, S., Nanzer, A., Ryanna, K., Hawrylowicz, C. (2010). Regulatory T cells, inflammation and the allergic response-The role of glucocorticoids and Vitamin D. *J. Steroid Biochem. Mol Biol., 120*, 86-95.

[23] Okada, H., Kuhn, C., Feillet H., Bach J.-F. (2010). The 'hygiene hypothesis' for autoimmune and allergic diseases: an update. *Clin. Exp. Immunol., 160*, 1–9.

[24] Belkaid,Y., Liesenfeld, O., Maizels, R.M. (2010). 99th Dahlem Conference on Infection, Inflammation and Chronic Inflammatory Disorders: Induction and control of regulatory T cells in the gastrointestinal tract: consequences for local and peripheral immune responses. *Clin. Exp. Immunol., 160*, 35–41.

[25] Rook, G. A.W. (2010). 99th Dahlem Conference on Infection, Inflammation and Chronic Inflammatory Disorders: Darwinian medicine and the 'hygiene' or 'old friends' hypothesis. *Clin. Exp. Immunol., 160*, 70–79.

[26] Bower, J.E., Ganz, P.A., Aziz, N. (2005). Altered cortisol response to psychologic stress in breast cancer survivors with persistent fatigue. *Psychosom. Med. 67*, 277-280.

[27] Gaab, J., Baumann, S., Budnoik, A., Gmunder, H., Hottinger, N., Ehlert, U. (2005). Reduced reactivity and enhanced negative feedback sensitivity of the hypothalamus-pituitary-adrenal axis in chronic whiplash-associated disorder. *Pain, 119*, 219-224.

[28] Negrini, S., Fenoglio, D., Balestra, P., Fravega, M., Filaci, G., Indiveri, F. (2006). Endocrine regulation of suppressor lymphocytes. Role of the glucocorticoids-induced TNF-like receptor. *Ann. N.Y. Acad. Sci., 1069*, 377-385.

[29] Trzonkovski, P., Szmit, E., Mysliwska, J., Mysliwski, A. (2006). $CD4^{+}CD25^{+}$ T regulatory cells inhibit cytotoxic activity of CTL and NK cells on humans-impact of immunosenescence. *Clin. Immunol, 119*, 307-316.

[30] Shurin, M.R, Shurin, G.V., and Chatta, G.S. (2007). Aging and the Dendritic Cell System: Implications for Cancer. *Crit. Rev. Oncol. Hematol., 64*, 90–105.

[31] Sandmand, M., Bruunsgaard, H., Kemp, K., Andersen-Ranberg, K., Pedersen, A.N., Skinhoj, P., Pedersen, B.K. (2002). Is ageing associated with a shift in the balance between Type 1 and Type 2 cytokines in humans? *Clin. Exp. Immunol., 127*, 107–114.

[32] Kato, H., Fujihashi, K., Kato, R., Dohi, T., Fujihashi, K., Hagiwara, Y., Kataoka, K., Kobayashi, R., McGhee, J.R. (2003). Lack of oral tolerance in aging is due to sequential loss of Peyer's patch cell interactions. *Int. Immunol., 15*, 145–158.

[33] Tesar, B.M., Du, W., Shirali, A.C., Walker, W.E., Shen, H., Goldstein, D.R. (2008). Aging augments IL-17 T-cell alloimmune responses. *Am. J. Transplant., 9*, 54-63.

[34] Zhou, L., Lopes, J.E., Chong, M.M., Ivanov, I.I., Min, R., Victora, G.D., Shen, Y., Du, J., Rubtsov, Y.P., Rudensky, A.Y., Ziegler, S.F., Littman, D.R. (2008). TGF-beta-induced Foxp3 inhibits T(H)17 cell differentiation by antagonizing RORgammat function. *Nature, 453,* 236–240.

[35] Ichiyama, K., Yoshida, H., Wakabayashi, Y., Chinen, T., Saeki, K., Nakaya, M., Takaesu, G., Hori, S., Yoshimura, A., Kobayashi, T. (2008). Foxp3 inhibits RORgammat-mediated IL-17A mRNA transcription through direct interaction with RORgammat. *J. Biol. Chem., 283*, 17003–17008.

[36] Koenen, H.J., Smeets, R.L., Vink, P.M., van Rijssen, E., Boots, A.M., Joosten, I. (2008). Human CD25highFoxp3pos regulatory T cells differentiate into IL-17-producing cells. *Blood, 112*, 2340–2352.

[37] Voo, K.S., Wang, Y.H., Santori, F.R., Boggiano, C., Wang, Y.H., Arima, K., Bover, L., Hanabuchi, S., Khalili, J., Marinova, E., Zheng, B., Littman, D.R., Liu, Y.J. (2009). Identification of IL-17-producing FOXP3+ regulatory T cells in humans. *Proc. Natl. Acad. Sci. USA , 106*, 4793–4798.

[38] Deknuydt, F., Bioley, G., Valmori, D., Ayyoub, M. (2009). IL-1β and IL-2 convert human Treg into T(H)17 cells. *Clin. Immunol., 131*, 298–307.

[39] Leveque, L., Deknuydt, F., Bioley, G., Old, L.J., Matsuzaki, J., Odunsi, K., Ayyoub, M., Valmori, D. (2009). Interleukin 2-mediated conversion of ovarian cancer-associated CD4+ regulatory T cells into proinflammatory interleukin 17-producing helper T cells. *J. Immunother., 32,* 101–108.

[40] Beriou, G., Costantino, C.M., Ashley, C.W., Yang, L., Kuchroo, V.K., Baecher-Allan, C., Hafler, D.A. (2009). IL-17 producing human peripheral regulatory T cells retain suppressive function. *Blood, 113*, 4240–4249.

[41] Elias, K.M., Laurence, A., Davidson, T.S., Stephens, G., Kanno, Y., Shevach, E.M., O'Shea, J.J. *(2008).* Retinoic acid inhibits Th17 polarization and enhances FoxP3 expression through a Stat-3/Stat-5 independent signaling pathway. *Blood, 111,* 1013-1020.

[42] Mucida, D., Pino-Lagos, K., Kim, G., Nowak, E., Benson, MJ., Kronenberg, M., Noelle, R.J., Cheroutre, H. (2009). Retinoic acid can directly promote TGF-beta-mediated Foxp3(+) Treg cell conversion of naive T cells. *Immunity, 30,* 471-472.

[43] Xiao, S. Jin, H., Korn, T., Liu, S.M., Oukka, M., Lim, B., Kuchroo, V.K. (2008). Retinoic acid increases Foxp3+ regulatory T cells and inhibits development of Th17 cells by enhancing TGF-beta-driven Smad3 signaling and inhibiting IL-6 and IL-23 receptor expression. *J. Immunol., 181,* 2277-2284.

[44] Quintana, F., Basso, A.S., Iglesias, A.H., Korn, T., Farez, M.F., Bettelli, E., Caccamo, M., Oukka, M., Weiner, H.L. (2008). Control of T(reg) and T(H)17 cell differentiation by the aryl hydrocarbon receptor. *Nature, 453,* 65-71.

[45] Veldhoen, M., Hirota, K., Christensen, J., O'Garra, A., Stockinger, B. (2009). Natural agonists for aryl hydrocarbon receptor in culture medium are essential for optimal differentiation of Th17 T cells. *J. Exp. Med., 206,* 43-49.

[46] Passerini, L., Allan, S.E., Battaglia, M., Di Nunzio, S., Alstad, A.N., Levings, M.K., Roncarolo, M.G., Bacchetta, R. (2008). STAT5-signaling cytokines regulate the expression of FOXP3 in CD4+CD25+ regulatory T cells and CD4+. *Int. Immunol.* , *20,* 421–31.

[47] Wuest, T.Y., Willette-Brown, J., Durum, S.K., Hurwitz, A.A. (2008). The influence of IL-2 family cytokines on activation and function of naturally occurring regulatory T cells. *J. Leukoc. Biol. 84,* 973–80.

[48] Yates, J., Rovis, F., Mitchell, P., Afzali, B., Tsang, J.Y., Garin, M., Lechler, R.I., Lombardi, G., Garden, O.A. (2007). The maintenance human CD4+CD25+ regulatory T cell function: IL-2, IL-4, IL-7 and IL-15 preserve optimal suppressive potency *in vitro*. *Int. Immunol. 19,* 785–99.

[49] Laurence, A., Tato, C.M., Davidson, T.S., Kanno, Y., Chen, Z., Yao, Z., Blank, R.B., Meylan, F., Siegel, R., Hennighausen, L., Shevach, E.M., O'shea, J.J. (2007) Interleukin-2 signaling via STAT5 constrains T helper 17 cell generation. *Immunity 26,* 371–81.

[50] Cohen, A.C., Nadeau, K.C., Tu, W., Hwa, V., Dionis, K., Bezrodnik, L., Teper, A., Gaillard, M., Heinrich, J., Krensky, A.M., Rosenfeld, R.G., Lewis, D.B. (2006) Cutting edge: decreased accumulation and regulatory function of CD4+ CD25(high) T cells in human STAT5b deficiency. *J. Immunol. 177,* 2770–2774.

[51] Viglietta, V., Baecher-Allan, C., Weiner, H.L., Hafler, D.A. (2004). Loss of functional suppression by CD4+CD25+ regulatory T cells in patients with multiple sclerosis. *J. Exp. Med. 199,* 971–979.

[52] Ehrenstein M.R., Evans, J.G., Singh, A., Moore, S., Warnes, G., Isenberg, D.A., Mauri, C. (2004). Compromised function of regulatory T cells in rheumatoid arthritis and reversal by anti-TNFα therapy. *J. Exp. Med., 200,* 277–285.

[53] Fujino, S., Andoh, A., Bamba, S., Ogawa, A., Hata, K., Araki, Y., Bamba, T., Fujiyama, Y. (2003). Increased expression of interleukin 17 in inflammatory bowel disease. *Gut, 52,* 65–70.

[54] Maloy, K.J. (2008). The interleukin-23/interleukin-17 axis in intestinal inflammation. *J. Intern. Med., 263,* 584–90.

[55] Sugiyama, H., Gyulai, R., Toichi, E., Garaczi, E., Shimada, S., Stevens, S.R., McCormick, T.S., Cooper, K.D. (2005). Dysfunctional blood and target tissue CD4+CD25high regulatory T cells in psoriasis: m.echanism underlying unrestrained pathogenic effector T cell proliferation. *J. Immunol., 174,* 164–73.

[56] Ma, C.S., Chew, G.Y., Simpson, N., Priyadarshi, A., Wong, M., Grimbacher, B., Fulcher, D.A., Tangye, S.G., Cook, M.C. (2008). Deficiency of Th17 cells in hyper IgE syndrome due to mutations in STAT3. *J. Exp. Med., 205,* 1551–7.

[57] Gabrilovich, D.I., Nagaraj, S. (2009). Myeloid-derived suppressor cells as regulators of the immune system. *Nat. Rev. Immunol., 9,* 162-174.

[58] Makarenkova, V.P., Bansal, V., Matta, B.M., Perez, L.A., Ochoa, J.B. (2006). CD11b+/Gr-1+ myeloid suppressor cells cause T cell dysfunction after traumatic stress. *Immunology, 176,* 2085-94.

[59] Kipnis, J., Schwartz, M. (2005). Controlled autoimmunity in CNS maintenance and repair: naturally occurring $CD4^{+}CD25^{+}$ regulatory T-Cells at the crossroads of health and disease. *Neuromolecular. Med. 7*, 197-206.

[60] Ford, J.D., Campbell, K.A., Storzbach, D., Binder, L.M., Anger, W.K., Rohlman, D.S. (2001). Posttraumatic stress symptomatology is associated with unexplained illness attributed to Persian Gulf War military service. *Psychosom. Med., 63,* 842-849.

[61] Cohen, H., Ziv, Y., Cardon, M., Kaplan, Z., Matar, M.A., Gidron, Y., Schwartz, M., Kipnis, J., (2006). Maladaptation to mental stress mitigated by the adaptive immune system via depletion of naturally occurring regulatory CD4+CD25+ cells. *J. Neurobiol. 66,* 552-563.

[62] Okada, H., Kuhn, C., Feillet, H., Bach, J.F. (2010). The 'hygiene hypothesis' for autoimmune and allergic diseases: an update. *Clin. Exp. Immunol. 160,* 1-9.

[63] Rook, G.A. (2009). 99th Dahlem conference on infection, inflammation and chronic inflammatory disorders: darwinian medicine and the 'hygiene' or 'old friends' hypothesis. *Clin. Exp. Immunol., 160,* 70-9.

[64] Whitlock, D.R., Feelisch, M. (2009). Soil bacteria, nitrite, and the skin. In: Rook GAW, ed. The hygiene hypothesis and Darwinian medicine. Basel: Birkhäuser, 103–116.

[65] van den Biggelaar, A,H,, Holt, P.G. (2010). 99th Dahlem conference on infection, inflammation and chronic inflammatory disorders: neonatal immune function and vaccine responses in children born in low-income versus high-income countries. *Clin. Exp. Immunol.,160,* 42-47.

[66] Rozy, A., Chorostowska-Wynimko, J. (2008). Bacterial immunostimulants--mechanism of action and clinical application in respiratory diseases. *Pneumonol. Alergol. Pol., 76,* 353-9.

[67] Camargo, C.A. Jr., Rifas-Shiman, S.L., Litonjua, A.A., Rich-Edwards, J.W., Weiss, S.T., Gold, D.R., Kleinman, K., Gillman, M.V. (2007). Maternal intake of vitamin D during pregnancy and risk of recurrent wheeze in children at 3 y of age. *Am. J. Clin. Nutr., 85,* 853-859.

[68] Devreux, G. Litonjua, A.A., Turner, S.W., Craig, L.C., McNeil, G., Martindala, S., Heims, P.J., Seaton, A., Wess, S.T. (2007). *Am. J. Clin. Nutr., 85,* 853-859.

[69] Kalliomaki, M., Salminen, S., Arvilommi, H., Kero, P., Koskinen, P., Isolauri, E. (2001). Probiotics in primary prevention of atopic disease: a randomised placebo-controlled trial. *Lancet, 357,* 1076–1079.

[70] Summers, R.W., Elliott, D.E., Urban, J.F., Jr, Thompson, R.A., Weinstock, J.V. (2005). Trichuris suis therapy for active ulcerative colitis: a randomized controlled trial. *Gastroenterology, 128,* 825–832.

[71] Holmén, N., Lundgren, A., Lundin, S., Bergin, A.M., Rudin, A., Sjövall, H., Ohman, L. (2006). Functional CD4+CD25high regulatory T cells are enriched in the colonic mucosa of patients with active ulcerative colitis and increase with disease activity. *Inflamm. Bowel Dis., 12,* 447-456.

[72] Lee, J.H., Yu, H.H., Wang, L.C., Yang, Y.H., Lin, Y.T., Chiang, B.L. (2007). The levels of CD4+CD25+ regulatory T cells in paediatric patients with allergic rhinitis and bronchial asthma. *Clin. Exp. Immunol., 148,* 53-63.

[73] Lane, N., Robins, R.A., Corne, J., Fairclough, L. (2010). Regulation in chronic obstructive pulmonary disease: the role of regulatory T-cells and Th17 cells. *Clin. Sci. (Lond), 119*, 75-86.

[74] Han, G.M, O'Neil-Andersen, N.J., Zurier, R.B., Lawrence, D.A. (2008) CD4+CD25high T cell numbers are enriched in the peripheral blood of patients with rheumatoid arthritis. *Cell. Immunol., 253,* 92-101.

[75] Schwartz, M., Kipnis, J. (2005). Therapeutic T cell-based vaccination for neurodegenerative disorders: the role of CD4+CD25+ regulatory T cells. *Ann. N.Y. Acad. Sci., 1051,* 701-708.

[76] Ojwang, J.O., Gruenberg, A., Loechlef, E.L. (1989). Synthesis of duplex oligonucleotide containing of nitrogen mustard inerstrand DNA-DNA cross link. *Cancer Res., 49,* 6529-6537.

[77] Colvin, M. (1993). Alkylating agents and platinum antitumor compounds. In Holland JF, Frei E III, Bast RCJ, Kufe DW, Morton DL, Weichselbaum RR (eds). *Cancer Medicine*. Philadelphia, Lea & Feiberg, p.p. 733-54.

[78] Mizumoto, K., Glascott, P.A., Farber, Jr., Farber, J.L. (1993). Role for oxidative stress and poly(ADP-ribosil)ation in the killing of cultured hepatocytes by methyl methane-sulfonate. *Biochem. Pharmacol., 46,* 1811-1818.

[79] Pukhalsky, A.L., Toptygina, A.P. (1993). The comparative study of mechanisms of antiproliferative effect mafosfamide and cyclosporine A applied in low doses. *Bull. Exp. Biol., 115,* 514-515.

[80] Pukhalsky, A.L., Toptygina, A.P., Viktorov, V.V. (1993). Immunosuppresive action of cyclophosphamide in mice: Contribution of some factors to determination of strain differences. *Int. J. Immunopharmacol., 15,* 509-514.

[81] Pukhalsky, A., Toptygina, A., Khaidukov, S. (1995). Interleukin-2 receptor β chain as a possible target for low doses of mafosfamide. *Med. Inflam., 4,* 175-180.

[82] Pukhalsky, A., Shmarina, G., Alioshkin, V., Sabelnikov, A. (2006). Alkylating drugs applied in non-cytotoxic doses as a novel compounds targeting inflammatory signal pathway. *Biochem. Pharmacol., 72,* 1432-1438.

[83] Shmarina, G., Pukhalsky, A., Alioshkin, V., Sabelnikov, A. (2007). Melphalan reduces the severity of experimental colitis in mice by blocking tumor necrosis factor-alpha signaling pathway. *Ann. N. Y. Acad. Sci., 1096,* 97-105.

[84] Stallmach, A., Witting, B.M., Moser, C., Fischinger, J., Duchmann, R., Zeitz, M. (2003). Safety and efficacy of intravenous pulse cyclophosphamide in acute steroid refractory inflammatory bowel disease. *Gut, 52,* 377-382.

[85] Sokolov, E.I., Zykov, K.A., Pukhalsky, A.L., Tsyplenkova, V.G., Shevelev, V.I. (2002). Inhalation of ultra-low doses of alkylating drugs in the treatment of asthma. *Pulmonology*, 12(3), 82-88 (in Russian).

[86] Pukhalsky, A.L., Shmarina, G.V., Alioshkin V.A., Sabelnikov A. (2008). HPA axis exhaustion and regulatory T cell accumulation in patients with a functional somatic syndrome: Recent view on the problem of Gulf War veterans. *J. Neuroimmunol., 196,* 133-138.

[87] Pukhalsky, A., Shmarina, G, Alioshkin, V. (2010). New approaches to cognitive disorder treatment in armed conflict participants. In: *Military Psychiatry: New Development*, Eds.: Dennis G. Stanton & Lawrence R. Castaneda, Nova Publisdhers.

[88] Tanaka, H., Tanaka, J., Kjaegaard, J., Shu, S. (2002). Depletion of CD4+CD25+ regulatory cells augments the generation of specific immune T cells in tumor-draining lymph nodes. *J. Immunother., 25,* 207-217.

[89] Prasad, S.J., Farrand, K.J., Matthews, S.A., Chang, J.H., McHugh, R.S., Ronchese, F. (2005). Dendritic cells loaded with stressed tumor cells elicit long-lasting protective tumor immunity in mice depleted of CD4+CD25+regulatory T cells. *J. Immunol., 174,* 90-98.

[90] Imai, H., Saio, M., Nonaka, K., Suwa, T., Umemura, N., Ouyang, G.F., Nakagawa, J., Tomita, H., Osada, S., Sugiyama, Y., Adachi, Y., Takami, T. (2007). Depletion of CD4+CD25+ regulatory T cells enhances interleukin-2-induced antitumor immunity in a mouse model of colon adenocarcinoma. *Cancer. Sci., 98,* 416-23.

[91] Matsushita, N., Pilon-Thomas, S.A., Martin, L.M., Riker, A.I. (2008). Comparative methodologies of regulatory T cell depletion in a murine melanoma model. *J. Immunol. Methods, 333,* 167-179.

[92] Dannull, J., Su, Z., Rizzieri, D., Yang, B.K., Coleman, D., Yancey, D., Zhang, A., Dahm, P., Chao, N., Gilboa, E., Vieweg, J. (2005). Enhancement of vaccine-mediated antitumor immunity in cancer patients after depletion of regulatory T cells. *J. Clin. Invest., 115,* 3623-3633.

[93] Radojcic, V., Bezak, K.B., Skarica, M., Pletneva, M.A., Yoshimura, K., Schulick, R.D., Luznik, L. (2010). Cyclophosphamide resets dendritic cell homeostasis and enhances antitumor immunity through effects that extend beyond regulatory T cell elimination. *Cancer Immunol. Immunother., 59,* 137-148.

[94] Salem, M.L., Al-Khami, A.A., El-Naggar, S.A., Díaz-Montero, C.M., Chen, Y., Cole, D.J. (2010). Cyclophosphamide induces dynamic alterations in the host microenvironments resulting in a Flt3 ligand-dependent expansion of dendritic cells. *J. Immunol., 184,* 1737-1747.

Index

A

D

H

I

J

K

L

M

N

O

P

Q

R

S

T

U

V

W

X

Y